# MEAT

* *These titles are now out of print*

# MEAT

PROCEEDINGS OF THE TWENTY-FIRST EASTER
SCHOOL IN AGRICULTURAL SCIENCE,
UNIVERSITY OF NOTTINGHAM, 1974

*Edited by*

D. J. A. COLE, B.SC., PH.D.

*Senior Lecturer in Animal Production, Department of Agriculture and Horticulture,
University of Nottingham School of Agriculture, Sutton Bonington, Loughborough,
Leicestershire*

*and*

R. A. LAWRIE, PH.D., D.SC., F.R.S.E.

*Professor of Food Science, Department of Applied Biochemistry and Nutrition,
University of Nottingham School of Agriculture, Sutton Bonington, Loughborough,
Leicestershire*

BUTTERWORTHS

THE BUTTERWORTH GROUP

ENGLAND
Butterworth & Co (Publishers) Ltd
London: 88 Kingsway, WC2B 6AB

AUSTRALIA
Butterworths Pty Ltd
Sydney: 586 Pacific Highway, Chatswood, NSW 2067
Melbourne: 343 Little Collins Street, 3000
Brisbane: 240 Queen Street, 4000

CANADA
Butterworth & Co (Canada) Ltd
Toronto: 2265 Midland Ave., Scarborough M1P4S1

NEW ZEALAND
Butterworths of New Zealand Ltd
Wellington: 26–28 Waring Taylor Street, 1

SOUTH AFRICA
Butterworth & Co (South Africa) (Pty) Ltd
Durban: 152–154 Gale Street

First published 1975

ISBN 0 408 70660 0

Printed and bound by R. J. Acford Ltd., Industrial Estate, Chichester.

# PREFACE

Meat continues to be a major food commodity. Despite the high cost of production of meat animals and their lower efficiency of protein synthesis compared with plants and micro-organisms, meat is likely to be important in the human diet for as long as can be foreseen in the future. It provides both satisfaction in eating and sound nourishment for a significant proportion of the world's population.

The contributions of the 21st Easter School were intended to emphasise the fact that the sequence of events, from the conception of meat animals to their incorporation in the human diet, is continuous. The properties of the commodity when eaten are influenced, in the nature and degree of their expression, by all the earlier components in this chain of circumstances. It was felt that the wide spectrum of disciplines thus involved would enable participants having expertise in a particular part of the food chain to assess its significance in relation to other equally important, if less familiar, phases.

With this intention, the papers, following an introductory review of the meat-eating habit in man, were arranged in sessions covering production, preservation, composition, eating quality, human nutrition and an assessment of the future role of meat.

It was also felt appropriate that the School should be organised jointly by the Department of Agriculture and Horticulture, representing animal production, and the Department of Applied Biochemistry and Nutrition, representing food science. This reflects the interdepartmental nature of many of the interests in the Faculty of Agricultural Science.

Nottingham 1974
D. J. A. Cole
R. A. Lawrie

# ACKNOWLEDGEMENTS

It is a pleasure to publicly acknowledge the efforts of those who contributed papers at the Easter School and thus ensured its success.

We are indebted to Professor D. Lewis, Dean of the Faculty of Agricultural Science who opened the conference and to those who kindly acted as Chairmen of the sessions: Professor J. C. Bowman, of the University of Reading, Professor E. F. Williams, O.B.E., of the University of Nottingham, Professor A. G. Ward, C.B.E., of the University of Leeds and Chairman of the Food Standards Committee, Professor M. Ingram, C.B.E., of the Meat Research Institute, Langford, Miss Dorothy Hollingsworth, O.B.E., Director of the British Nutrition Foundation Ltd., and Dr. P. N. Wilson of BOCM-Silcock Ltd., Basingstoke.

We are also grateful to Mr. Roland Moyle, M.P., Parliamentary Secretary to the Ministry of Agriculture, who gave an address at the Dinner marking the 21st Anniversary of the Easter Schools; and to Professor J. D. Ivins, Deputy Vice-Chancellor, who responded on behalf of the University of Nottingham and the School of Agriculture.

The University of Nottingham wishes to express its appreciation to the following organisations whose generosity made possible the contributions from overseas speakers:

Armour & Co. Ltd., (Deltec Foods Ltd.)
Birds Eye Foods Ltd.
The Boots Company Ltd.
British Bacon Curers Federation
The Distillers Company Ltd./The United Yeast Company Ltd.
Fatstock Marketing Corporation Ltd.
Griffith Laboratories (UK) Ltd.
H. J. Heinz Co. Ltd.
F. Hoffmann—La Roche & Co. Ltd. (of Basle, Switzerland)
Pedigree Petfoods Ltd.
Pork Farms Ltd.
J. Sainsbury Centenary Grant

In conclusion we should like to thank warmly all those members of staff and students who gave their time in the interests of the School. The help of Miss G. E. Fox and Mrs. D. M. Borrows was particularly appreciated.

# CONTENTS

## INTRODUCTION

## I MEAT PRODUCTION

## II PREPARATION AND STORAGE

# CONTENTS

# CONTENTS

# INTRODUCTION

# 1

# THE MEAT-EATING HABIT IN MAN

JOHN YUDKIN

*University of London*

## INTRODUCTION

A useful approach to the subject of meat eating is to
consider first the subject of non-meat eating.
Vegetarianism has a long and distinguished history.
Its adherents and ardent proponents make up a roll of
honour that includes eminent names such as Pythagoras,
Plato, Socrates, Ovid, Leonardo da Vinci, Milton, Isaac
Newton, Isaac Pitman, Tolstoy and Bernard Shaw.

It is likely that the abstention from meat, either
permanently or on particular days during the year, that
is common to several religions, owes its origin to
economic pressures, though these may be long forgotten.
Often, as individuals from such religious groups become
more affluent, they become back-sliders and forsake
their strict vegetarianism.

The adoption of vegetarianism by members of meat-
eating cultures is usually motivated by considerations
of compassion. It is true that to these is added closely
argued evidence relating to man's anatomy and physiology,
and to the hazards to health that it is claimed is the
price to be paid for the indulgence in meat. However,
often these arguments tend to be rationalisations to
support an instinctive repugnance to the slaughter of
animals solely in order to gratify man's appetite.
Shelley's *A Vindication of Natural Diet* of 1813 was
written mainly, according to the preface in a later

reprint, '...to show that a vegetable diet is the most
natural, and therefore the best for mankind.  It is not
an appeal to humanitarian sentiment.'  Yet here are some
of Shelley's lines that show that compassion plays by no
means a small part in his adoption and advocacy of
vegetarianism.  He speaks of the time when he who ceases
to eat meat '...will hate the brutal pleasures of the
chase by instinct; it will be a contemplation full of
horror and disappointment to his mind, that beings
capable of the gentlest and most admirable sympathies,
should take delight in the death-pangs and last
convulsions of dying animals.'


## VEGETARIAN MAN


Without discussing whether compassion is a unique
characteristic that distinguishes man from the lower
animals, we would all agree that it is an important
human characteristic.  My firm belief is that compassion,
like all other instincts, had an important biological
purpose in the preservation of the species, but that in
some of its manifestations it now conflicts with the
sophistications of modern life.  This is only one
example of the dissociation of wants and needs that has
produced many of the dilemmas of affluence.  Man wants
to rest from his labours, thus satisfying the need to
reverse the biochemical processes that produce fatigue,
in order that he can resume his labours.  However, he can
now satisfy his want for rest far beyond his need, by the
use of a vast range of devices from mechanical means of
transportation to labour-saving inventions at home and
at work, to a degree that makes sedentary man a prey to
obesity, coronary thrombosis, and a range of other
diseases of affluence.  Man wants to copulate - and here,
'man' embraces 'woman' - and this serves his need to
perpetuate his kind.  However in today's conditions,
affluent man, unlike primitive man, has been able to feed
and to protect his young so that few die, and as a
result the uncontrolled or inadequately controlled
satisfaction of his sexual wants has resulted in the
population explosion that understandably causes so much
concern.  Similarly, man's intuitive compassion,
especially for his own family, was of importance in
making and maintaining the human race against the animate

and inanimate forces that seemed to be bent on the destruction of his species.

With the increase in affluence that gave him the technical knowledge to provide security, food, shelter, public health measures and medical care, he could afford the extension of his instinctive compassion beyond his family to his tribe, and then to his fellow-countrymen, then to mankind as a whole, and finally to the animal kingdom. Leaving aside those communities, such as the Brahmins or the Jains, whose concern for animals is often less of positive kindness than of refraining from deliberate or even accidental killing of animals, animal welfare is a luxury that is rare in populations that are poor, or have only recently ceased being poor.

This extension of the instinct of compassion to animals, and thus the reluctance or refusal to kill animals for food, is, I contend, another example of the conflicts that arise between wants and needs, for there is a great deal of evidence that meat has for millions of years played an important role, even an essential role, in man's diet. Let me, however, make clear at the outset that I am not intending to show that man is by nature strictly carnivorous, but that he is not by nature herbivorous. Man belongs to the few species that are able to eat from an extremely wide range of foods of animal and plant origin. This is in extreme contrast to the animals that have a very limited diet, such as the giraffe that lives almost exclusively on acacia leaves, or the koala bear that lives almost exclusively on eucalyptus leaves and from only some eight or ten out of the 400 or more species that exist in Australia. Man shares this omnivorous habit with the pig and the rat, a habit that has enabled these species, and man, to inhabit virtually every part of the globe.

Much of the discussion about what are the proper foods of man has concerned itself with anatomy, especially the teeth, the jaw and the alimentary canal. It would be tedious to repeat this discussion in detail; it is, however, worth pointing out that anatomy has more relevance to distant evolutionary history than to current eating patterns. Both the brown bear and the panda belong anatomically to the carnivores, yet the former is omnivorous and the panda almost exclusively herbivorous. It seems to me that the evidence as a whole points to an anatomical adaptation to a diet for man that was neither entirely herbivorous nor entirely carnivorous, but

omnivorous.  However, rather than pursue this ancient
discussion I intend to concentrate on two other sorts
of evidence - from palaeontology and from nutrition -
that meat is a part of man's natural diet.

I have hesitated to use the word 'natural' since it
has recently become so debased.  We are told about
'natural' tobacco, as if man was evolved as a smoking
animal.  We see in health food stores a range of highly
sophisticated preparations labelled 'natural', and
produced by elaborate technological procedures that
include molecular distillation, selective absorption and
elution, and a range of other physical, chemical and
microbiological procedures.  These techniques are often
more sophisticated than anything used in the production
by food manufacturers of what have been called 'deficient
and devitaminised garbage which they have the temerity
to label food.'

Yet although the word 'natural' has been debased in
this way, I would still like to use it in its original
meaning.  Thus, I mean by the natural foods of man the
sort of foods he was eating before he became, uniquely,
a food producer.  Just as the 'natural' food of a panda
is bamboo shoots, of the chameleon insects, of the lion
carcasses, so it seems that man and his ancestors for at
least 2 million years and perhaps much longer ate meat
as a large part of his natural diet.


## CARNIVOROUS MAN


Certainly the primates from which man descended were
herbivorous.  They emerged some 70 million years ago,
and continued as vegetarians up to about 20 million years
ago.  However, then the rainfall began to decrease, and
the earth entered on a 12 million year period of drought.
The forests shrank and their place was taken by the
ever-increasing areas of open savannah.

At this time, the man-ape from which we are descended
emerged.  He forsook the vegetarian and fruitarian
existence of his immediate ancestors, and changed to a
predatory and hunting existence that was largely
carnivorous.  He could do this because he had adopted a
completely erect posture, his arms and hands were freed
from the need to be used for locomotion, and he could
use tools and weapons.  His earliest weapons were bones;

only later did he begin to use stones, and still later
the axe. Thus, it appears that for several million
years man's ancestors ate meat as an important part of
their diet.

One great advantage of a meat-rich diet is that it is
far more concentrated than are the leaves and shoots and
fruits that made up a large part of man's earlier diet.
Animals that live on this sort of diet have to spend an
inordinate amount of time in finding and consuming their
bulky food; the adoption of the meat-eating habit allowed
man, if he was lucky in his kill, more time for other
pursuits and skills, such as devising new tools and
learning the beginnings of pictorial art.

The importance of meat in man's diet is indicated by
the way in which made-up vegetarian foods are so often
prepared with hydrolysed vegetable protein so as to give
a meaty flavour, or given meat-like names such as soya
steak or nut cutlets.

Until fairly recently, in evolutionary terms, man like
all other animals depended for his food on hunting,
scavenging or gathering from the animal and vegetable
kingdoms in his environment. It was less than 10,000
years ago, compared with the 2 million years or more of
his largely carnivorous ancestry, that man became
uniquely a food producer. The domestication of grasses
became the cereals that are now the staple food of the
majority of present-day mankind, and this was followed
or accompanied by the domestication of root crops and
of wild animals that were used not only as animals of
burden but also for food. His new foods were to a large
extent plant storage organs - seeds and roots - so that,
though vegetable in origin, they were more concentrated
than were the leaves and shoots that made up much of the
diet of the primates from which he was descended. This,
together with the greater reliability of food production
compared with food gathering, often still allowed him
some degree of leisure, and thus the continuation and
extension of his creative abilities in the field of
invention and art.

The results of the discovery of agriculture, the
neolithic revolution, were many and far reaching.
Compared with hunting and foraging, agriculture usually
yielded more food. It also allowed man to cultivate
areas where existing stocks of food would have been
inadequate. Thus the human population grew, partly
because fewer died of food shortage and partly because

man spread into increasing areas of the earth's surface. However, in due course the limits of food production again became the limits of the numbers that could be fed. The inevitable pressure of population on food supplies tended to perpetuate a type of diet quite different from that of man's hunting ancestors. It is much easier to produce vegetable foods than to produce animal foods. For a given area of land, some ten times as many calories can be produced in the form of cereals or root crops than in the form of meat, eggs or milk.

## THE FIRST DIETARY REVOLUTION

Thus, the effect of the neolithic revolution was to alter the components of the diet. Instead of being rich in protein and moderately rich in fat, it was now rich in carbohydrate - mostly starch - and poor both in protein and in fat. It is likely that both protein deficiency, and deficiency of many of the vitamins, began to affect large sections of the human species only after man became a food producer.

We have accepted that natural foods are those animal or plant materials that can be found by hunting or gathering. We can then state that, in choosing from such foods, palatability is an excellent guide to nutrition. That is to say, a species survives because, by eating the foods that it wants, it is choosing the foods that it needs. This proposition becomes even more evident when stated in the reverse: if the foods chosen do not satisfy the needs of a species, that species cannot survive during the natural selection that determines the course of evolution. Today man's preferred foods, other than manufactured foods, are meat and fruit, and there is no reason to believe that there has been a fundamental change in food preferences. If one remembers that, for early man, 'meat' implied not only the muscular tissue, but all the viscera and soft bones of his kill, then his two preferred foods are capable of satisfying entirely his nutritional requirements.

As we have seen, man's potential as an omnivore allowed him to survive on a wide range of other foods, such as seeds, leaves and roots, when his preferred meat and fruit were not available. However, throughout history those individuals that, through power or purse, were able to eat meat and fruit did so.

We saw too that the neolithic revolution enormously increased man's supply of food, and thus his ability to survive in a wide variety of climates. However, the utilisation of the starch-rich cereals and root crops that soon became his staple foods was, I suggest, greatly affected by two other discoveries. The first, without which I believe these foods would have remained only minor items of man's diet, was the discovery of fire. It seems that the first controlled use of natural fire began some 500,000 years ago, to be followed much later by the ability to make fire at will. Food could now be cooked, and as a result, the new flavours of cooked meats and a few vegetable foods were discovered and enjoyed. Cooking became an essential part of man's eating pattern with the cultivation of the starch-rich cereals and roots, since he has only a limited capacity to digest uncooked starch. Without the knowledge of how to cook these foods, they could not have become his staples, until in many populations they soon made up 80% or more of his total diet.

## BREAD WITH SALT

There was a second discovery that I believe affected the change of the diet associated with the neolithic revolution. This was the introduction of salt, sodium chloride. The date of this is not known, but it was certainly not very long after the beginning of food cultivation. Salt is found in much higher concentration in meat than it is in vegetable foods; these on the other hand have a higher concentration of potassium. It has been suggested that man, by nature carnivorous, requires salt only when he adopts a diet that is mainly vegetarian. Certainly it is true that animal products are palatable without salt, whereas vegetable products - bread and potatoes, for example - are quite unpalatable without salt. Thus, the use of salt, whether from the evaporation of sea water or from salt mines, must have helped early man to make the change to the vegetable-rich diet that accompanied his transformation from a food hunter and gatherer to a food producer. However, it seems that salt is one of man's discoveries that is now taken largely for pleasure rather than for need. In this, its

consumption joins a range of other of man's habits -
sugar, tobacco, alcohol, tea and coffee. Nevertheless,
it differs from these in that in quite small quantities,
perhaps one-tenth of the amount we usually consume, there
is a physiological requirement for salt, although this
is certainly met on a meat-rich diet without the need to
add sodium chloride itself.

The inclusion of a high proportion of vegetable foods
at the expense of some of the meat in man's diet has
several nutritional disadvantages. Partly, they derive
from the fact that vegetable foods are deficient in some
essential nutrients. The most important example is
vitamin $B_{12}$, which is found in no vegetable or fruit
ordinarily consumed by man, although it is a product of
fermentation by some micro-organisms. In addition, the
proteins of particular vegetable foods are short of
some of the essential amino acids; cereals lack lysine
and pulses lack the sulphur-containing amino acids.
This does not matter if the diet contains a mixture of
vegetables, in which case the proteins will supplement
one another.

As well as having a low content of some nutrients,
many vegetable foods are able to reduce the availability
of nutrients that may be present in apparently adequate
amounts in the diet. The danger of this secondary or
conditioned sort of deficiency is especially present for
those populations - the vast majority - for whom
agriculture resulted in the consumption of large amounts
of cereals. Chiefly because of their content of phytate
and other phosphates, cereals have the property of
combining with several of the mineral elements of the diet.
This is particularly true for whole grain cereals, since
milling in most instances reduces the content of phytate.
As a result, it turns out that the iron present in the
vegetable foods, or even iron salts added to bread, are
badly absorbed by the human intestine, whereas the iron
of meat is readily absorbed when adequate calcium is
present, for example from carcass bones.

An indication of the relative unavailability of dietary
iron from vegetable foods came to our notice in a study
we recently made of the diets of the 1860s. That was a
period when anaemia was common, especially amongst
working-class adolescent girls. It was often so severe
that the victims' pallor led to the disease being called
'the green sickness' or chlorosis. We found that the
poor working-class diet of those days contained if

anything more iron than does our present-day diet, when chlorosis has disappeared and even moderate degrees of anaemia are uncommon. It seems very likely that it was the lack of meat, coupled with a lack of calcium, that was responsible for the chlorosis of those days.

The absorption of other mineral elements as well as that of iron is affected by the phytate in cereals. Since phytate is partly destroyed during yeast fermentation, this danger is greater when cereals such as wheat are eaten in unleavened products. Some of the recent cases of rickets in Britain that have occurred in Asian immigrants may owe their origin to the consumption of large quantities of chapattis. More importantly, it now seems that the widespread zinc deficiency reported from Iran and Egypt is also a secondary deficiency caused by the binding of dietary zinc by the phytate in the unleavened bread customarily consumed in those countries.

The view that 'bread is the staff of life' and represents man's major natural dietary component is of course widespread. This is why we are still frequently told that we should be much better and much healthier if we would go back to the bread made from whole wheat. As we have seen, man has consumed bread for less than one-half of one per cent of the period of his existence as a separate species. At the other extreme, the view is expressed, often by the same food reformers, that man is by nature a vegetarian; this in turn is to go back in evolution several millions of years before the emergence of our species.

We have considered the situation of man's early diet, in which meat and probably fruit played a large part both quantitatively and nutritionally. We have also considered the profound dietary changes that followed the neolithic revolution, in which meat consumption fell, and was to a large extent replaced by starchy foods such as cereals and root crops. Thus the neolithic revolution was also a dietary revolution. We saw that there nevertheless was and is an instinctive desire for meat and for fruit, so that those individuals - and much more recently those populations - that were able to do so always ate more of these foods, and to some extent reduced their consumption of the less palatable starch-rich foods.

## THE SECOND DIETARY REVOLUTION

We must now touch on the second dietary revolution.
This has come about during the past 200-300 years, and
coincided with the industrial revolution.  The rapid and
accelerating increase in science and technology made it
possible for man to satisfy his wants for palatable
foods by eating more of the natural foods meat and
fruit.  He also became increasingly able to manufacture
completely new and quite different palatable foods.  We
are now able to separate palatability from nutritional
value, so that we have many foods in which we have
dissociated wants from needs.  And we can do this in
both directions.  We now have many high protein foods
from plant or microbial sources, of good nutritional
value but of such low palatability that without exception
they have failed to be accepted with anything like the
enthusiasm shown by those that produce them.  Conversely,
we can make a variety and ever-expanding range of foods
and drinks, largely by the use of sugar, that are almost
irresistibly attractive, but which at best contribute
little but calories to nutrition, and at worst contribute
to a range of diseases from caries to coronaries.

Sugar, the quintessential sweetness that our ancestors
enjoyed virtually only when they ate fruits, came with
several nutrients, particularly vitamin C.  The total
consumption of sugar from fruits by pre-agricultural man
is unlikely to have exceeded a few grams a day; the
average consumption in the industrialised countries of
the world of sugar, largely manufactured from the cane
and the beet, now exceeds one hundred grams a day.  This
nutrient-free sugar, together with the low-nutrient fat,
cocoa powder and other constituents that accompany sugar
in confectionery, cakes, biscuits, ice cream and soft
drinks, makes up over 25% of the calories of our total
diet.

One of the effects of the existence of these highly
palatable, calorie-rich foods is that they tend to make
us overeat, so that they constitute a major cause of the
obesity that afflicts about one-third of our population.
A second effect is that since not everybody eats these
foods *in addition* to their ordinary foods, they must at
least to some extent be taken *instead* of other foods.
It is accepted as axiomatic by both economists and
nutritionists that the consumption of meat and fruit

shows a high income elasticity, so that we would expect
that there has been a continuing increase in consumption
as incomes have risen in such countries as Britain and
the USA. Some time ago, we predicted that this tendency
would however be counteracted by the parallel increase
in the consumption of the new highly palatable foods
rich in calories. From the few relevant figures that
are available, we were able to confirm this prediction
for meat in Britain and for fruit in the United States.
Whilst the consumption of meat in the lowest income group
in Britain had risen, consumption in the wealthiest group
had fallen. Indeed, it is not necessary to seek these
elusive and precise figures; it is known both that
average meat consumption has remained virtually unchanged
since before the war, and that the poorer sections of our
population are certainly eating more meat than they did
35 or 40 years ago. It follows that the wealthier
section are eating less meat, and the most reasonable
explanation is that other palatable foods have partly
replaced meat in these diets.


## PALATABILITY AND THE NOVEL FOODS

The role of palatability in nutrition is an excellent
instance of an attitude of the layman being more
realistic than that of the physiologist or nutritionist.
The man in the street, and even more the housewife in
the kitchen, know that people eat foods that they like,
and do not eat foods that they do not like. That the
expert has still not properly appreciated this is shown
by the failure, as I mentioned, of virtually all of the
high protein foods that have been produced to make any
real impact to relieve the protein shortage - or the
alleged shortage - in the poorer countries. It seems
quite illogical to expect the inhabitants of these
countries to eat unappetising foods, however much we
tell them that the foods are highly nutritious, any more
than we can be expected to eat foods that we do not like
just because they are said to be good for us. The wise
food manufacturer who, in this country, will spend months
devising a new product and carrying out elaborate
organoleptic tests before launching it on the market,
seems entirely to suspend this wisdom when devising the
new high-protein products designed for the third world.

Partly, the confusion arises because of our
unthinking failure properly to distinguish between foods
and nutrients.  The nutritionist has been at fault so
long as he describes human requirements solely in terms
of nutrients, forgetting that people eat food and not
nutrients.  The layman on the other hand has been at
fault in so far as he often talks of protein, or
carbohydrate, or vitamins, when he is really referring
to meat, or bread, or orange juice.  However, it is not
only the layman who falls into this latter trap.  Many
of the elaborate official calculations on the extent of
world protein needs are in fact nothing of the sort;
they are calculations of the extent of world demand for
meat.  Even if, in the varied circumstances existing in
the different populations in which this demand has been
assessed, the assessed demand really represents a
nutritional need, how do we know it is for the protein
of the meat, and not for the nicotinamide, or the
vitamin $B_{12}$, or the highly assimilable iron?

As I have suggested, I believe it is possible, with
our modern technology, to make foods that are very
attractive and palatable, so that, once we recognise
the importance of palatability, we should be able to
produce novel proteins in forms that might be accepted
as reasonable substitutes for meat.  Indeed, the first
examples of attempts to do this are already being
marketed in a small way.  However, we should look very
carefully at the nutrient content of these products,
not so much when they are used to extend meat products,
but in situations where they may be taken as complete
substitutes for meat, for example in institutions with
long-stay inmates in this country, or in the under-
nourished populations abroad.

It may be that we are wrong to think of these new
protein-based foods as substitutes for meat; perhaps
we should consider them as new foods in their own right
If their object is really to help improve the nutrition
of the poorer peoples of the world, we shall have to
know very much more about their real nutritional needs,
the extent to which the new food will be taken into the
diet, and - most importantly - what other foods will be
consumed in smaller amounts, or perhaps even greater
amounts, because of the consumption of the new food.
We have not even begun to study the interdependence of
food in our diets, let alone in the diets of those for
whom these foods are supposed to have been so
assiduously developed.

## CONCLUSION

My main thesis has been that man has evolved as an omnivorous animal, with a preference for meat and fruits which, between them, could satisfy his nutritional needs. It is, however, quite impossible now to contemplate a situation in which he could revert to a diet composed largely of fruits and of meat – with the bones and the viscera as well as the muscles of animals, as he used to consume them. We have already been made aware that the demand for meat is rapidly outstripping the possible supply. If we say that the diet of our pre-agricultural ancestors would be likely to be ideal for our own healthy sustenance, we must now contemplate ways in which present diets can be complemented and extended. Even the most wealthy, for example, will not be able to ensure adequate calcium intake in a culture in which the chewing of meat bones is considered uncouth and unacceptable. Fortunately, we have an excellent alternative for our calcium, namely milk, which in addition is a good source of some of the nutrients we would otherwise be getting from viscera.

Increasingly, it appears that, in whatever way we adapt our consumption of traditional foods, we shall also need to eat quite new foods. For the first time, mankind is proposing to produce on a large scale foods that are destined to satisfy his nutritional needs, and not simply to satisfy his gastronomic wants. If the survey I have briefly made is correct, even only in broad outline, then it teaches us three important lessons. The first is that the instinctive desire for meat helped man for a very long period of his existence to provide some of his essential nutrient requirements. The second is that we must try and bring together both wants and needs if we are to succeed in producing substitutes for meat in the future. Thirdly, we must put a much greater effort into studying the factors that determine what foods people will eat and what they will reject. In particular, we need to know much more about what effects the acceptance of one new food can have upon the consumption of the other foods in the diet. We constantly forget that nutrition is as much to do with people as it is with food; it is as much a social science as it is a natura.. science.

# I

## MEAT PRODUCTION

# 2

# GROWTH OF MEAT ANIMALS

R. T. BERG
*University of Alberta, Canada*

R. M. BUTTERFIELD
*University of Sydney, Australia*

## INTRODUCTION

Growth, according to Fowler (1968), has two aspects. The first is measured as an increase in mass (weight) per unit time. The second involves changes in form and composition resulting from differential growth of the component parts of the body. In appraising growth of meat animals, there is a trend towards greater emphasis on those components of the body which have most commercial value. Often the carcass is the unit of trade and carcass weight has been an acceptable end point for measurement of growth in meat animals. Use of carcass weight eliminates the influence of certain variables which affect liveweight such as gut fill and other non-carcass components. The ultimate use of the carcass as edible components has shifted emphasis to studies of growth of the edible, high-value regions.

Much emphasis in meat animal research has been placed on the proportion of high-priced cuts, which has meant looking at the problem from the consumer's point of view. It was perhaps assumed that the animal would and could conform to any changes which the consumer's desires might dictate. Perhaps the problem has been approached from the wrong end. If the requirements from the animal's side were known, there might be a better understanding of how far the animal could be changed to

meet consumer requirements.  Emphasis on high-priced
regions or cuts has often resulted in the shifting of
fat around the carcass with the result of apparent
increases in high-priced regions but with little or no
real improvement.  For example, increases in kidney and
channel fat usually increase the proportion of the
hindquarter while increases in subcutaneous fat over the
loin may result in a higher proportion of loin.  As the
consumer has evolved a preference for lean meat (muscle)
rather than fat, such changes do not improve the
desirability of the product.

The alternative is to look at growth from the animal
end; to find out how animals are made, what they are made
of and what are the basic functional requirements in
terms of the various components.  Then, within the
limits imposed by functional requirements, one could
attempt to alter the component systems to better satisfy
the needs or desires of consumers.

How then should the problem of discovering more about
how animals are made be approached?  Instead of
considering an animal as various butcher's cuts - loin,
round or ham, flank and brisket - the functional systems
might be investigated, how they grow, how they develop
and what possibilities there are of adapting these
systems to our desires.  To the consumer the important
systems, in animals used for meat production, are the
muscle system, the bone system and the fat system, which
together make up most of the components of the carcass.
Other systems, such as the circulatory, nervous etc.,
although important functionally, do not constitute an
important part in the use of animals for meat production.
Thus, the approach in this chapter will be to try to come
to a proper understanding of the functional requirements
of these important systems and of what course of action
might be taken to alter them to best fit our needs.

## FUNCTIONAL BASES FOR GROWTH

The three major tissues with which we are concerned
have certain characteristics which help them to play
their functional roles.  *Bone* forms the skeleton.  It
is the framework which provides support for the soft
tissues, protection for the vital organs, and levers
for the action of muscles.  The skeleton is formed in a
very regular pattern which varies little with respect

to the numbers of bones and the size of bones relative
to one another within a species.  Bones grow in length
and in circumference which results in some change in
shape and relationships with maturity.

*Muscles* constitute the labour force of the body, being
able to convert energy from food into work.  Each muscle
is capable of just three types of activity: it can
contract and pull its ends together, it can relax and
let its ends go further apart, or it can just hold steady
and keep its ends fixed.  To achieve any particular
movement requires the combined action of many muscles
which are contracting, relaxing or just remaining fixed.
Each muscle with a particular piece of work to do is
formed in a way which best allows it to do that job.
Each muscle has the same basic structure, being made up
of fibres which can contract, with each muscle fibre
being attached at each end to a fine strand of tendon
which is in turn attached to something else, usually a
bone.  Hard-working, powerful muscles like those located
distally on the leg have many fibres in complex arrays
terminating in strong, heavy tendons.  The connective
tissue or supportive tissue form a high proportion of
such muscles and they are basically less tender as a
result.  Other muscles, like the *psoas*, do not have to
be very powerful, but contract over a long distance and
the connective tissue attachments do not need to be as
strong, thus they have a low proportion of connective
tissue and are a more tender muscle.

Each muscle is attached to two or more points of the
skeleton so that the bones act as levers.  These points
of attachment are fixed.  They do not change from animal
to animal but a particular muscle always runs between
the same points.   Muscles have a function, which at
this stage involves fixed patterns and points of attachment that result from the evolutionary experimentation
of the species.

*Fat*, the third important carcass tissue, does not have
neat boundaries like muscles and bones.  Body fat is a
reasonably fluid substance, and is laid down in depots,
the amount in any depot being influenced partially by
physical pressure.  In thin animals little resistance
would be encountered and fat would be deposited in
different depots in fairly definable patterns or
sequences.  As an animal becomes very fat and some of
the larger, looser depots such as the body cavity and
the flank are physically full, other depots which offer

more resistance such as intramuscular areas will
increase.  Of course one depot does not completely fill
before others begin but one could consider the whole
process as a fluid (fat) flowing around and filling
spaces; as pressure (fattening) increases, spaces
offering more resistance are filled.

Thus, the carcass part of a meat animal has a pattern
of a very rigid skeleton, a softer series of active
muscles with attachments specifically fixed relative to
the skeleton, and relatively fluid fat which, although
following certain patterns of deposition, is much less
fixed with respect to location than the other tissues.

A dramatic change in functional requirements of
component tissues of an animal occur at birth.  The dark,
wet, protected environment within the mother gives way
to the exposure to the dry, brightly lit world.  The
new-born animal has to participate in its own nutrition
and locomotion in order to survive.  The change in
function which occurs at birth is accompanied by changes
in size and shape - aspects of the phenomenon of growth.
A most striking example is provided by the kangaroo
which, when it emerges from its mother's womb, is very
small and in a very early stage of development.  It is
born with a large head, large front legs and no back
legs at all because at that time its needs for survival
are front legs to drag itself up to its mother's pouch
and a powerful set of jaws to enable it to suck.
However, it will eventually be an animal with massive
hind legs and relatively small front legs and head,
which equip it well to thrive and survive in the harsh
environment of its particular habitat.

Changes are not so extreme in our domestic meat
animals but they still occur.  An examination of the
three systems in turn, bone, muscle and fat, will show
what happens to them as a result of functional demands.


BONE

Bone is relatively well developed at birth and the
skeleton at this time bears a strong resemblance to
that of the adult.  However, the bones lower down on
the limbs are relatively much longer in the newborn
animal than in the adult.  Long legs assist the newborn
to follow its mother and to run from its enemies.  As
the animal grows and develops, bones adjust, becoming

thicker to carry a greater load. They adjust also to
the greater demands of muscles which become more powerful
putting greater stress on bones to which they are
attached.


MUSCLE

Muscles most clearly show the influence of function on
growth. For example, when the calf is born, it has
little need for some muscles and extreme need for
others. The relative development of the various
muscles at birth is determined by the evolutionary
experience of the species to meet the immediate needs
of the newborn animal. Immediately after birth, the
muscles of the distal limbs are essential so that the
calf may stand, move around and obtain food in its new
environment. Through evolution Nature has provided
well-developed leg muscles and jaw muscles in anticipation
of the immediate post-natal functional demands. Other
muscles, not so important at birth respond later to the
demands imposed on them by use.
As the calf grows its needs change and a transition
takes place in the relative weights of muscles as their
overall size increases in the growing animal. For
example, the muscles of the belly wall have little work
to do in the day-old calf and are thus very light. As
the animal grows and becomes more independent, it starts
to consume bulky feed which is carried around in its huge
developing gut supported by the muscles of the belly wall.
Thus, these muscles in the adult animal comprise a much
greater proportion of total muscle than in the calf.
Another early change of function for the growing calf
relates to its increased mobility which is accommodated
by an increase in the more powerful muscles of the back
leg.
Later in life another change in function occurs if the
calf is a male, which results in a relative increase in
certain muscle groups. As the bull reaches sexual
maturity, it becomes involved in courtship behaviour
with females and engages other males of its species in
battle for this privilege; so the muscles of the neck,
which are specifically needed for fighting, start to
grow much faster than the other muscles.
The stimuli and mechanisms which produce different
rates of growth in different muscles are largely unknown.

However, they probably result from a combination of
evolutionarily-acquired genetic patterns conditioned
post-natally by response to use.  The relative development of different muscles at birth have to be a result
of the evolutionary pattern but the relative importance
of the genetic template in post-natal changes would be
difficult to assess.  Would the neck muscles of the
young male develop to the same extent if he did not have
the opportunity to do battle with his peers?  It is
known that the greatest adjustment in muscle-weight
distribution takes place soon after birth; in the
adolescent phase little further change takes place and
proportions of muscles in the heifer or steer do not
seem to differentiate further.  The young adult male
enters a maturing phase which adapts the musculature
to the dual role of survival and reproduction with its
attendant battles.

There is some knowledge of why muscles grow in the way
they do but a rather extreme example can be provided of
how muscles respond to functional needs.  Bryden (1969)
dissected a number of elephant seals which spend part
of their time on land and part in the sea, with the
result that muscle-weight distribution changed quite
markedly as muscles used for swimming were well developed
in that environment and those used for 'humping' about
on land responded to that demand.


## FAT

It is possible that fat growth also responds to
functional demands although our knowledge in this area
is more limited.  Fat acts as an energy store.  At
birth, the new-born calf has very little fat storage in
carcass tissues.  Its mother and internal fat depots
will provide its energy needs.  In the growing animal,
fat storage is a survival mechanism to get animals through
periods of low food availability, e.g. droughts or
winters.  If a species has evolved in an environment
which provides a continuous food supply, its need for
fat storage would not be critical; species evolving in a
cyclical environment of plenty and scarcity would
probably have evolved a greater capacity for fat storage
during times of plenty.

Most of the fat stores are energy stores and their
locations around the body are probably a result of

expediency, the depots developing in less restricted
'free-space' areas.  There are discernible patterns of
fat deposition among the depots and fat can be shifted
around by selection.  Man with some of his domestic
animals has had a great influence on how early animals
fatten, how much they fatten and on where the fat is
stored in or on the body.  Nature has developed fat
stores in animals to fill functional needs but this
storage does not follow the precise architecture seen
for the skeleton and skeletal muscles.

## FACTORS INFLUENCING TISSUE GROWTH PATTERNS

Some changes which occur within each of the three
carcass tissues have been described and now the way in
which each grows in relation to the others is discussed.
As an average animal grows the proportions of bone,
muscle and fat change.  This is illustrated in *Figure
2.1* where the vertical axis represents the weight of a

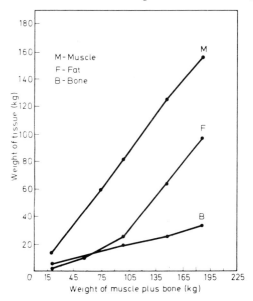

*Figure 2.1   Tissue growth relative to muscle plus bone
growth in steers.   (Adapted from Berg and Butterfield,
1968)*

given tissue and the horizontal axis represents a
general increase in size as an animal grows. This graph
has been developed from dissection of many carcasses
from animals of several species as they have grown and
can be used to depict an average relative growth pattern,
for example, of an average beef steer.

Bone represents a relatively large part of the carcass
at birth, as was previously shown, and it grows slowly
and steadily as the animal increases in size. The total
musculature is a relatively small part of the carcass at
birth and muscle growth proceeds more rapidly than bone
as is shown by the steeper line in *Figure 2.1*. Muscle-
weight increases compared to that of bone as growth
progresses. At birth a calf carcass has about twice as
much weight in muscle as in bone whereas the adult ratio
may be about five to one.

The third line in *Figure 2.1* represents the growth of
fat associated with the carcass. As pointed out earlier
there is not much need for carcass fat at birth and
there is therefore relatively less at that stage. Fat
increases slowly until the animal reaches a certain
stage of maturity and then commences to grow at an
increasing rate. Normally it seems that when muscle
and bone growth slow down, available energy is converted
to fat and what might be called the fattening phase sets
in.

*Figure 2.1* depicts the general situation of relative
tissue growth. The factors which alter this general
pattern resulting in some changes in patterns of growth
or proportions of tissues will now be examined. For
any given animal of a particular breed or sex, fed on
an adequate level of nutrition, the growth patterns in
*Figure 2.1* tend to be weight dependent, that is each
tissue tends to reach a specific weight at a given
carcass weight. Age on the other hand, within limits,
has very little influence on the level each tissue
reaches.

Genetic differences influence the patterns of relative
tissue growth. In cattle some breeds are considered to
be early-maturing, which means that they reach their
mature carcass proportions at light weights. Early
maturity can be illustrated by superimposing what happens
on the relative growth graph as shown in *Figure 2.2*. In
early-maturing breeds the line for fat growth is shifted
to the left, for late-maturing breeds the line is shifted
to the right. Thus we might call early-maturing breeds,

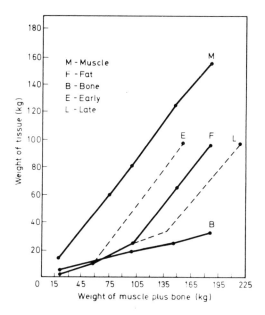

*Figure 2.2  Influence of early or late maturity on
tissue growth patterns in steers*

early fatteners since they fatten at lighter weights;
late-maturing breeds are late fatteners.   Early fattening
in some breeds of cattle is a result of man's selection.
Not only is earliness of fattening under genetic
influence; rate of fat deposition as well as ultimate
proportion of fat at maturity are also affected.
   Genetic differences also exist in muscle growth
relative to bone growth.   Breeds selected for body
thickness or for draught usage generally exceed those
selected for dairy character in muscle:bone ratio.
Heavy muscling is manifested early in the post-natal
period and breeds with high muscle:bone ratios retain
these higher ratios throughout life barring periods of
weight loss.   In *Figure 2.1,* the line for muscle growth
of a heavy-muscled breed would be above that shown for
the average animal; the line for bone growth would be
lower.
   The extreme of what can be accomplished in altering
tissue growth patterns and body composition by genetic

means is illustrated by the 'double-muscled' animals.
Their bone growth is a little slower than average, their
muscle growth is considerably faster and they seem to
have almost lost the ability to store carcass fat.  This
type of animal occurs in many breeds as a genetic
deviant and illustrates well our original thesis of the
relationship of growth and development and function.
The simple function of walking is impaired; the packing
of large masses of muscle behind the stifle upsets the
angulation of the joints of the hind leg causing
'knuckling' at the fetlock joints.  Reproduction is also
impaired, partly as a result of altered physical
relationships.  Some success has been achieved in some
breeds where a modified form of double muscling is
retained which is more compatible with function.

Sex also affects tissue growth patterns.  Compared to
steers, the fattening curve of heifers would be shifted
to the left on *Figure 2.1* and it may be steeper; the
fattening curve for bulls would be shifted to the right
and would not be as steep as that for steers.  At given
weights of muscle plus bone, the lines depicting growth
of muscle and of bone do not differ among the sexes.
Bulls may have higher muscle:bone ratios than steers at
slaughter but this is caused by bulls being larger and
having a greater total amount of muscle at slaughter.
Bulls have a more prolonged impetus for muscle growth
than steers whereas steers slow down in muscle growth
at lighter weights and an upsurge in fattening takes
place.

Nutrition, particularly the level of intake of
digestible energy, can affect carcass composition.
Again the major effect is on the fat curve, a high plane
of nutrition shifting the fattening line to the left and
a low plane shifting it to the right.  It should be
remembered that there is a strong tendency for an animal
to reach a certain carcass composition at a given weight
and that it seems to be able to accomplish this over
reasonably wide ranges in nutrition.  This tendency has
led some researchers to conclude that body composition
is completely dependent on body size; however we believe
that there is sufficient evidence that carcass fatness
at a given body weight can be influenced by plane of
nutrition.  A low plane of nutrition before the fattening
phase has little or no effect on ultimate carcass
composition provided that the animal is finished on an
adequate plane.

A period of weight loss depletes all tissues but the relative effect on fat is greater than on muscle, while bone resists depletion to a greater extent than muscle and fat. Re-alimentation following weight loss tends to restore normal body composition.

In general, partition of nutrients for tissue growth priorities are relative rather than absolute. In animals on a positive energy balance, bone and muscle growth proceed together maintaining a genetically determined ratio. The amount of fat deposited will depend on how much surplus energy is available over maintenance and bone and muscle growth requirements. Under starvation (negative energy balance) the reverse process starts with fat and muscle reserves, and to some extent bone reserves, being depleted to satisfy vital functions.


## FACTORS INFLUENCING MUSCLE-WEIGHT DISTRIBUTION

Muscle growth is accompanied by large developmental changes some of which were mentioned earlier. The animal at birth is equipped with a set of muscles which have grown during ante-natal life in such a way that they play their role in ensuring survival. The new-born animal is able to walk and to suck because muscles of the distal parts of the limbs are well developed as are those of the jaws. Such muscles have been called 'early develop-ing' in that they completed relatively more of their total growth at birth. In contrast there are others which have little immediate function at birth and these have completed only a small part of their total growth. The muscles of the abdominal wall fall into this category and they are termed 'late developing'. Between these extremes there is a spectrum of development which reflects the sequences of priorities of use which enable the animal to adjust and survive in the early post-natal period. Normally after the major adjustments have been made soon after birth, muscle groups as proportions of the total do not change much until sex-influenced differentiation takes place in the maturing male. The female and castrate remain adolescent in this respect.

It should be obvious that normal development, responding to functional demands, brings about changes

in muscle-weight distribution. Any comparisons which
are made between groups of animals with respect to their
muscle-weight distribution should take into account
their stage of compositional development. Some factors
which might be expected to influence muscle-weight
distribution, such as growth rate, breed and sex, will
be examined, and then data on species differences will
be presented and an attempt will be made to relate these
to function.

Rate of growth does not seem to affect the amount of
muscle relative to bone at any given weight of muscle
plus bone. It also does not seem to affect the relative
weights of individual muscles when comparison is made at
the same total muscle weight. Therefore it is expected
that in young animals (calves), during the immediate
post-natal period featuring maximum differential muscle
growth, comparisons of fast- and slow-grown animals at
the same age will reveal differences in muscle-weight
distribution.

Breed has been thought to influence muscle-weight
distribution. However, data from precise anatomical
dissections do not reveal any marked differences between
breeds which differ a great deal in physical appearance.
*Table 2.1* gives results from dissections of carcasses of
several breed groups of bulls, steers and heifers.
Within each sex group the breed groups are remarkably
similar in the proportions of muscles in each anatomical
region. The only statistically significant differences
between breeds were found in the abdominal wall region
and in the neck to thorax region. Interestingly, Jersey
bulls although lighter, show the greatest percentage in
the latter, sex-influenced, muscle group indicating
greater compositional maturity. The Holsteins' development in this region had not reached the same degree of
maturity as the other breeds.

*Table 2.1* also provides muscle-weight distribution
comparisons between sexes (bulls, steers and heifers).
Differences reflect differential growth of neck muscles
in the bull, steers show a faint image of the same
effect and heifers are not affected.

PIGS

*Table 2.2*, which presents data from 109 pigs at four
slaughter weights, three breeds and two sexes, confirms
that the concept of uniformity within species is equally
true in pigs as in cattle.  The differences between
breeds are confined to only one muscle group and are so
small as to be of no economic importance, as demonstrated
in the similar yields of 'expensive' muscle weights.  The
weight range of slaughter in this work covered a period
when differential muscle growth of muscle groups was
almost completely absent.  It can be presumed that this
also applied to the vast majority of individual muscles.
The only exception to this uniform relative growth rate
is shown by the intrinsic muscles of the thorax and neck,
which grew a little faster than total muscle over the
range of weights.  This aligns with the findings of very
late development in steers and bulls, however, the barrows
are not significantly different from the gilts which may
have been expected if the barrows were weakly following
the path to crest development that may be exhibited by
mature boars, similarly to bulls.

Recently available data from dissection of several
species have been compiled (Berg and Butterfield,
unpublished observations).  Muscle-weight distribution
of eight species are compared on an index basis to that
of cattle in *Table 2.3*.

The major difference between the pig and the ox is
that the pig has a much greater proportion of its total
muscle weight surrounding the spinal column.  The
muscles in the two groups located distally in the limbs
of the pigs are relatively small and this seems to
align with the reduced agility of pigs compared with
the other species.  The muscles connecting the thorax
to the thoracic limb are also reduced in the pig and
indeed, it will be seen that all the other species
(except the elephant seal) are less developed here than
the ox.


SHEEP

The sheep, as a small domestic ruminant, is probably
expected to be more like the ox than any other animal,
and indeed, its muscle-weight distribution features only
three major departures from that of the ox.  The muscles

*Table 2.1* Distribution of muscle weight in breed groups

| Breed[1] | Bulls | | | | |
|---|---|---|---|---|---|
| | He | Shx | XB | Ho | Je |
| Number of animals | 13 | 12 | 22 | 8 | 8 |
| Age (days) | 461 | 361 | 430 | 386 | 407 |
| Liveweight (lb) | 1026 | 850 | 1079 | 915 | 648 |
| Hot carcass weight (lb) | 615 | 515 | 652 | 511 | 334 |
| ANATOMICAL MUSCLE GROUPS AS % OF TOTAL MUSCLE[2] | | | | | |
| 1. Proximal hind leg | 28.4 | 28.4 | 28.3 | 28.8 | 28.4 |
| 2. Distal hind leg | 4.5 | 4.1 | 4.4 | 4.3 | 4.0 |
| 3. Around backbone | 12.4 | 12.4 | 12.6 | 12.1 | 12.1 |
| 4. Abdominal region | 9.8 | 10.0 | 8.8 | 10.9 | 10.6** |
| 5. Proximal front leg | 12.4 | 12.5 | 12.3 | 12.8 | 12.6 |
| 6. Distal front leg | 2.2 | 2.3 | 2.2 | 2.5 | 2.4 |
| 7. Thorax to front leg | 10.3 | 10.2 | 10.7 | 10.7 | 10.4 |
| 8. Neck to front leg | 5.6 | 5.4 | 5.7 | 5.0 | 5.3 |
| 9. Neck to thorax | 12.4 | 12.1 | 12.7 | 11.1 | 13.0** |
| Expensive muscles[3] | 40.8 | 40.8 | 40.9 | 40.9 | 40.5 |
| Expensive muscles[4] | 53.2 | 53.3 | 53.2 | 53.7 | 53.2 |
| Hind quarter | 46.7 | 47.8 | 47.6 | 48.7 | 47.6 |
| Front quarter | 53.3 | 52.2 | 52.4 | 51.3 | 52.4 |

[1] He = Hereford, Shx = Shorthorn cross, XB = Hybrid and Ho = Holstein, Je = Jersey.
[2] The totals do not sum to 100% since some muscle was anatomical groups.
[3] Sum of muscle groups 1 plus 3.
[4] Sum of muscle groups 1 plus 3 plus 5.
**Differences statistically significant at the 1% level.

of bulls, steers and heifers (after Berg and Mukhoty, 1970)

| | Steers | | | | | Heifers | | Sex Sig. |
|---|---|---|---|---|---|---|---|---|
| He | Shx | XB | BSx | Ho | | He | Shx | |
| 11 | 22 | 32 | 14 | 6 | | 10 | 12 | |
| 402 | 383 | 434 | 404 | 480 | | 365 | 398 | |
| 823 | 830 | 1016 | 1005 | 1027 | | 672 | 745 | |
| 486 | 526 | 602 | 574 | 589 | | 391 | 461 | |
| 29.5 | 29.6 | 29.5 | 29.6 | 29.6 | | 31.3 | 31.6 | ** |
| 4.3 | 4.3 | 4.4 | 4.4 | 4.2 | | 4.2 | 4.4 | |
| 12.3 | 12.3 | 12.3 | 12.2 | 12.3 | | 12.2 | 12.1 | |
| 10.8 | 11.8 | 10.4 | 10.6 | 11.3** | | 11.7 | 11.4 | ** |
| 12.5 | 12.3 | 12.6 | 12.9 | 12.5 | | 12.5 | 12.4 | |
| 2.4 | 2.4 | 2.4 | 2.4 | 2.5 | | 2.1 | 2.4 | |
| 10.0 | 10.6 | 10.8 | 10.7 | 10.4 | | 10.5 | 10.2 | |
| 5.2 | 5.1 | 5.8 | 5.2 | 5.1 | | 5.1 | 5.2 | |
| 10.4 | 9.5 | 10.4 | 10.3 | 9.4** | | 9.1 | 9.0 | ** |
| 41.8 | 41.9 | 41.8 | 41.8 | 41.9 | | 43.5 | 43.7 | ** |
| 54.3 | 54.2 | 54.4 | 54.7 | 54.4 | | 56.0 | 56.1 | ** |
| 50.2 | 49.9 | 49.5 | 49.7 | 49.5 | | 50.5 | 50.3 | ** |
| 49.8 | 50.1 | 50.5 | 50.3 | 50.5 | | 49.5 | 49.7 | ** |

other crossbreds, BSx = Brown Swiss crossbreds,

weighed as scrap muscle and was not included in the

Table 2.2 'Standard muscle groups' as percentage of weight of total side muscle in 109 barrows and gilts of three breeds and slaughtered at four liveweights (after Richmond and Berg, 1971)

| Muscle group | Liveweight (kg) | | | | Breeds[1] | | | Sex | |
|---|---|---|---|---|---|---|---|---|---|
| | 23[2] | 68 | 91 | 114 | D | H | Y | B | G |
| 1. Proximal pelvic limb | 26.56 | 28.40 | 28.25 | 28.67 | 28.39 | 28.42 | 28.50 | 28.67 | 28.21 |
| 2. Distal pelvic limb | 3.99 | 3.96 | 3.84 | 3.87 | 3.86 | 3.84 | 3.97 | 3.84 | 3.95 |
| 3. Spinal | 16.83 | 17.01 | 17.42 | 17.44 | 17.69a[4] | 17.17ab | 17.01b | 17.54a | 17.05b |
| 4. Abdominal | 12.41 | 11.32 | 10.98 | 11.16 | 11.16 | 11.21 | 11.10 | 11.22 | 11.09 |
| 5. Proximal thoracic limb | 12.35 | 12.29 | 12.05 | 11.79 | 11.87 | 11.94 | 12.32 | 11.90 | 12.18 |
| 6. Distal thoracic limb | 2.15 | 1.94 | 1.89 | 1.85 | 1.88 | 1.89 | 1.91 | 1.84a | 1.95b |
| 7. Thorax to thoracic limb | 7.35 | 7.56 | 7.64 | 7.38 | 7.42 | 7.57 | 7.58 | 7.48 | 7.57 |
| 8. Neck to thoracic limb | 4.39 | 4.90 | 4.84 | 4.97 | 4.88 | 4.91 | 4.92 | 4.77 | 5.03 |
| 9. Neck and thorax | 9.28 | 9.39a | 10.02b | 9.76ab | 9.66 | 9.81 | 9.69 | 9.66 | 9.78 |
| 10. Scrap | 4.69 | 3.21 | 3.06 | 3.08 | 3.15 | 3.20 | 3.01 | 3.07 | 3.17 |
| Expensive Groups[3] A | 30.61 | 32.36 | 32.09 | 32.54 | 32.25 | 32.27 | 32.48 | 32.50 | 32.16 |
| B | 47.46 | 49.38 | 49.51 | 49.98 | 49.95 | 49.44 | 49.48 | 50.04a | 49.20b |
| C | 59.81 | 61.67 | 61.56 | 61.78 | 61.82 | 61.38 | 61.80 | 61.94 | 61.39 |

[1] D = Duroc x Yorkshire; H = Hampshire x Yorkshire; Y = Yorkshire.
[2] 23 kg group not tested statistically against other weight groups.
[3] A (Group 1 + Group 2); B (Group 1 + Group 2 + Group 3); C (Group 1 + Group 2 + Group 3 + Group 5).
[4] a, b, c, means within the same classification followed by different letters differ significantly at P < 0.05.

*Table 2.3* Muscle-weight distribution of several species expressed relative to *Bos taurus*. Weight of standard muscle groups as percentage of total muscle of each species compared with similar value from appropriate *Bos taurus* animals = 100

| Muscle groups | Bos taurus | Pig | Sheep | Water buffalo | Banteng | Moose | Deer | Bison | Elephant seal |
|---|---|---|---|---|---|---|---|---|---|
| Proximal pelvic limb | 100 | 97 | 94 | 103 | 123 | 117 | 111 | 108 ⎤ | 20 |
| Distal pelvic limb | 100 | 91 | 109 | 105 | 142 | 155 | 132 | 98 ⎦ | |
| Around spinal column | 100 | 141 | 139 | 79 | 103 | 96 | 122 | 102 | 190 |
| Abdominal wall | 100 | 103 | 97 | 79 | 76 | 58 | 56 | 71 | 280 |
| Proximal thoracic limb | 100 | 94 | 90 | 117 | 90 | 133 | 89 | 104 ⎤ | 30 |
| Distal thoracic limb | 100 | 79 | 130 | 107 | 133 | 158 | 115 | 100 ⎦ | |
| Thorax to thoracic limb | 100 | 70 | 72 | 96 | 71 | 81 | 74 | 84 | 120 |
| Neck to thoracic limb | 100 | 93 | 109 | 120 | 106 | 89 | 82 | 158 | 53 |
| Neck and thorax | 100 | 97 | 90 | 116 | 73 | 64 | 105 | 91 | 157 |

surrounding the spinal column are a much higher
proportion of total muscle than in the ox and are
comparable with the pig. In keeping with most of the
other species, except the pig, the distal muscles of
the thoracic limb are better developed. Again the
muscles connecting the thorax to the thoracic limb are
less well developed.


WATER BUFFALO

Data for the water buffalo were obtained from just two
bulls and could therefore be misleading. However the
buffalo seems to have a greatly reduced proportion of
total muscle surrounding the spinal column. This may
be associated with lumbar vertebrae which are different
from cattle (Butterfield, 1964). The muscles of the
limbs comprise a higher proportion of total muscle than
in the ox which may reflect the greater agility of the
buffalo in swampy ground.


BANTENG

Banteng data came from a single banteng steer. This
animal has a much higher proportion of its muscle weight
in the hindlimb and around the spinal column, and
relatively less muscle weight in the forequarter and
abdominal wall. The marked increase in development of
the distal muscles of all limbs is probably an indication
of the agility of the animal; while the decrease in the
weight of muscles in the abdominal wall is more typical
of non-domesticated species studied. These cattle are
somewhat similar in external appearance to deer and, in
fact, resemble them perhaps more than they do other
cattle. It is not surprising that the general picture
of its muscle-weight distribution is more like the deer
than the ox, with the notable exception of the muscles
surrounding the spinal column, in which the Banteng
resemble *Bos taurus*.

MOOSE

Data from three bull moose were available.  It is
perhaps a surprise, and certainly a sobering thought,
to those who believe it is simple to assess relative
muscle development in the live animal, to find that the
moose is relatively better developed in the hindquarter muscles than the ox.  This is a good illustration
of the influence of length of bones in altering the
appearance of an animal relative to muscle development.
The most outstanding feature of the moose is the massive
development of the muscles of the distal part of the
limbs, and, in fact, all four intrinsic groups of the
limbs are well developed.  This is no doubt associated
with the special wading habit of the species, which
demands individual action from the limbs rather than
a co-ordinated effort involving the whole musculature.
The three forequarter groups associated with the trunk
and attachment of the limb are relatively smaller than
in the ox and have the smallest index figures for any
species studied for the intrinsic muscles of the thorax
and neck.  In common with the other wild species the
abdominal wall is very light.


DEER

Deer show a greater proportion of their muscle weight
in the hindlimbs and around the spinal column than
cattle.  They also show a decrease in the abdominal wall,
the proximal part of the thoracic limb and the muscles
connecting the thorax to the thoracic limb.  However,
the mature male deer showed a slightly greater response
in his development of forequarter groups than the bull.


BISON

Bison data came from a single male.  Surprisingly, this
bull showed less variation from cattle than any species
with which comparison was made.  Apart from the lighter
thorax to thoracic limb muscles, consistent with all
except the buffalo, the only major change was a
reduction in abdominal wall and a very marked increase
in the muscles connecting the neck to the thoracic limb.
It was interesting that this increase in forequarter

muscles was not associated with an increase in the
intrinsic muscles of the neck and thorax, because in
most other species sex-induced differences are seen in
the intrinsic neck muscles. However, the hump of
*Bos indicus* bulls is developed in the rhomboid muscle
in either its cervical part (Brahman) or thoracic part
(Boran). The rhomboid muscle is associated with the
attachment of the forelimb to the neck and thorax, and
hence this enlargement in *Bos indicus* cattle is in the
same group as the enlargement in bison. It therefore
appears that *Bos indicus* is more like the bison than it
is like *Bos taurus* in this characteristic.


ELEPHANT SEAL

A very extensive study of growth and development of the
elephant seal by Bryden (1969) included a study of
muscle-weight distribution which produced a great deal
of substantial evidence on the effect of function on
relative muscle growth. The muscle-weight distribution
of mature male seals is compared with that of a mature
*Bos taurus* bull. Owing to the extreme differences of
shape, the standard muscle groups were modified by
Bryden and an attempt is made to align the data as well
as possible. In order to do this certain groups are
combined. In line with the reduced function of legs in
this species there is a marked shift of muscle weight
to the muscles of the trunk. Of particular interest is
the heavy abdominal wall which plays a large part in
locomotion. No other very sensible comparisons seem
possible.
  In general it appears that the smaller the species
the more likely it is that an increased proportion of
its muscle weight will be concentrated around its spinal
column. The more agile the animal, the greater is the
development of the muscles in the distal parts of the
limbs; the more mobile, the greater is the proportion of
the whole of the limbs. The domestication of animals
appears to bring about an increase in the relative
weight of the muscles of the abdominal wall to cope
with the more continuously fully-loaded digestive tract.
The more aggressive animals have relatively larger
muscles in the neck region. Muscular structure of each
species has evolved to suit its needs. Muscle-weight
distribution within a species has a regular, constant

development pattern and differences between species
reflect functional differences of agility, mobility,
weight support etc. The possibility of changing muscle-
weight distribution in any domestic species does not
seem too promising as it would seem necessary to alter
functional relationships to accomplish this end.


## SUMMARY

Growth is the manifestation of increase in mass and of
differential increases of the components which comprise
that mass. Man's requirements from meat animals are the
edible components obtainable from their carcasses and
thus growth studies of meat animals emphasise these
components. The major components of a carcass are
muscle, bone and fat and these tissues adapt and grow
in such a way that they play their functional role in
the survival and reproduction of the species. The role
as meat producers is a secondary imposition added to
the animals' normal functional requirements. In
attempting to alter the composition of carcasses of
meat animals one must be cognisant of possible adverse
effects on functional requirements of the animal.
The carcass of a meat animal emerges as a pattern of
a very rigid skeleton, a softer series of active muscles
with attachments specifically fixed relative to the
skeleton and relatively fluid fat which is much less
fixed with respect to location. A pattern of relative
tissue growth can be observed within a species. Bone
is fairly well developed at birth as are some essential
muscle groups but the total musculature is relatively
less well developed than bone at birth. Little fat is
normally present at birth and fat deposition is more
pronounced as maturity approaches. Major differences in
the pattern of fat deposition can be observed among
breeds, sexes and under the influence of differing
planes of nutrition. Breed differences exist in the
amount of muscle relative to bone at any stage of
development but sex and nutrition have little effect on
this ratio at equal weights of lean body mass.
Muscle growth follows evolutionarily-acquired genetic
patterns conditioned by response to use. A change in
function at birth results in a great adjustment in
muscle-weight distribution. In the adolescent phase

little further differential change takes place and the
female and castrate do not emerge from this phase.
Muscles of the young adult male enter a maturing phase
which adapts the muscles to the dual role of survival
and reproduction with its attendant battles. Within a
species of a given sex and maturity, little variation is
found in muscle-weight distribution. Between species
muscle-weight distribution reflects differences in
functional adaptations. In smaller species an increased
proportion of muscle weight is concentrated around the
spinal column. The more agile species have greater
muscle development in the distal limbs; the more mobile
the greater the development on the whole of the limbs.
Domestication appears to have brought about an increase
in the relative weight of the muscles of the abdominal
wall. The more aggressive species have relatively
larger muscles in the neck region. Muscle-weight
distribution within a species reflects a rather regular
developmental pattern and between species reflects
functional differences in agility, mobility, weight
support and aggressiveness. Changing muscle-weight
distribution within a species may be quite difficult
because of functional influences observed to be present.

REFERENCES

BERG, R.T. (1968). 'Genetic and environmental
   influences on growth in beef cattle', in *Growth and
   Development of Mammals*. Ed. G.A. Lodge and G.E.
   Lamming. Butterworths, London
BERG, R.T. and BUTTERFIELD, R.M. (1968). *J. Anim. Sci.*,
   27, 611
BERG, R.T. and MUKHOTY, H.M. (1970). *49th Annual
   Feeders' Day Report*, p.40. University of Alberta,
   Edmonton
BRYDEN, M.M. (1969). Ph.D. Thesis, University of Sydney,
   Australia
BUTTERFIELD, R.M. (1964). Report to Northern Territory
   Administration, Darwin, Australia
BUTTERFIELD, R.M. (1965). *Proc. N.Z. Soc. Anim. Prod.*,
   25, 152
BUTTERFIELD, R.M. and BERG, R.T. (1972). *Proc. Brit.
   Soc. Anim. Prod.*, p.109
FOWLER, V.R. (1968). 'Body development and some
   problems of its evaluation', in *Growth and Development*

*of Mammals.* Ed. G.A. Lodge and G.E. Lamming. Butterworths, London

RICHMOND, R.J. and BERG, R.T. (1971). *Can. J. Anim. Sci.,* 51, 41

# 3

# BREEDING AND MEAT PRODUCTION

P. GLODECK

*Institut für Tierzucht und Haustiergenetik,
University of Göttingen, West Germany*

## INTRODUCTION

In this chapter breeding is taken to mean the
utilisation of genetic variability within and between
farm animal populations for the improvement of meat
production. The overall objective is very complex,
consisting of a number of genetically and economically
very different component traits such as:

1. Reproductive rate.
2. Growth rate.
3. Feed efficiency.
4. Carcass composition.
5. Specific meat quality.
6. Postweaning losses (liveability).

Reproduction is important for the dams only whereas all
the others apply to the meat producing progeny more than
to the parents. All of these are complex biological
functions in themselves whose genetic properties are
still insufficiently studied, so that breeders have to
work with fairly imprecise parameter estimates for many
of them. Nevertheless, construction of selection
indices using the best parameter estimates available and
straightforward selection will be the most likely way of

making genetic progress in most cases. Apart from
that, heterosis may play an important role in component
traits related to reproduction, liveability and early
growth, and therefore has particular interest for
populations with low reproductive performance.

The following is restricted to the three main meat
producing species of farm animals: cattle, sheep and
pigs.


## RELATIVE ECONOMIC IMPORTANCE OF COMPONENT TRAITS OF MEAT PRODUCTION

The three species chosen differ considerably in their
reproductive rates, their efficiency of production and
the importance of specific quality traits. Therefore,
it appears to be useful to get some idea of the relative
economic importance of various characters within species.
These will be discussed separately because of some
completely different features of dual-purpose (that is
milk and meat producers) and beef cattle.


### DUAL-PURPOSE CATTLE

Recent estimates of the relative economic importance of
various traits in German dual-purpose cattle have been
published by Böckenhoff et al. (1967), Langholz (1970),
Haring (1972), Adelhelm et al. (1972) and Zeddies (1974).
It has been known for a long time that milk yield is
the most, and carcass quality the least, important
trait in dairy and dual-purpose cattle, with growth
characters intermediate. Adelhelm et al. (1972) have
looked for some components of reproductive performance
(*Table 3.1*). Under European Economic Community (EEC)
price assumptions (their Alternative 4) they calculated
for a 40 cow herd the economic weights per phenotypic
standard deviation including correlated effects.

It can be concluded that milk production is more than
twice as important as growth rate and feed efficiency.
Liveability, regular reproduction and carcass characters
are only a tenth as important as these. Growth capacity
or the weight of the cow becomes important when labour
is of high value.

*Table 3.1* Relative economic weights per phenotypic
standard deviation in Deutschmarks (including correlated
effects) in German dual-purpose cattle (after Adelhelm
*et al.*, 1972)

| Trait | | $\sigma_P$ | $I^1$ | $II^2$ |
|---|---|---|---|---|
| Fat free milk yield | (kg) | 600 | 601 | 606 |
| Butterfat yield | (kg) | 27 | 673 | 678 |
| Feed efficiency | (SE) | 250 | 348 | 359 |
| Daily gain | (g) | 100 | 323 | 361 |
| Growth capacity | (kg) | 50 | 98 | 304 |
| Primal cuts in carcass | (%) | 1.0 | 24 | 24 |
| Calving interval | (days) | 40 | 269 | 437 |
| Production life | (years) | 2 | 502 | 404 |

[1] I = situation where feed is lacking.

[2] II = situation where labour is lacking.


BEEF CATTLE

It is quite surprising that no recent calculations of
relative importance of production characters in beef
cattle can be cited from the literature. A general
account of the suitability of various traits for
selection has been given in the Meat and Livestock
Commission (MLC), *Scientific Study Group Report on Beef
Improvement* (1971). Extrapolating from dual-purpose
cattle (which to some extent are kept as suckler cows),
weaning weight of calves per cow and per year must be
the dominating trait. The next will be growth capacity
and efficiency of feed utilisation of the meat animal,
carcass quality which is mainly lean percentage will
follow only at a fair distance if it is slaughtered at
a proper time.

An older study by Lindholm and Stonaker (1957)
indicated that 56% of the costs of producing a steer
of 7-900 lb liveweight were production costs of the
feeder calf and only 44% were actual feed-lot costs of
the steer. The fixed costs of keeping the mother cow
amounted to 41% of the cost of a feeder calf or about
23% of the total costs. Compared with the total cost
of a fat steer the difference in returns between choice

and medium grade was only 20%. In their selection
index the standard deviation of each trait was weighted
as follows:

| | |
|---|---|
| Weaning weight | 39.1 |
| Days to finish | 17.6 |
| Daily liveweight gain | 4.0 |
| Feed per lb liveweight gain | 3.9 |

## SHEEP

British conditions are reflected in the MLC *Scientific
Study Group Report on Sheep Improvement* (1972a) which
shows that lamb sales account for 85-90% and wool sales
for only 10% of the gross returns in MLC recorded
flocks. On the cost side, in the same 400 flocks, feed
and forage accounted for 50-65% of total costs and
flock replacement 25-30%. From these figures the number
of lambs reared per ewe and per year is by far the most
important trait followed by growth rate and feed
utilisation efficiency of the lambs.

Similar weights are reported from Norway by Gjedrem
(1966) where in a selection index, number of lambs
reared, wool (kg) and weaning weight of lamb (kg) were
given the relative factors of 26:3:1, respectively. In
Germany Nitter and Jacubec (1970) have calculated
relative economic weights for various component traits
of lamb production for specified German conditions which
are given in *Table 3.2*. By far the most important trait
is reproductive performance of the ewe, ahead of growth,
efficiency of food utilisation and market grading of the
fat lambs. Wool accounts for less than 10% of the lamb
sale returns.

## PIGS

Many investigations have dealt with relative economic
weights in pig production traits. In Britain, MLC has
worked for years with selection indices which are
adapted to British consumer and producer demands. On
the continent, other conditions prevail and particular
indices are in use in Scandinavia, Holland and Germany.
The methods commonly used in developing economic weights
for pig selection indices are described among others by

*Table 3.2* Relative economic importance of component
traits in German sheep production (after Nitter and
Jacubec, 1970)

| Component trait | Profit per unit change in component trait (Deutschmarks) |
|---|---|
| No. lambs reared (per ewe and year) | 5.20 |
| Liveweight at optimal finish (kg) | 2.00 |
| Price (per kg lamb according to quality) (DM) | 1.95 |
| Post-weaning feed consumption (1000 SE/kg) | 1.74 |
| Ewes liveweight deviation (kg) | 1.56 |
| Pre-weaning feed consumption (1000 SE/kg) | 1.46 |
| Birthweight (kg) | 0.80 |
| Losses (birth to market) (%) | 0.69 |
| Wool (per ewe and year) (kg) | 0.63 |
| Price (per kg greasy wool) (DM) | 0.52 |
| Weaning weight (kg) | 0.05 |

Pease *et al.* (1967), Böckenhoff, Fewson and Bischoff
(1967) and Glodek (1969/1970). In Britain for seven
traits included in the MLC *Revised Combined-Test-Index*
(1972b) the economic weights given in *Table 3.3* are
used.

For Germany the weights in *Table 3.4* of Fewson,
Böckenhoff and Bischoff (1967) and Glodek (1969/1970)
may indicate representative ranges.

These figures show that in pigs litter performance
is not as dominating as in cattle and sheep, but it
assumes about equal relative economic merit to
fattening and carcass performance.

*Table 3.3* Relative economic weights of MLC-Combined-Test-Index for pigs in Britain (after MLC, 1972b)

| Trait | Value in pence | |
|---|---|---|
| | per unit | per $\sigma_P$ |
| Daily liveweight gain (lb) | 120 | 14.4 |
| Feed conversion ratio | 280 | 56.0 |
| Killing out percentage (%) | 31 | 56.0 |
| Trimming percentage (%) | 21 | 14.0 |
| Loin and shoulder (%) | 25 | 55.5 |
| Eye muscle area (cm$^2$) | 3 | 11.8 |
| EEL value (quality) (Pts) | 1 | 8.5 |

*Table 3.4* Relative economic importance for pig production traits in Germany (after Fewson *et al.*, 1967; Glodek, 1969/1970)

| Trait | Value per $\sigma_P$ in DM | |
|---|---|---|
| | Fewson et al. | Glodek |
| No. pigs weaned per sow | 12.00 | 14.58 |
| Daily liveweight gain (g) | 1.22 | 2.40 |
| Feed conversion ratio | 8.10 | 8.75/15.52 |
| Backfat thickness (cm) | – | 4.80/10.45 |
| Lean:fat ratio (loin) (1:...) | – | 6.55 |
| % Ham, loin, shoulder | 12.00 | 9.40 |
| Göfo value (quality) (Pts) | – | 2.37 |

[1]Expected prices of 1977/78 assumed.

[2]Higher figures for supposedly higher $\sigma_P$ in commercial herds.

# GENETIC VARIABILITY FOR COMPONENT TRAITS
# OF MEAT PRODUCTION

## DUAL-PURPOSE CATTLE

The number of publications in this field is very large
particularly in European countries but recently American
workers appear to have become interested in dual-purpose
cattle. Only recent investigations with sufficient data
to provide reliable estimates will be used. More complete
summaries have been given by Preston and Willis (1970),
Glaner (1970), Haring (1972) and Langholz (1974).

As far as important reproductive traits of dual-purpose
as well as beef cows are concerned one can follow Preston
and Willis (1970) when they state that heritabilities are
too low to justify intra-population selection programmes.
These findings have been confirmed by Cloppenburg (1966)
and Baptist and Gravert (1973) in the German 'Schwarzbunte'
(SB) breed, when they estimated heritabilities between 0
and 0.05 for various traits connected with regular
fertility of cows.

Milk and butter-fat performance are only of indirect
importance here because they may affect meat production
through their genetic relationship with growth and
carcass characters. For the main German breeds
'Schwarzbunte' (SB), 'Rotbunte' (RB) and 'Fleckvieh'
(FV), Langholz (1974) has summarised four recent studies
and his estimates of heritabilities and percentage
additive genetic standard deviation are given together
with the findings of Mason, Vial and Thompson (1972) for
British Friesian cattle in *Table 3.5*.

From these results all growth and efficiency characters
show about 5% additive standard deviation and medium to
high heritabilities so that intra-population selection
should be quite useful. Of the carcass traits, however,
only internal fat weights but neither percentage of lean
cuts nor dressing percentage show sufficient genetic
properties to be used as selection criteria. Much better
parameters with values of 5-15% $\sigma_A$ and heritabilities of
0.3-0.6 were found in many studies for measurements and
subjective scores on live animals and carcasses. However,
the correlations with real carcass values are usually too
unreliable to qualify as selection criteria.

Detailed genetic studies on meat quality characters
were published by Gravert (1963) on 24 progeny groups of

Table 3.5 Estimates of heritability and genetic standard deviation for meat production traits in German dual-purpose cattle (after Langholz, 1974)

| Breed (source) trait | I (SB) | | II (SB) | | III (SB) | | III (RB) | | IV (FV) | | V (BF) | |
|---|---|---|---|---|---|---|---|---|---|---|---|---|
| | $\sigma_A$ % | $h^2$ | $\sigma_A$ % | $h^2$ | $\sigma_A$ % | $h^2$ | $\sigma_A$ % | $h^2$ | $\sigma_A$ % | $h^2$ | $\sigma_A$ % | $h^2$ |
| Daily liveweight gain | 4.6 | 0.45 | 4.5 | 0.57 | 4.7 | 0.52 | 4.6 | 0.63 | 6.4 | 0.59 | 3.2 | 0.34 |
| Daily carcass gain | 5.9 | 0.66 | 5.0 | 0.56 | 4.4 | 0.46 | 5.0 | 0.66 | 4.2 | 0.33 | 3.2 | 0.18 |
| KSTE/kg liveweight gain | 6.2 | 0.68 | 4.0 | 0.40 | 3.7 | 0.20 | 5.3 | 0.45 | - | - | - | - |
| Dressing percentage | - | - | 1.7 | 0.42 | 1.2 | 0.32 | 1.6 | 0.51 | 2.2 | 0.88 | 2.0 | 0.53[1] |
| % Primal cuts | - | - | 1.3 | 0.32 | 1.3 | 0.26 | 1.6 | 0.46 | 1.2 | 0.22 | 5.4 | 0.58[1] |
| Weight internal fat | 13.1 | 0.34 | 16.5 | 0.63 | 12.2 | 0.23 | 16.2 | 0.48 | 32.5 | 0.80 | 10.8 | 0.46 |
| Weight of round | - | - | 2.9 | 0.54 | - | - | - | - | 2.3 | 0.44 | - | - |

I. Langlet et al. (1967): bull progeny groups 28 days to 350 kg liveweight
II. Langholz and Jongeling (1972): 77 bull progeny groups 28 days to 350 kg liveweight
III. Trappmann (1972): 100 bull progeny groups 70-450 kg liveweight
IV. Kräusslich et al. (1970): 34 bull progeny groups 112-500 days
V. Mason et al. (1972): 32-53 bull progeny groups, 390-609 steers to 18 months

[1] % muscle and fat in rib joint dissection.

the SB breed. Glaner (1970) later reported on 40
progeny groups of the RB breed and some of his figures
are given in *Table 3.6*.

*Table 3.6* Estimates of heritability and genetic
standard deviation for meat quality traits in the German
RB breed: 383 bulls of 350 kg (after Glaner, 1970)

| Trait | $\sigma_A$ % | $h^2$ | $\pm$ | $s_{h^2}$ |
|---|---|---|---|---|
| Dry matter in meat (%) | 0.7 | 0.05 | $\pm$ | 0.10 |
| Fat in meat (%) | 28.6 | 0.41 | $\pm$ | 0.16 |
| Protein in meat (%) | 1.8 | 0.13 | $\pm$ | 0.12 |
| Fat distribution | 27.6 | 0.43 | $\pm$ | 0.16 |

BEEF CATTLE

Summaries of older estimates of heritability are given
by Preston and Willis (1970) and median values from
there have been extracted for the MLC report (1971).
Reproductive traits have already been cited and new
results by Dearborn *et al.* (1973), including conception
rates and foetal mortality, fit into the general picture
of very low heritability estimates. The MLC median
values together with some recent estimates are given in
*Table 3.7*.
Although figures presented in *Table 3.7* sometimes
conflict and are often based on too few data, one could
draw general conclusions very similar to those for dual-
purpose cattle. Selection for the various growth
characteristics and internal or subcutaneous fat
measurements will be efficient and if meat quality
becomes important then marbling, firmness and colour
could be used. Selection for fertility in beef cattle
is not advisable and the possibility of utilising
heterotic effects must be considered.

SHEEP

A summary of heritabilities and repeatabilities for
various reproductive traits in sheep has been given by
Turner (1969). She demonstrated that sufficient
heritabilities exist only for lifetime or repeated

Table 3.7 Estimates of heritability and additive genetic standard deviation for beef production traits in the literature

| Trait | I $h^2$ (Median) | II $\sigma_A$ % | II $h^2$ | III $\sigma_A$ % | III $h^2$ | IV $\sigma_A$ % | IV $h^2$ |
|---|---|---|---|---|---|---|---|
| Daily liveweight gain | 0.5 | - | - | | | 6.4 | 0.55 |
| Adjusted live/carcass weight | 0.5 | 7.6 | 0.56 | | | 6.6 | 0.85 |
| Cutability percentage | 0.3 | 1.6 | 0.35 | 7.3 | 0.53 | 2.0 | 0.66 |
| Subcutaneous fat thickness | 0.4 | 24.5 | 0.51 | | | | |
| Percentage kidney fat | 0.6 | | | 13.6 | 0.50 | 10.5 | 0.39 |
| L. dorsi area | 0.4 | 5.1 | 0.32 | 7.4 | 0.40 | 4.4 | 0.25 |
| Meat colour | 0.2 | | | | | 6.5 | 0.19 |
| Marbling | 0.3 | 25.0 | 0.33 | | | 10.2 | 0.31 |
| Conformation score | | | | | | 7.6 | 0.45 |

I. MLC (1971) and Preston-Willis (1970)
II. Cundiff et al. (1970): 75 sire groups, 503 steers, 3 breeds
III. Brackelsberg et al. (1970): 46 sire groups, 247 steers, 3 breeds
IV. Dinkel et al. (1973): 70 sire groups, 679 steers, 1 breed

records, whereas first or one year records, or traits,
like failure to lamb etc., show only very low genetic
variation. Growth and carcass traits have been more
frequently analysed and a detailed discussion of results
with particular respect to British conditions was
published by Bowman (1966). The MLC report (1972a) has
put together 'probable value' ranges for the important
characteristics which are given together with some
recent estimates in *Table 3.8*.

In accord with the estimates of Ercanbrack and Price
(1972) for the fattening performance of four breeds of
lambs on grazing or concentrate feeding, *Table 3.8* shows
genetic standard deviations of 5-10%. The information
on carcass and meat quality traits, however, is very
limited and once more fat measurements on the carcass
show high, but weight of meat cuts very low genetic
variation.


PIGS

Since selection indices are in common use in Britain as
well as the rest of Europe and in the USA, the traits
which are included in such indices will be discussed.

The first obvious fact in European indices is that
usually no litter production traits are included, which
is different from the situation under the more extensive
production schemes in the USA. Considering the
economic importance of litter production this can only
be explained by the very poor additive genetic properties
of such traits (Glodek, 1970b; Strang and King, 1970).
There would probably be some potential in using lifetime litter production records (Biedermann, 1971) but
competitive breeding programmes could hardly afford the
prolonged generation intervals that would be caused.

No doubt seems to remain in European countries that
growth rate and/or efficiency of feed utilisation,
carcass composition (e.g. amount of lean cuts and lean:
fat ratio) as well as specific meat quality have to be
included in a present day selection index.

Slight differences between countries do exist, however,
in the preference of traits to measure carcass
composition and meat quality. In *Table 3.9* some
parameters for components of the aggregate breeding
values of current selection indices are given.

Table 3.8 Estimates of heritability and genetic standard deviation for growth, carcass and wool characters in sheep

| Source | I $h^2$ | II $h^2$ | III $\sigma_A$ % | III $h^2$ | IV $\sigma_A$ % | IV $h^2$ |
|---|---|---|---|---|---|---|
| Final weight | 0.15-0.55 | 0.11 | 5.7 | 0.53 | 2.8 | 0.43 |
| Daily gain | 0.10-0.30 | 0.26 | 13.8 | 0.55 | 4.3 | 0.32 |
| Eye muscle area | – | 0.14 | – | – | 7.0 | 0.70 |
| Carcass conformation | 0.25-0.35 | – | 7.1 | 0.16 | – | – |
| % Loin | – | 0.32 | – | >1 | – | – |
| % Leg | – | 0.16 | 6.2 | 0.25 | – | – |
| % Lean cuts | – | – | – | – | 2.0 | 0.68 |
| % Kidney fat | – | – | – | – | 17.5 | 0.62 |
| Fleece weight | 0.30-0.50 | – | 4.9 | 0.21 | – | – |

I. MLC-Scientific Study Group Report (1972a)
II. Bowman and Hendy (1972): 18 sire groups, 360 Dorset Horn castrated male progeny
III. Weniger et al. (1968): 96-112 sire groups, 1102-1510 lambs, Merino
IV. Bradford and Spurlock (1971): 10 sire groups, 167 $F_1$ lambs, Suffolk x Corriedale

Table 3.9 Genetic parameters used in aggregate breeding values of selection indices for pigs in European countries

| Trait | I | | II | | III | | IV | |
|---|---|---|---|---|---|---|---|---|
| | $\sigma_A$ % | $h^2$ | $\sigma_A$ % | $h^2$ | $\sigma_A$ % | $h^2$ | $\sigma_A$ % | $h^2$ |
| Daily gain | 2.5 | 0.18 | 4.3 | 0.45 | 6.8 | 0.46 | 4.7 | 0.31 |
| Feed efficiency | 4.5 | 0.33 | 4.5 | 0.52 | 4.6 | 0.47 | 5.8 | 0.56 |
| Backfat thickness | | | 8.3 | 0.52 | 7.2 | 0.36 | 7.9 | 0.42 |
| Loin eye area | 4.1 | 0.36 | 7.8 | 0.56 | 7.2 | 0.59 | 6.4 | 0.41 |
| Lean:fat ratio | | 0.39 | – | | 14.4 | 0.30 | 4.5 | 0.28 |
| Ham weight % | | | | | 0.8 | 0.05 | – | |
| Lean cuts % | 5.0 | 0.51 | | | – | – | – | |
| Trimming % | 4.1 | 0.25 | | | – | – | – | |

I. MLC (1972b):Combined test results Large White and Landrace 1972/73
II. Flock (1968): 17,000 German Landrace test pigs 1960-67
III. Pfleiderer (1973): 10,800 German Landrace test pigs 1968-70
IV. Glodek (1970b):5,700 German Landrace test pigs 1964-67

## IMPORTANT GENETIC CORRELATIONS AMONG
## COMPONENT TRAITS OF MEAT PRODUCTION

It is not intended to present complete genetic
correlation matrices for all traits and species described.
Since all favourable or non-significant correlations do
not affect intrapopulation selection procedures to a
large extent, only significant unfavourable genetic
correlations among economically important traits in the
chosen species will be discussed.

### DUAL-PURPOSE CATTLE

The most important relationship for dual-purpose cattle
is the genetic correlation between milk yield and beef
production traits. For economic reasons selection has
to concentrate primarily upon milk production and any
unfavourable effects upon amount and quality of the
secondary product, beef, must largely be tolerated. If
this becomes impossible, the end of dual-purpose cattle
would be the consequence. Although this situation may
eventually be reached with further genetic progress in
selection for milk production, at present, most analyses
have shown non-significant or favourable genetic
correlations among milk and beef production traits
(*Table 3.10*).

The same authors reported generally favourable geneti
correlations between growth characters and carcass
composition. Correlations between growth rate and
amount of lean meat to such meat quality characters as
dry matter, water-binding capacity, fibre diameter, pH
value and others tend to be slightly unfavourable,
although at the moment no real meat quality problems,
as exist in pigs, are known in dual-purpose cattle
(Gravert, 1963; Herbst, 1964; Glaner, 1970). The very
important relationship between fertility and milk or
meat production is not sufficiently analysed but when
fertility traits show very small genetic variability,
they are not expected to exhibit strong genetic
correlations with other traits. However, as very high
specific performance (milk or growth) could weaken the
animals health and adaptability under intensive
management conditions it could be that liveability and
reproductive fitness will be affected without any direct
genetic relationship (Dannenberg, 1967; Weseloh, 1968).

*Table 3.10* Genetic correlations among milk and meat production traits in dual-purpose cattle

| Combination of traits | $r_g$ | Source | Amount of data |
|---|---|---|---|
| kg milk: gain, feed efficiency | 0.11 ± 0.20 | I | 354 mother/son pairs (Schwarzbunte) |
| kg milk: meatiness score | 0.12 ± 0.23 | I | |
| kg milk: fattening traits | 0.41 ± 0.25 | I | 37 sire progeny groups |
| kg milk: carcass traits | 0.48 ± 0.28 | I | with 344 sons (Schwarzbunte) |
| kg butterfat: carcass weight gain | 0.44 | II | 63 sire progeny groups of the |
| kg butterfat: feed efficiency (STE) | -0.37 | II | German Schwarzbunte |
| kg butterfat: % primal cuts | 0.24 | II | |
| kg butterfat: gain 365-500 days | 0.37 | III | 65 sire progeny groups (Fleckvieh) |
| kg milk: gain 6-12 month | -0.47 | IV | 53 sire progeny groups with |
| kg milk: fleshing index | -0.59 | IV | 609 steers |
| kg milk: % lean in rib | -0.23 | IV | British Friesian |

I. Langlet *et al.* (1967)
II. Langholz and Jongeling (1972)
III. Kräusslich *et al.* (1973)
IV. Mason *et al.* (1972)

BEEF CATTLE

In beef cattle most correlations between growth and
carcass characteristics are favourable and fairly close,
so that selection for growth rate also improves lean
meat production.  Particular meat quality problems are
of minor importance, although real breed differences are
known in growth capacity and beef quality.

Selection for carcass desirability is very often based
upon conformation scores and, at least in some breeds,
has led to problems in regular reproduction and meat
quality.  Of particular interest in this respect are
calving difficulties in heavy muscled breeds, where
double muscled calves are frequent (Vissac, 1971).

SHEEP

Again reproduction and meat performance are the main
variables and there are large breed differences in both
traits but within breeds Turner (1969) reports
consistently positive genetic correlations between body
weight and reproductive rate.  An open question in this
respect is the effect of feed consumption of the ewe
and the conversion of feed into lambs by ewes of varying
body weight.  Somewhat more conflicting but generally
slightly negative are the correlations between fleece
weight and reproduction in different breeds.  Although
no serious problems are expected in the near future more
information is clearly needed.

PIGS

In highly selected meat-type pig populations under
intensive production regimes several unfavourable
correlations were detected in recent years.  Pig breeders
in Western Europe have increased the amount of lean in
the carcass so rapidly that the most extreme purebred
stock can now hardly be called well adapted to even
normal production conditions.  The consequences are
increased losses due mainly to leg weakness and heart
failure, serious meat quality problems as indicated in
rising PSE frequency and more reproductive problems in
males and females.  Rapid selection for such traits as
loin eye area and body conformation may also result in

decreased growth rate similar to the situation in typical
sire breeds such as Pietrain or Belgian Landrace.

Unfortunately litter performance records from pedigree
breeders do not allow calculations of genetic correlations
with growth and carcass data because they include only
litters above a minimum size, and no figure is known
about the litters below that minimum. However, practical
breeders admit rising culling rates among gilts and
increasing complaints about boars with poor sexual
activity and fertility. Investigations of Teuscher
(1972) have clearly shown that incidence of leg weakness
is correlated with weight of loin and particularly ham,
and it is probable that the favoured ham shape even more
than ham weight is the main obstacle. Ham weight is a
poor selection criterion because it seems to be difficult
to measure accurately on large numbers and on the other
hand appears to have little to do with what is paid for
in European markets, namely ham shape. However, ham
shape can be judged even on the live animal and has given
better heritabilities than weights taken in our routine
sib testing.

Relations between amount of lean meat and its quality
are being tested, at least in all Landrace populations
and here again the worst findings are between loin eye
area and meat quality, which is usually measured at the
loin. It does not seem to be important which criterion
(pH, water-binding capacity, EEL, Göfo etc.) is taken
(Weniger, Steinhauf and Glodek, 1970) as all have their
disadvantages but all express about the same correlations.
In the *Table 3.11* some genetic correlations estimated
recently are given. In a recent publication Richter,
Flock and Bickhardt (1973) calculated a genetic
correlation of $r = -0.60 \pm 0.22$ between ultrasonic backfat
thickness of boars and their logarithmic creatine
kinase values (LCK), determined in blood samples taken
after a standardised physical exercise. The genetic
correlation between liveweight gain and LCK values was
found to be $-0.17 \pm 0.33$. Although the data are few
and the repeatability of the method (0.68 to 0.81) is
fairly low, it can be applied in testing live animals
and therefore deserves a lot of interest.

Table 3.11 Genetic correlations among meat quality and other production characteristics in European selection indices in pigs

| | I | II | III | IV | V | VI |
|---|---|---|---|---|---|---|
| Daily gain | 0.10 | 0.00 | | -0.30 | -0.20 | -0.04 |
| Feed efficiency | -0.17 | 0.00 | 0.27 | 0.28 | 0.28 | -0.05 |
| Backfat thickness | | 0.20 | | 0.00 | 0.18 | |
| Loin eye area | 0.02 | -0.25 | -0.50 | -0.30 | -0.56 | -0.43 |
| Lean:fat ratio | | 0.25 | 0.21 | 0.19 | 0.46 | 0.47 |
| Lean % (R, B) | 0.12 | | | | | |
| Ham % | | | -0.20 | | | |

I. MLC sib index (EEL value)
II. Averdunk (1974) unpublished results (Göfo value)
III. Breloh and Schmitten (1970), (subjective score 1-5)
IV. Pfleiderer (1974), (Göfo value)
V. Flock (1973), (Göfo value)
VI. Glodek, cited in Weniger et al. (1970), (subjective score 1-5)

SOME GENERAL CONCLUSIONS ABOUT APPROPRIATE
BREEDING METHODS FOR GENETIC IMPROVEMENT
OF MEAT PRODUCTION

In all simple cases with sufficient genetic variability,
selection within purebred or crossbred populations will
be the first choice. It can be direct or indirect
selection, it usually will have to be index selection
because more than one trait must be considered. It
always requires performance testing which can be done on
the breeding animal or its relatives or both as a basis
for estimating the animals breeding value.

In more complicated cases, where very little additive
genetic variance is present or where high unfavourable
correlations with other important components of overall
performance exist, various crossbreeding methods can be
tried in order to utilise non-additive genetic variance
as well.

INTRA-POPULATION SELECTION (IPS)

All traits with intermediate and high heritabilities can
be improved by IPS, and this includes all growth and
feed efficiency characters, and many carcass composition
and some meat quality traits of the three species
discussed here. None of the important fertility,
reproductive and liveability characters in the three
species can be improved with competitive speed by such
methods. As these traits are of major importance in
all species IPS alone cannot be the best breeding method
for any of them. An exception for the time being may be
dual-purpose cattle, because in this case milk production is of such overwhelming importance and generation
interval is so long that for a good time to come IPS
will be required to guarantee a large proportion of
possible genetic progress. In all other species however,
IPS will be used most efficiently as a method to improve
parent or grandparent populations which then are taken
to produce the commercial crossbreds with maximum overall performance.

In *dual-purpose cattle* selection programmes based on
average daughter milk yield of the bulls and bull dam's
own milk performance are in common use. Improvement of
beef production is usually incorporated as individual

mass selection of young potential test bulls from
élïte matings for their own growth rate and sometimes
body composition judged subjectively or measured by
ultrasonics or X-ray methods.  Selection must not be
too strong in order not to harm progress in milk
production.  Most schemes also involve a progeny test
of proven élite bulls for growth and carcass performance
of son groups, but the cost of this is usually higher
than the value of additional progress made.  It might
serve some purpose to watch how genetic correlations
between milk and meat production and quality are
changing with the rising milk production.  It will alsc
provide accurate growth and carcass information for
breed comparisons.

In *beef cattle* production under European conditions
IPS is a useful tool for improving growth rate and
carcass desirability of pure breeds, whereas all productivity traits of the dams, apart from some cow culling,
should be improved by crossbreeding.  As long as
genotype/environment interactions are not important, an
intensive station performance test of the best young
bulls, from weaning (at the latest) to the breed-
specific market weight or age with an assessment of live
fat measurement and conformation judgement, will
probably offer the best selection intensities.  Females
can be selected within herds according to a less accurate
field index just before their first mating and the dam's
reproductive performance may be included.  *Elite* bulls
can also be picked after the first progeny generation
has been recorded in the herd but reliable progeny tests
can only be achieved in large herds with many bulls per
generation or with artificial insemination.  If, in the
future, meat quality problems have to be incorporated,
some sort of progeny or sib test with detailed carcass
evaluation may be necessary.

In *sheep* the general situation is very similar to
that in beef cattle but as quite a high proportion of
sheep is kept on marginal pastures where cattle cannot
live, specialisation of populations is much greater.
This leads to more differentiated selection criteria
and, with genotype/environment interactions being more
important, the typical British type of stratification
schemes develop.  In marginal hill areas nature leaves
very little for artificial mass selection, whereas in
intensively kept ram breeds, growth and carcass traits
are of major importance and can be best improved by

mass selection among ram lambs based on station
performance test records as with beef bulls. Wool
characteristics have so little economic worth that
within-herd field records provide enough information
for the justified culling.

In *pigs*, selection schemes are more complicated
because carcass composition is of relatively higher
economic importance and meat quality problems are
already far too great to be ignored in selection.
Nevertheless in Britain and the rest of Europe mass
selection among young boars and gilts on their own
performance index which includes gain and efficiency
as well as ultrasonic fat and sometimes muscle measure
ments has brought about the greatest improvement and will
do so for some generations. In Britain, the MLC Combined
Test provides very accurate information on the boars own
breeding value and detailed carcass and meat quality
information on their sibs. In the rest of Europe, station
testing of boars has not yet become very popular although
it has often been recommended (Hartmann and Fewson, 1967).
The main objection has been the large disease risk if
boars are returned to breeding herds after test. This
can be avoided if boars are reared at home and only sibs,
to be slaughtered after test, go to the station.
Although slightly less accurate than with station
testing of boars, selection indices have been developed
(Glodek, 1970a) to include weight for age, ultrasonic
backfat thickness of the boar and efficiency and carcass
information of his sibs. In this way station capacity
can be utilised very effectively and within-herd gilt
selection could also gain from station records of sibs.
A very serious problem is that some breeders still use
such station records as a progeny test of their herd
boars and sows and used in that way it becomes very
misleading information because most sib groups come
from deliberately selected matings. With a good sib
testing scheme no progeny tests are required the only
exception could be for the top class of AI boars which
are permitted unlimited insemination. If methods to
measure meat quality and stress susceptibility on live
animals, such as described by Richter, Flock and
Bickhardt (1973), become practicable for large scale
use, and if health control schemes can be improved
markedly, then boar station tests will be boosted
considerably because they also provide the great
advantage of applying automatic selection pressure upon

leg weakness problems.  At present individual pedigree
breeders tend to avoid carefully any losses when they
rear young boars at home by providing unusually
comfortable conditions which are not generally met in
practice, so that these boars are then a problem in
commercial herds.


COMMERCIAL CROSSBREEDING (CC)

There are three main advantages of commercial crossbreeding.  It can:

1. Utilise specialised sire and dam populations as
   particularly Smith (1964) and Moav (1966) have
   shown in theory and application to various species.
2. Utilise considerable heterosis effects in low
   heritability traits such as reproduction,
   liveability and early growth.
3. Make selection criteria within specialised parent
   populations less complicated and thereby more
   effective than in completely purebred populations.

Certain present and likely future applications of CC in
the three species are reviewed here and references are
given to the literature which provides more detailed
information.

In *dual-purpose cattle*, commercial crossbreeding is
not at all common in Europe because milk production is
the dominating trait and heterosis effects are not
large enough (Pearson, Lucia and McDowell, 1968) to
compensate for the more complicated organisational
situation that would be required.  However, CC has a
large potential for beef production from the dairy herd
in that cows not good enough to breed purebred replacement stock can be inseminated by beef bulls to produce
better beef calves.  In Britain, this is already a very
practical scheme where as many purebred Friesian and
Ayrshire cows are mated with beef bulls as with
their respective dairy bulls.  A major breakthrough
for the future could be reached when sex
determination through biotechnical manipulations
in sperm becomes practicable.  On the other hand, the use
of CC for utilising direct heterosis effects is still
unlikely (Mason, 1966) and will only become attractive
when extreme milk production levels lead to serious

reproductive and liveability problems under normal intensive management conditions.

In *beef cattle*, commercial crossbreeding seems to offer much more because the dominating aim, (weaning weight per cow and per year) contains various heterotic component traits. The amount of heterosis varies very much among traits and breeds as the summaries of Mason (1966) and Preston and Willis (1970) show. While growth heterosis does not seem to be competitive with the effects of using dual-purpose crossing partners (Pahnisch *et al.* 1971), Mason reports 18-25% heterosis for the ratio of calves weaned to cows mated. How much specific combining ability could be expected from the generally small experiments is still disputed. Lasley *et al.* (1973) found between 0 and 6.8%, whereas McDonald and Turner (1972) in crosses including Brahman Zebus calculated 84.8% for general, 6.6% for maternal and 8.6% for specific combining effects in weaning weight. Crossbred cows seem to be advantageous but as specific crosses provide organisational problems, rotational crossbreeding systems are preferred although they do not utilise differential breed effects or maximum heterosis. Whether large co-operative breeding schemes or new biotechnical developments in oestrus synchronisation and artificial insemination could change this in the future, will depend on accurate experiments to show what gains could be made.

In *sheep*, as the first step for commercial cross-breeding, an accurate breed evaluation is in progress at many centres (Dickerson *et al.* 1972; Sidwell and Miller, 1971; MLC, 1972a). From such data test cross experiments can be designed in the way Nitter and Jacubec (1970) have suggested and they will show the relative importance of general, maternal and specific combining ability under various management conditions. Traditional production schemes in Britain and the USA, and experimental results in south Germany clearly demonstrate that specific breed or line crosses will have a future in intensive lamb production. The very large differentiation among existing breeds in additive performance level calls for specific crosses and excludes rotation schemes in order to maximise profits. Information on heterosis effects is limited and conflicting but purebreeding will only have an opportunity for improving parent or grand parent populations.

In *pigs*, various types of commercial 'hybrids' are
already on the market and it is predicted that their
development will follow that of broilers.  For such
predictions astonishingly little is known about the
relative importance of general, maternal and specific
combining effects.  The classical American experiments
of Henderson (1949) and Hetzer *et al.* (1961) had
insufficient data and conflicting results.  Worldwide
literature reviews on heterosis effects in pigs (Glodek.
1970a; Sellier, 1970; Kalm, 1973), however, indicate
average figures in litter performance of the order of
10-15% and around 5% in growth characteristics, whereas
carcass quality is not affected.  There is also some
very interesting information in recent experiments
(e.g. Glodek, 1973) that stress susceptibility and meat
quality are improved markedly in crossbred pigs.  Our
own experience with a large test cross programme
including nine European and American breeds also
indicated that, apart from heterosis, a good deal of
additive combination effects can be utilised if one
decides to produce specific crosses.  At the present
time most competitors in the hybrid market quite
justifiably seem to concentrate upon these because no
one will be able to claim that his 'lines' are already
differentiated and homogeneous (that is highly selected)
enough to attack the most specific hybrid effects with
a great chance of success.  Findings of some American
experiments (Stanislaw *et al.* 1967; Louca and Robinson,
1967; Wong, Boylan and Rempel, 1971) should warn us not
to start with very sophisticated methods of heterosis
utilisation (e.g. reciprocal recurrent selection) before
enough genetic differentiation among parent populations
has been reached and mass selection response within
parent populations has become diminishing.  On the
other hand it has been shown that even in many classical
pedigree purebreeding schemes in Europe fairly quick
genetic progress can be made in highly heritable traits
by very simple mass selection and that correlated
complications follow just as quickly.  From a genetic
point of view I am therefore firmly convinced that a
large percentage of our present so called hybrid
programmes on the pig market do not have sound enough
foundations to survive the next 5-10 pig generations
until the real hybrid competition starts.  I am also
aware of the fact that many non-genetic factors influence
the survival of pig breeding schemes, but to discuss
these would have exceeded the scope of this paper.

ACKNOWLEDGEMENTS

I wish to acknowledge the help of Mr. D.E. Steane and
Mr. A.J. Kempster of the Meat and Livestock Commission
for reading the manuscript and making valuable
suggestions.

REFERENCES

ADELHELM, R., BÖCKENHOFF, E., BISCHOFF, T., FEWSON, D.
and RITTLER, A. (1972). Die Leistungsmerkmale beim
Rind - wirtschaftliche Bedeutung und Selektions-
würdigkeit, Hohenheimer Arbeiten 64, Ulmer Stuttgart.
AVERDUNK, G. (1974). Revised Selection Index in
Bavaria, personal communication
BAPTIST, R. and GRAVERT, H.O. (1973). Züchtungskunde,
45, 399
BIEDERMANN, G. (1971). Züchtungskunde, 43, 346
BÖCKENHOFF, E., FEWSON, D. and BISCHOFF, T. (1967).
Züchtungskunde, 39, 270
BOWMAN, J.C. (1966). 'Meat from Sheep', Anim. Breeding
Abstr., 34, No. 3, 1
BOWMAN, J.C. and HENDY, L.R.C. (1972). Anim. Prod., 14,
189
BRACKELSBERG, P.O., KLINE, E.A., WILLHAM, R.L. and
HAZEL, L.N. (1971). J. Anim. Sci., 33, 13
BRADFORD, G.E. and SPURLOCK, G.M. (1972). J. Anim. Sci.,
34, 737
BRELOH, B. and SCHMITTEN, F. (1970). Züchtungskunde,
42, 93
CLOPPENBURG, R. (1966). 'Geburtsverlauf bei Nachkommen
von schwarzbunten Bullen einer westfälischen
Besamungsstation,' Dissertation, University of
Göttingen
CUNDIFF, L.V., GREGORY, K.E., KOCH, R.M. and DICKERSON,
G.E. (1971). J. Anim. Sci., 33, 550
DANNENBERG, K. (1967). 'Ist die Konzeptionsbereitschaft
beim Rind leistungsabhängig?' Dissertation, University
of Göttingen
DEARBORN, D.D., KOCH, R.M., CUNDIFF, L.V., GREGORY, K.E.
and DICKERSON, G.E. (1973). J. Anim. Sci., 36, 1032
DICKERSON, G.E., GLIMP, H.H., TUMA, H.J. and GREGORY,
K.E. (1972). J. Anim. Sci., 34, 940

DINKEL, C.A. and BUSH, D.A. (1973). *J. Anim. Sci.*, 36, 832

ERCANBRACK, S.K. and PRICE, D.A. (1972). *J. Anim. Sci.*, 34, 713

FEWSON, D., BOCKENHOFF, E. and BISCHOFF, T. (1967). *Züchtungskunde*, 39, 324

GJEDREM, T. (1966). *Acta Agric. Scand.*, 16, 21

GLANER, H.D. (1970). 'Untersuchungen über die Vererbung der Fleischleistung bei Rindern der Rasse Deutsche Rotbunte in der Nachkommen prüfstation Futterkamp/ Schleswig-Holstein', Dissertation, University of Kiel

GLODEK, P. (1969). *Züchtungskunde*, 41, 174

GLODEK, P. (1969/1970). *Zübiol.*, 86, 127 and 273

GLODEK, P. (1970a). 'Konstruktion und Eigenschaften von Selektionsindices für mehrere Merkmale und Informationsquellen', Habilitationsschrift, University of Göttingen

GLODEK, P. (1970b). 'Neue genetische Parameter für Zuchtleistungseigenschaften beim Schwein', *DGfZ-GfT-Jahrestagung*, 61

GLODEK, P. (1973). 'Halbzeitergebnisse im Bundeshybridzuchtprogramm', *SuS*, 21, 332

GRAVERT, H.O. (1962/1963). *Zübiol.*, 78, 43 and 139

HARING, H.J.F. (1972). 'Zuchtplanung in der Rinderzucht aus ökonomischer Sicht', Dissertation, University of Göttingen

HARTMANN, W. and FEWSON, D. (1967). *Züchtungskunde*, 39, 1

HENDERSON, C.R.H. (1949). 'Estimation of general, specific and maternal combining abilities in crosses among inbred lines of swine', Dissertation, Iowa State College Library

HERBST, K. (1964). 'Die Prüfung der Fleischbeschaffenheit an Nachkommengruppen des schwarzbunten Rindes', Dissertation, University of Göttingen

HETZER, H.O., COMSTOCK, R.E., ZELLER, J.H., HINER, R.L. and HARVEY, W.R. (1961). *USDA Techn. Bull.*, 1237

KALM, E. (1973). 'Bisherige Ergebnisse aus Gebrauchskreuzungsversuchen, *Proc. Pig Breeding Workshop, Göttingen*, 89

KRÄUSSLICH, H., AVERDUNK, G., GOTTSCHALK, Al, SCHMITTER, W., SCHUMANN, H. and SCHWARZ, E. (1970). *Bay. Ldw. Jb.*, 47, 1

LANGHOLZ, H.J. (1970). *Züchtungskunde*, 42, 454

LANGHOLZ, H.J. (1974). 'Die Nachkommenprufung auf
Fleischleistung beim Zweinutzungsrind in der
Bundesrepublik Deutschland', *Proc. 1st World Congr.
Appl. Gen., Madrid*
LANGHOLZ, H.J. and JONGELING, C. (1972). *Züchtungskunde,*
__44__, 368
LANGLET, J., GRAVERT, H.O. and ROSENHAHN, E. *Zübiol.,*
__83__, 358
LASLEY, I.F., SIBBIT, B., LANGFORD, L., COMFORT, J.E.,
DYER, A.J., KRAUSE, G.F. and HEDRICK, H.B. (1973).
*J. Anim. Sci.,* __36__, 1044
LINDHOLM, H.B. and STONAKER, H.H. (1957). *J. Anim. Sci.,*
__16__, 998
LOUCA, A. and ROBINSON, O.W. (1967). *J. Anim. Sci.,*
__26__, 67
MASON, I.L. (1966). 'Hybrid vigour in beef cattle',
*Anim. Breeding Abstr.,* __34__, No. 4, 28
MASON, I.L., VIAL, V.E. and THOMPSON, R. (1972).
*Anim. Prod.,* __14__, 135
MLC (1971). Scientific study group report on beef
production
MLC (1972a). Scientific study group report on sheep
production
MLC (1972b). Combined test selection index revision
1972, MLC-Stat. Dep., 3/72
MCDONALD, R.P. and TURNER, J.W. (1972). *J. Anim. Sci.,*
__35__, 1146
MOAV, R. (1966). *Anim. Prod.,* __8__, 193, 203 and 365
NITTER and JACUBEC, (1970). *Züchtungskunde,* __42__, 436
ÖZTAN, A. (1971). 'Untersuchungen zur Gewichtsentwicklung
beim Deutschen Merinolandschaf und schwarzköpfigen
Fleischschaf', Dissertation, University of Göttingen
PAHNISCH, O.F., KNAPP, B.W., URICK, J.J., BRINKS, J.S.
and WILSON, F.S. (1971). *J. Anim. Sci.,* __33__, 14
PEARSON, LUCIA and MCDOWELL, R.E. (1968). 'Crossbreeding
of dairy cattle in temperate zones: A review of recent
studies', *Anim. Breeding Abstr.,* __36__, No. 1, 1
PEASE, A.H.R., COOK, G.L., GREIG and CUTHBERTSON, M.A.
(1967). *Combined testing,* PIDA - Report DA 188
PRESTON, T.R. and WILLIS, M.B. (1970). *Intensive Beef
Production,* Pergamon Press, Oxford
RICHTER, L., FLOCK, D.K. and BICKHARDT, K. (1973).
*Züchtungskunde,* __45__, 429
SELLIER, P. (1970). *Ann. Genet. Sel. Anim.,* __2__, 145
SIDWELL, G.M. and MILLER, L.R. (1971). Production in
Some Breeds of Sheep and Their Crosses (I-III)

SMITH, C. (1964). *Anim. Prod.*, 6, 337
STANISLAW, C.M., OMTVEDT, I.T., WILLHAM, R.L. and
    WHATLEY, A.J. (1967). *J. Anim. Sci.*, 26, 16
STRANG, G.S. and KING, J.W.B. (1970). *Anim. Prod.*, 12,
    235
TEUSCHER, T. (1972). 'Untersuchungen uber das
    Beinschwächesyndrom an einer Zuchtpopulation des
    Hausschweines (Deutsche Landrasse)', Dissertation,
    University of Berlin
TRAPPMANN, W. (1972). *Züchtungskunde*, 44, 17
TURNER, Helen Newton (1969). 'Genetic improvement of
    reproduction rate in sheep', *Anim. Breeding Abstr.*,
    37, No. 4, 545
VISSAC (1971). 'L'hypertrophie musculaire d'origine
    genetique on caractere culard', *Proc. 22. EAAP - Conf.*,
    *Paris*
WENIGER, J.H., GLODEK, P., ASSADI-MOGHADDAM, R., SCHMIDT,
    L. and BURGHART, M. (1968). *Züchtungskunde*, 40, 34
WENIGER, J.H., STEINHAUF, D. and GLODEK, P. (1970).
    *Zübiol.*, 87, 230
WESELOH, H. (1968). 'Zusammenhänge zwischen
    Milchleistung, Fruchtbarkeit und Lebensdauer beim
    Rind', Dissertation, University of Göttingen
WONG, W.C., BOYLAN, W.J. and REMPEL, W.E. (1971).
    *J. Anim. Sci.*, 32, 605
ZEDDIES, J. (1974). Zur ökonomischen Bewertung, Planung
    und Beurteilung von Rinderzuchtprogrammen
    Habilitationsschrift, University of Göttingen

# 4

# NUTRITION OF FARM ANIMALS

H. SWAN and D. J. A. COLE

*Department of Agriculture and Horticulture,*
*University of Nottingham*

```
'From Christmas to May
weak cattle decay...'
                    Thomas Tusser
                    1524? - 1580
```

To state that nutrition is an important aspect of meat animal production is a grave understatement. The poet of Elizabethan rural life knew well the importance of nutrition in the growth and development of the meat animal. Robert Bakewell, of Dishley Grange who has been given credit by historians as the architect of the high producing animal, through his pioneer work on animal breeding, would have freely admitted his debt of gratitude to the Enclosure Act and the Norfolk four-course rotation. As with many historical legends the conclusion may well be wrong. The development of the Norfolk four-course rotation allowed, for the first time, a reasonable availability of animal feed between Christmas and May. Thus, adequate levels of nutrition were presented to farm animals. In response the average weight of cattle slaughtered at Smithfield Market increased by 48 per cent between 1724 and 1774.

In the United Kingdom, feed costs are the largest item in any cattle, sheep or pig production system. Clearly the relationship between feed cost and carcass value is of great importance. Given standards for carcass composition and quality, the aim must be to

minimise food costs commensurate with the optimum rate
of liveweight gain. In the longer term one further
constraint is clear, nutritional regimes must be
devised, which minimise the inclusion of animal and
vegetable materials, that are of adequate nutritional
value for direct human consumption in one form or
another. This means maximising forage and by-product
utilisation in ruminants and minimising the animal and
vegetable protein consumption of the pig. In order to
achieve this situation several questions must be
answered.

1. How does the nutrient density of the diet
   influence feed intake, efficiency of feed
   utilisation and carcass composition?
2. What are the minimum levels of supplementary
   protein required in the diet of meat animals.
3. Is there a basis for channelling available feed
   resources to the appropriate livestock types?

It is generally assumed that the differences between
pigs and ruminants result in widely different nutritional
patterns. For example, the voluntary feed intake
responses are often different but this may be a
reflection more of the diet than of inter-specific
differences between animals. When pigs and cattle are
offered diets of similar nutrient density (i.e. they do
not impose a physical limitation) a similar intake
response may be obtained. *Figure 4.1* indicates that
voluntary intakes of dry matter and digestible energy
per unit of metabolic size of pigs and cattle, are of
a similar pattern and similar order of magnitude.
Thus, cattle and pigs may fit the same general
relationship when fed similar diets. Cole, Hardy and
Lewis (1972) have suggested a schematic representation
of voluntary feed and digestible energy intake under
conditions of varying nutrient density (*Figure 4.2*) and
this might be extended to include cattle. However,
under typical production systems cattle are fed diets
of low nutrient density which impose a physical
limitation to intake while pigs are given diets of
higher nutrient density, the intake of which is
controlled physiologically under *ad libitum* feeding
conditions.

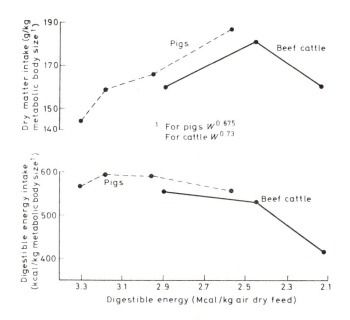

*Figure 4.1    Relationship between the voluntary intake
of food dry matter, digestible energy and nutrient
density of the diet in pigs (Cole, Duckworth and Holmes,
1967) and beef cattle (Swan and Lamming, 1974)*

## NUTRIENT INTAKE AND CARCASS COMPOSITION

With normal production systems there are apparent
differences in the deposition of tissues in pigs and
ruminants.  In considering carcass composition care must
be taken to ensure that dietary treatments are evaluated
fairly in terms of carcass parameters.  It is important
that comparisons are made within a breed and that
slaughter weights are chosen that lead to equal carcass
weights.  When these criteria are met for beef cattle
the conclusion can be drawn that within a range of
dietary digestible energy concentration of 3.48 Mcal
DE/kg DM to 2.57 Mcal DE/kg DM, digestible energy
concentration had no effect on carcass composition
(Swan and Lamming, 1974).

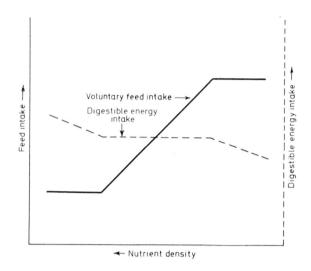

*Figure 4.2 Schematic representation of voluntary feed and digestible energy intake under conditions of varying nutrient density (Cole, Hardy and Lewis, 1972)*

As part of a large balance experiment involving 64 steers over the liveweight range 300–450 kg, three main treatments were selected to investigate the relationship between dietary energy concentration and carcass composition. Beef cattle were offered the diets shown in *Table 4.1*.

Analysis of the diets showed a large difference in apparently digestible energy and estimated metabolisable energy between the diets. This was reflected in differences in the dry matter intake and daily liveweight gain (*Table 4.2*).

Analysis of the carcass (*Table 4.2*) showed little variation in the chemical components as a result of dietary treatment. However, there was a significant correlation between the variation in chemical components of the carcass and carcass weight.

In this experiment it was evident that the major factor influencing carcass composition was carcass weight which was significantly related to carcass fatness. Carcass protein and ash were unaffected by crude protein, crude fibre or energy concentration in

*Table 4.1* Composition of diets

| | Dietary energy concentration | | |
| | High | Medium | Low |
|---|---|---|---|
| INGREDIENT (%) | | | |
| Ground barley straw | 10 | 30 | 50 |
| Ground barley | 54 | 31 | 8.2 |
| Soya bean meal | 13 | 16 | 18.8 |
| Molasses | 20 | 20 | 20 |
| Minerals and vitamins | 2 | 2 | 2 |
| Cod liver oil | 1 | 1 | 1 |
| | 100 | 100 | 100 |
| ANALYSIS | | | |
| Dry matter (%) | 83.3 | 83.3 | 83.5 |
| Ash (% DM) | 6.8 | 7.3 | 8.1 |
| Crude protein (% DM) | 14.2 | 13.1 | 12.3 |
| Ether extract (% DM) | 2.02 | 1.93 | 1.77 |
| Acid detergent fibre (% DM) | 10.5 | 20.3 | 29.1 |
| Gross energy (Mcal/kg) | 4.4 | 4.5 | 4.5 |
| Apparent digestibility (%) | 79 | 66.4 | 57.6 |
| Apparent digestible energy (Mcal/kg) | 3.48 | 2.95 | 2.57 |
| Metabolisable energy (DE x 0.82, Mcal/kg) | 2.86 | 2.42 | 2.11 |

*Table 4.2* Animal performance and chemical composition of carcass

| | Dietary energy concentration | | | |
| | High | Medium | Low | |
|---|---|---|---|---|
| INGREDIENT (%) | | | | |
| Gain/day (kg) | 1.24 | 1.32 | 0.90 | ±0.07 |
| Dry matter intake (kg/day) | 8.71 | 9.89 | 8.82 | ±0.39 |
| Efficiency of food utilisation (kg DM/kg liveweight gain) | 7.02 | 7.50 | 9.8 | ±0.43 |
| Efficiency of utilisation of concentrate (kg conc. DM/kg liveweight gain) | 6.32 | 5.24 | 4.9 | – |
| Carcass weight (kg) | 262 | 257 | 245 | ±3.81 |
| Killing out percentage | 57.2 | 54.7 | 53.7 | ±0.60 |
| Carcass gain (kg/day) | 0.71 | 0.72 | 0.48 | – |
| CARCASS | | | | |
| Carcass weight (kg) | 262 | 257 | 245 | ±3.81 |
| Side weight (kg) | 130 | 129 | 122 | |
| WEIGHT OF CHEMICAL COMPONENTS IN DRY MATTER OF DISSECTED SIDE: | | | | |
| Ash (kg) | 5.85 | 5.78 | 5.72 | ±0.09 |
| Fat (kg) | 36.3 | 33.4 | 33 | ±1.74 |
| Protein (kg) | 20.2 | 19.9 | 19.1 | ±0.35 |
| Water (kg) | 64.9 | 64.5 | 61.5 | ±1.15 |
| Total | 127.3 | 123.6 | 119.2 | |

the diet.  This is important, as the practical
implication is that diets containing a high proportion
of forage may be used in cattle production systems
without adverse effects on carcass composition.  Given
that slaughter weight is fixed, dietary composition can
be determined on the basis of liveweight gain.

The situation in the pig is quite different, there
being abundant evidence that both digestible energy
concentration and amino-acid composition of diets have
a marked effect on carcass composition.  Thus, increases
in dietary digestible energy will serve to increase the
proportion of fat in the carcass and increases in crude
protein, or in the limiting amino acid, will raise the
lean content (e.g. Cooke, Lodge and Lewis, 1972).  Thus,
the much more precise nutritional work carried out with
the pig indicates that carcass composition is highly
dependent on the balance of dietary nutrients achieved.
If amino-acid supply is the limiting factor in protein
synthesis then an excess of digestible energy will be
deposited as fat.  While such effects may occur in
cattle they are of such small magnitude as to be
difficult to detect in practice.  The reasons why
carcass growth in the pig is much more sensitive to
dietary nutrient balance than in cattle are probably
related to:

> Digestive tract function.
> Nutrient density of the diet.


DIGESTIVE TRACT FUNCTION

The major sources of energy in the pig diet are starch
and fat.  The processes of digestion in the pig lead
to most of the starch being converted to glucose by
enzymic action in the mouth, stomach and small
intestine.  Dietary fat would be absorbed in micelle
form.  If the dietary supply of glucose and lipid
exceeds the energy requirement for the maintenance of
essential body function and muscle protein synthesis,
the remainder is readily and efficiently transformed
into storage lipid.

The process of ruminant digestion is the factor which
probably accounts for the greatest difference between
pigs on the one hand and cattle and sheep on the other.
That the feed conversion efficiency of ruminants is

poorer than that of pigs is a basic fact of animal
production. The reason is that food is fermented in
the reticulo-rumen before being digested in the abomasum
(or stomach). In experiments covering a wide range of
diets it has been demonstrated that up to 90% of dietary
starch disappears in the reticulo-rumen although a larger
proportion of cellulose may be fermented in the hind gut,
particularly the caecum. Of the energy lost in the rumen,
up to 15% is accounted for by the heat of fermentation
and methane. The remainder is accounted for by microbial
cell yield and the products of microbial anaerobic
fermentation, namely volatile fatty acids. Microbial
cells are protein rich and are subsequently digested by
the host animal. Thus, given that nitrogen is not a
limiting factor, the process of rumen fermentation may
influence the energy:protein ratio to a large extent.
Furthermore, only small quantities of starch escape
fermentation to be directly absorbed as glucose. The
animal depends largely on propionic acid and gluconeo-
genesis for its glucose supply.


NUTRIENT DENSITY OF THE DIET

There is a great difference in the range of nutrient
densities offered to pigs on one hand and cattle on the
other. It has been demonstrated (Cole *et al.*, 1968)
that a high nutrient density diet led to a greater level
of carcass fat than a low nutrient density diet (which
was physically limiting) under *ad libitum* feed conditions,
owing to a greater intake of digestible energy.
    The beef animal would generally consume a diet of much
lower digestible energy, in which a balance between
energy and protein was being effected by the rumen
microflora. The exceptional case is the barley beef
animal which is offered a high nutrient density diet.
The effect of high nutrient density on carcass fat is
demonstrated when a beef animal of low growth potential
is offered such a diet.
    Thus, there is a difference between pigs and ruminants
in terms of the relationship between diet and carcass
quality. As pigs are sensitive to variations in energy
and nutrient intake, they are given allowances of these
rather than being offered food *ad libitum* in situations
where carcass quality is of importance. On the other
hand, within the normal range of beef cattle diets,

nutrient density, and consequently nutrient intake, does not have a direct effect on carcass composition.

## DIETARY PROTEIN REQUIREMENT AND CARCASS COMPOSITION

Nutrition of the meat animal is, as in the case of Man, a question of balance. Economic constraints are more stringent for farm livestock than appears to be the case in the Hotel and Catering Industry. The balance to be achieved is between nutrient intake and requirement. There is no evidence to indicate that the nutrient requirement of the pig is different from that of the beef animal. It has been demonstrated that the voluntary food intake response to dietary nutrient density is very similar in both cases. It is probable that the beef animals' requirement for essential amino acid is similar to the pig. In the discussion of response to dietary protein, it must be remembered that the major difference is the modifying effect of the rumen.

Moderate increases in protein level have shown improvements in carcass quality (e.g. 14 to 18%) for pigs of 23 to 55 kg liveweight (Robinson, Morgan and Lewis, 1964), and 11 to 19% for pigs of 46 to 92 kg liveweight (Robinson and Lewis, 1964). Much higher ranges of crude protein levels (15 to 27.5%) at the same digestible energy content (3.6 Mcal/kg) have been fed to gilts over the liveweight range 23 to 60 kg. While lean content increased over the range of protein levels there was no response in terms of feed utilisation or growth above 18% crude protein.

As protein content of the diet is increased it is necessary to establish whether or not the response is also influenced by the amount of energy supplied. In order to study the concept of energy:protein ratio in diets of relatively high nutrient density, Cooke (1969) used four digestible energy levels at each of four crude protein levels at a common scale of feeding. There were no significant interactions between energy and protein, which is in line with other work at this centre and indicates that digestible energy and crude protein act independently in their effect on growth rate, feed utilisation and carcass quality with diets of high nutrient density.

Although diets are usually prepared by the use of naturally occurring complete proteins, it is recognised that the major function of dietary crude protein is to supply the limiting amino acids.
It is generally accepted that lysine is the first limiting amino acid in diets prepared from the readily available ingredients in Britain. For the growing pig (25 to 55 kg liveweight) this is often provided by formulating diets of about 17.5% crude protein where the major protein-rich ingredients are fish meal and soya bean meal. The extent to which total protein is used to supply one essential amino acid, namely lysine, is illustrated by an experiment reported by Taylor, Cole and Lewis (1973). Eight diets were formulated to contain 17% down to 10% crude protein, by decrements of 1%. Synthetic L-lysine monohydrochloride was used to maintain levels of lysine in all diets at the same level as that in the 17% crude protein diet. Growth rate, feed utilisation (*Figure 4.3*) and carcass composition did not deteriorate until below 14% crude protein. The results indicated firstly, that the addition of crude protein above 14% was solely to provide lysine which was the first limiting amino acid and secondly that below 14% crude protein one or more essential amino acids became limiting.

Subsequently (Taylor, Cole and Lewis, 1974), it was established that threonine was the amino acid which was limiting at this point. However the best response for growth rate and carcass composition was obtained in the presence of adequate methionine plus cystine and trypophan, as well as lysine.

Clearly, the dietary needs of the pig in terms of individual essential amino acids and non-essential nitrogen can be established. Situations could arise where it would be more useful to formulate to these requirements rather than to meet them by the supply of total crude protein.

In the case of the ruminant animal, the influence of dietary crude protein level is much more difficult to evaluate because of the modifying influence of the reticulo-rumen. Microbial fermentation is a non-selective process, protein is likely to be fermented with the same frequency as starch or other carbohydrate sources, the amino nitrogen forming ammonia, a proportion of which will be incorporated into microbial cell protein. The question of the extent of microbial protein synthesis

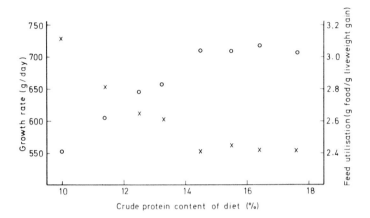

Figure 4.3 The growth rate (o) and feed utilisation
(x) of pigs fed diets of different crude protein level
but the same lysine content (Taylor, Cole and Lewis,
1973)

has been reviewed by Purser (1970). It was calculated
that the theoretical maximum yield of microbial protein
was 18.3 g digestible protein/Mcal DE. Other estimates
range from 21 up to 46 g digestible protein/Mcal DE.
It is apparent that the products of rumen fermentation
are capable of supplying the ruminant with sufficient
protein if the ratio is near to 45 g digestible protein/
Mcal DE. If the ratio is at the low level of 18 g
digestible protein/Mcal DE the supply of microbial
protein would be less than the requirement of the
productive animal. In addition to the supply of protein
there is the question of energy:protein ratio.

An evaluation of two beef cattle experiments recently
carried out at Nottingham University (Owers, Swan and
Wilton, 1974) showed that there may be a relationship
between energy:protein ratio, feed intake and growth
rate. Both feed intake and growth rate were maximised
at approximately 48 kcal (200 kJ) per 100 g digestible
protein per kg liveweight$^{0.7}$ (Figure 4.4). In these
experiments there was no relationship between digestible
crude protein intake and carcass protein content. It is
likely that dietary crude protein level and in particular
energy:protein ratio was influencing growth rate through

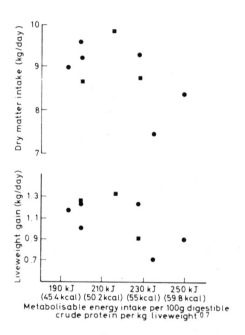

*Figure 4.4  Relationship between dietary energy:protein ratio, growth rate and dry matter intake in beef cattle (● , Owers, Swan and Wilton, 1974; ■ , Swan and Lamming, 1974)*

an effect on feed intake. Clearly under circumstances where feed intake is not subject to limitation because of indigestibility, there may be a deleterious effect of excess dietary nitrogen on voluntary food intake. It is striking that in many experiments in which urea has been compared with vegetable protein, there has been a depressant effect of the urea treatment on feed intake. This may have been due to excess concentration of nitrogen in the diet leading to a reduction of dry matter intake.

# IS THERE A BASIS FOR CHANNELLING APPROPRIATE FEED RESOURCES TO THE AVAILABLE LIVESTOCK TYPES?

Livestock production is used as a method of supplementing and to some extent upgrading the vegetable diet otherwise available to Man. Energy is readily available to Man as carbohydrate; livestock products cannot be considered as an exclusive source of energy. The value of meat is its enhanced amino-acid composition. The aim of policy-makers and livestock producers should be to concentrate the available resources within the area of most effective utilisation. The message in the production of pig meat is to minimise the inputs of animal and vegetable protein around optimum growth rate and carcass composition. Thus work involving the identification and supply of limiting amino acids must be of the highest priority.

In the case of ruminants both protein supplements and cereal grains should be minimised around optimum growth rates. There is no evidence of a specific relationship between nutrient density and carcass composition in the beef animal. Thus diets of low nutrient density, fed *ad libitum*, should be used, the protein level being adjusted to maximise food intake. There can be no justification in presenting to the pig a diet acceptable to Man and to the ruminant that which can be utilised by the pig.

REFERENCES

COLE, D.J.A., DUCKWORTH, J.E. and HOLMES, W. (1967). *Anim. Prod.*, 9, 141-148
COLE, D.J.A., DUCKWORTH, J.E., HOLMES, W. and CUTHBERTSON, A. (1968). *Anim. Prod.*, 10, 345-357
COLE, D.J.A., HARDY, B. and LEWIS, D. (1972). In *Pig Production*, p.243. Ed. D.J.A. Cole. Butterworths, London
COOKE, R. (1969). Ph.D. thesis, University of Nottingham
COOKE, R., LODGE, G.A. and LEWIS, D. (1972). *Anim. Prod.*, 14, 35-46
OWERS, M., SWAN, H., and WILTON, B. (1974). *Livestock Prod. Sci.* (in the press)

PURSER, D.B. (1970). *J. Anim. Sci.*, 30, 988-1001

ROBINSON, D.W., MORGAN, T.J. and LEWIS, D. (1964).
*J. agric. Sci., Camb*, 62, 369

ROBINSON, D.W. and LEWIS, D. (1964). *J. agric. Sci., Camb*, 63, 185

SWAN, H. and LAMMING, G.E. (1974). Unpublished data

TAYLOR, A.J., COLE, D.J.A. and LEWIS, D. (1973).
*Proc. Brit. Soc. Anim. Prod.*, 2, 87

TAYLOR, A.J., COLE, D.J.A. and LEWIS, D. (1974).
*Proc. Brit. Soc. Anim. Prod.*, 3, (in the press)

# 5

# THE INFLUENCE OF SEX ON MEAT PRODUCTION

M. KAY and R. HOUSEMAN

*The Rowett Research Institute,*
*Bucksburn, Aberdeen*

In animal production one of the prime objectives is the economic production of lean meat, and for many years research in nutrition, genetics and animal husbandry has been directed towards the identification and selection of the most efficient convertors of feed into edible end-products.

It has been recognised for some time that the sex of an animal has a considerable influence on its pattern of growth and development and is undoubtedly one of the main influences of the carcass composition of meat-producing animals.

The results of numerous experiments have shown that there are differences in growth rate, feed efficiency and carcass composition between the different sex types. These aspects have been particularly well documented for the comparison between the castrated male and its entire counterpart (Brannang, 1966; Turton, 1968; Walstra and Kroeske, 1968; Wismer-Pedersen, 1968; Field, 1971; Plumpton and Teague, 1972).

There is, however, a lack of information on the mechanisms by which the effects of sex are exerted at the cellular level and the consequent effect on the performance of the animal as a whole. Much of the information which exists has been derived from experiments in which exogenous hormones have been administered. The elucidation of the response of animals to hormonal treatment has been seriously handicapped by the lack of a quick, reliable method of hormone assay.

In addition very few experiments have provided sufficient data for a comparison of the overall efficiency of the different sexes, in terms of output of lean meat for a given input of energy.

This chapter is concerned firstly with the way in which sex manifests itself in meat animals with particular reference to growth rate, feed utilisation and carcass characteristics, and finally an attempt has been made to compare the efficiencies of the different sex types with regard to carcass production for a given energy input.

## HORMONAL EFFECTS ON METABOLISM

Before puberty growth hormone (GH) is possibly the most important endocrine regulator of growth, since androgenic and oestrogenic hormones are not found in significant amounts before the onset of sexual maturity. Growth, being a complicated process, is influenced in a number of ways by GH. It stimulates the growth of bone and cartilage, the oxidation of fat, the retention of nitrogen in the body and the synthesis of protein. At adolescence, the adrenal cortex increases its secretion of sex hormones as do the gonads. Androgens, in general, stimulate the growth of body tissues as a whole as well as organs specific to the male mammal (Brody, 1945; Gaunt, 1954). The fusion of the epiphyses in the long bones is brought about by the increase in androgen secretion at the time of puberty, and therefore castration should lead to an increase in the length of the bones of the limbs because of the delayed closure of the epiphyses. Tandler and Keller reported as early as 1920-1 that the castrated animal is taller than the entire male or female.

Androgens are also known to stimulate certain muscle groups particularly those in the neck and head regions. This accounts for the characteristic growth of the entire male with the greater development of the forequarter. In addition, comparisons between entire and castrated males have shown that androgens not only promote development of the early-maturing parts of the body, but also the development of the early-maturing tissues. Thus in a male, or an animal under the influence of androgens, there is a corresponding increase in the proportion of bone and muscle in the carcass compared with the castrated animal.

It is considered that androgens induce the synthesis of protein by regulation of the ribonucleic acids and the

protein biosynthesis system at the microsomal level
(Wilson, 1962; Kochanian, 1966). The response of the
muscles is not uniform for the different animal species
and the endocrine status may modify the effect of androgens. Oestrogens also appear to exhibit anabolic effects.
Wilkinson, Carter and Copenhaver (1955a), Wilkinson *et al.*
(1955b) and Preston and Burroughs (1958) have shown that
there is an increase in the growth of the early-maturing
bones and muscle in both ruminants and pigs. More
recently, *in vitro* studies have shown that oestradiol
causes an accumulation of phospholipids, ribonucleic acids
and protein in the rat uterus (Aisawa and Muller, 1961;
Noteboom and Gorski, 1963; Talwar and Segal, 1963).

The effects of androgens and oestrogens on basal metabolism have not been fully explained but the evidence
suggests that these hormones have a marked effect. Bugbee
and Simond (1926) and Ptajzek (1928) reported a reduced
basal metabolic rate in dogs after castration. Aude (1927)
found that basal metabolism was reduced by about 30% in
castrated cockerels. The basal metabolism of sheep and
pigs was measured by Ritzman, Colovos and Benedict (1936),
and on the basis of the results obtained from one ram and
one boar, they suggested that heat production was lower
after castration than before. In a subsequent experiment
five castrated lambs gave 5-10% lower values than five
entire males.

It has been well documented that castration brings about
changes in temperament. Ritzman, Colovos and Benedict
(1936) observed that after castration the animals reclined
more but stood up and laid down less than they did before
castration.

## ASPECTS OF GROWTH AND DEVELOPMENT

One of the most important factors affecting the production
of meat is that of nutrition, and a definition of the
response of the sexes to variations in two of the most
important constituents of the diet - dietary protein and
energy, is important in evaluating optimum dietary
conditions for the production of meat from the different
sexes. However, very few experiments have been reported
in which the differential response of the sexes to various
dietary factors has been investigated.

The changes in the composition of an animal from birth
to maturity have been well-defined and can be best understood if the body is considered to consist of two compartments: a lean body mass of constant composition and fat

which is more variable in amount with respect to age and weight.

It would seem on the basis of evidence from Elsley, McDonald and Fowler (1964) that it is not possible to alter to any great extent the proportions of essential tissues by normal nutritional means, but fat deposition seems to be solely dependent on the nutritional status of the animal. The amount of fat deposited depends on the amount of energy available in excess of that required for the growth of essential tissue.

From evidence in the literature it would seem that the entire male has a greater heat production associated with maintenance (Bugbee and Simond, 1926; Aude, 1927; Ptajzek, 1928; Ritzman, Colovos and Benedict, 1936) and a greater potential for lean tissue growth than the castrated male.

It would be expected therefore that on a given energy intake, the entire male would have less energy available for fat deposition than its castrated counterpart resulting in a carcass with a larger content of lean tissue.

## EFFECT OF SEX ON GROWTH RATE

The published results which are available for pigs differ from those obtained for cattle, particularly with regard to the difference between entire males and castrates. For example, under conditions of *ad libitum* feeding, castrated male pigs grow faster than boars. This is due mainly to the castrates' larger appetite which raises the daily food intake to a higher level than that of the boar. When boars and castrates are given the same amount of food the responses which are observed are conflicting, and this may be due to differences in the protein content of the diets used.

Work by Prescott and Lamming (1964) at Nottingham together with recent Canadian work has demonstrated the possibility of sex x dietary protein interaction and Fowler (1968) suggested that there was a different response to dietary protein in the growth rate of lean tissue between boars and castrates. These results have been confirmed under *ad libitum* feeding in so far as the response of boars to dietary protein has been found to be much greater than that of either castrates or gilts (Houseman, 1973).

In *Figures 5.1(a)* and *(b)*, a summary of the results
for the effect of sex on growth rate in pigs is given.
Only those experiments in which feed intake and growth
rate for any two sexes were measured, have been used for
this analysis. Details of results for *ad libitum* feeding
and fixed scale systems are given separately.
Generally, on *ad libitum* feeding, castrates have a
2.1% higher growth rate than boars, although in some
situations owing to a sex x protein interaction, the
boar becomes superior at high levels of dietary protein.
The results with boars and gilts on *ad libitum* feeding
are quite variable; on average, boars grow 3.8% faster
than gilts. Again a sex x protein interaction is
evident.

Experiments with cattle at the beginning of this
century highlighted the superior weight for age of bulls
compared with steers (Schuppli, 1911). Since that time
and in particular over the period 1950-74, numerous
experiments have been undertaken and the more important
of these will be referred to later. In general, the
castration of both male and female cattle has a retarding
effect on growth rate. In a review on the effects of
castration, Brannang (1966) suggested that the reduction
in daily gain through the castration of males is on
average 10% (range 5-15%) when the whole period from
calfhood to slaughter is included. Curran, Crowley and
McGloughlin (1965) found that spaying heifers weighing
320 kg while at pasture, reduced growth rate by 15%.
Many other workers have shown that the growth rate of
spayed heifers in feed lot conditions is 10% less than
that of intact heifers (Hart, Guilbert and Cole, 1950;
Dinusson, Andrews and Beeson, 1950; Kercher *et al.*, 1960;
Nygoard and Embry, 1966; Whetzal, Embry and Dye, 1966;
Keith *et al.*, 1967). On the other hand, Rako *et al.*
(1963) suggested that spaying heifers at 3.5 months old
did not reduce their growth rate up to a slaughter weight
of 370 kg at 14 months of age. The results of some
experiments have been inconclusive, possibly because the
ovaries are not the only site of oestrogen production
and one can conclude that the stress of the operation
itself may have militated against better performance.
Certainly the practice has not been accepted commercially.

Under grazing conditions, Turton (1962) suggested that
the growth of bulls was no better than that of steers.
This observation was subsequently confirmed with animals
reared with their dams at pasture (Watson, 1969). The
bulls and steers in Watson's experiment were slaughtered

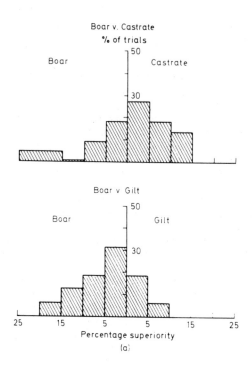

Boar v. Castrate

% of trials

*Figure 5.1(a)* *Growth rate (g/day):* ad libitum *feeding*

at 10 months of age when they weighed approximately
300 kg, so that testosterone secretion was unlikely to
have been pronounced until the months just before
slaughter and it is possible that the superiority of the
bulls may have become apparent at a heavier slaughter
weight (Lindner and Mann, 1960). Harte (1969) summarised
the results of a series of experiments made in Ireland
with bulls at pasture and found that the growth rate of
the bulls was 10% greater than that of the steers. It
is possible that the increased growth rate of the bulls
was brought about by differences in energy intake
associated with the use of better quality pastures.
Brannang (1966) when reporting 15% better gains for
bulls over steers at pasture, stressed that all animals
had 'abundant pasture of good quality'. Observations
on the behaviour of bulls at pasture have shown that
they spend 20% less time grazing and more time resting
than steers.

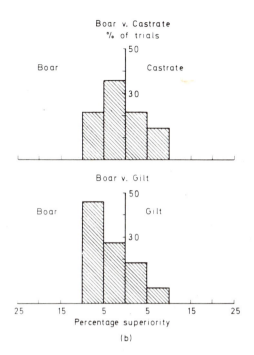

Figure 5.1(b)   *Growth rate (g/day): restricted feeding*

In most of the published experiments that have compared
differences in growth rate caused by sex, the animals
have been slaughtered at similar weights or ages.  This
inevitably introduces difficulties associated with
comparing the sexes at different stages of maturity as
outlined by Guilbert and Gregory (1952).  These arise
on account of the heavier mature weight for the bull as
opposed to the cow.  In experimental design such effects
are best catered for by using more than one slaughter
weight.

## EFFECT OF TIME OF CASTRATION ON GROWTH RATE

The effect of castration at different ages on
performance and carcass composition is important in that
it enables decisions to be made about the optimal time
for the castration of entire animals, thus maximising
feed utilisation and minimising the characteristic odour
of entire animals.

In experiments carried out with pigs and reported by
Clausen (1960), Bratzler *et al.* (1954), Charette (1961)
and Kroeske (1968), it has been shown that castration
carried out at birth and up to about 80 kg liveweight
has had little effect on growth rate to slaughter.

It is important to emphasise, however, that for
optimum feed utilisation and carcass quality, castration
should be delayed; this will increase the weight of lean
tissue in the carcass. Boar odour can be detected at
different ages and at different liveweights so that no
firm recommendation can be given with regard to the
optimum time for castration. It has been reported that
boar odour may be detected at 55 kg liveweight by Staun
(1965), while Prescott and Lamming (1964) found no taint
difference between boars and castrates at much heavier
weights. Similarly, it has been reported that boar odour
is present in the carcass up to 22 days after castration
(Lerche, 1936) while Bratzler *et al.* (1954) suggested
that it was eliminated 21 to 44 days after castration
at 81 kg liveweight. The castration of heavy pigs is
not a practical proposition at the present time.

With cattle, several experiments have found no
difference in growth rate to slaughter resulting from
early or late castration prior to puberty (Champagne
*et al.*, 1969; Kocenov, 1962). The differences in post-
pubertal growth rate between the sexes will be greatest
in situations in which diets containing a high
concentration of metabolisable energy are offered. Plane
of nutrition exerts a strong influence on the age at
which sexual maturity is attained; the higher the level
of nutrition, the earlier the onset of puberty. In a
heifer reared on a diet containing predominantly milk,
Amir, Kali and Volcani (1967) reported that first
oestrus appeared at 138 days when the animal weighed
220 kg. This was 122 days earlier than for a heifer
grown more slowly on a conventional diet. Thus it seems
that size or body development rather than chronological
age determines onset of puberty. Lassiter *et al.* (1972),
however, reported large differences in weight at puberty

among different breed crosses. In experiments with young
bulls offered diets containing predominantly barley,
Robertson, Wilson and Morris (1967) reported that
spermatozoa were first seen in the semen of one bull at
5½ months (approximately 220 kg liveweight). They were
present in samples from all bulls at 10½ months (420 kg
liveweight). It is likely that in well-grown males,
puberty is reached soon after the animal achieves a
weight of 220 kg.

The growth curves for bulls and steers in the experiment
reported by Brannang *et al.*(1970) and for 80 animals from
an experiment made at the Rowett Institute (Kay,
unpublished data) have been plotted in *Figure 5.2*. The
higher growth rate of the Swedish Red and White bull
occurs much earlier and at a lighter weight than with the
British Friesian. Preston *et al.* (1968) made an
experiment with 15 pairs of dizygotic British Friesian
twins and found that the difference in the growth rate
between bulls and steers was not manifest until after
they reached 250 kg liveweight.

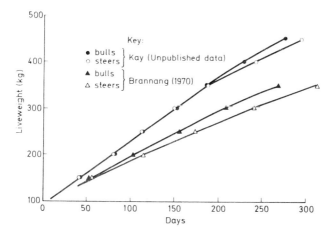

*Figure 5.2   Growth curve for bulls and steers fed
ad libitum*

## THE EFFECT OF SEX ON FEED INTAKE AND
## FEED UTILISATION

The effects of sex on feed intake can only be determined
in experiments in which the appetite potential is
allowed to express itself, i.e. *ad libitum* feeding.  In
experiments with pigs the results are quite clear in
that castrates eat more feed relative to time or body
weight than do either gilts or boars.  Boars eat the
least in this respect.  A summary of the experiments
showing the effect of sex on feed intake is given in
*Figure 5.3*.

Feed utilisation by the boar is superior to that of
the castrate and gilt at all levels of feeding.  This is
more apparent on *ad libitum* feeding where differences in
feed intake are larger.  This is possibly due to two
effects.  Firstly, with the boar the energy costs of
maintenance are higher than in the castrate and
consequently there is less energy available for tissue
deposition.  A consequence of this is that with boars a
greater lean tissue growth rate can be found.  The second
factor, possibly a result of this phenomenon, is mediated

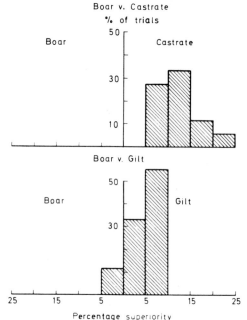

*Figure 5.3  Feed intake (kg/day)*

through the increased nitrogen retention which has been observed in boars. German work has shown that the average daily nitrogen retention from 30 to 110 kg liveweight was 28% greater in boars than in castrates (Piatowski and Jung, 1966). The experiments in which feed utilisation data have been recorded are shown in *Figure 5.4*. In the majority of comparisons, the boar shows a superior feed utilisation to the gilt or castrate, the average figure being 6.2% and 8.7% over gilts and castrates, respectively.

In experiments with cattle in which diets containing a high concentration of metabolisable energy have been offered *ad libitum* and the animals slaughtered at similar liveweights, feed intake by steers was on average 1.5% higher than that of bulls (Wickens and Ball, 1967; Prescott and Lamming, 1964; Nicholls *et al.*, 1964; Bailey, Probert and Bohman, 1966; Robertson, Wilson and Morris, 1967; Brannang *et al.*, 1970). However, Bailey and Hironka (1969) fattened yearlings weighing

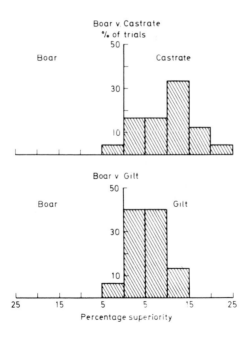

*Figure 5.4  Feed utilisation (kg feed/kg gain)*

approximately 170 kg on diets containing a large
proportion of barley and found that daily feed intake
was 9.01 kg for steers and 9.96 kg for bulls, an increase
of 10% in favour of bulls.  The difference was
significant and may have arisen as a result of the low
weight for age of the animals at the start of the
experiment.  According to information published by
Brannang *et al.* (1970) castration does not seem to affect
appetite as no differences in energy intake were noted
between bulls and steers of the same age.


## THE EFFECT OF SEX ON CARCASS CHARACTERISTICS

From experiments reported in the literature it appears
that boars are superior to both gilts and castrates in
lean meat production at most slaughter weights.  They
also appear to have a lower killing-out percentage than
castrates, probably as a result of differences in the
weight of genitals, the greater weight of internal
organs, offal, head and skin.  They also have a smaller
backfat thickness and, in general, features which are
associated with increased carcass leanness are evident;
these include greater area of *l.dorsi* muscle and a
higher proportion of muscle and bone in the carcass and
in various commercial cuts.  With respect to muscle:bone
ratios it must be emphasised that the lean tissue of
boars contains only a slightly higher proportion of
bone than those of castrates or gilts (Prescott and
Lamming, 1964; Fowler, Taylor and Livingstone, 1969).

There is also a greater development of forequarter
musculature in boars and thus in some investigations,
a greater weight of lean cuts have been obtained from
this part of the animal.  The results for one carcass
parameter, backfat thickness, are summarised in *Figure
5.5.*

For cattle, the bull carcass contains approximately
8% more muscle and 38% less fat than that of the steer.
The additional fat in the steer is .not evenly distributed
among the various fat depots, there being an increase in
mesenteric fat, a very large increase in subcutaneous fat
and a smaller increase in intermuscular fat (Brannang,
1966).  In spite of the higher fat content and the
resulting increase in the energy density of the tissues
of the steer, the differences in killing-out percentage
between bulls and steers are small.  No important
differences in the relative weights of the internal

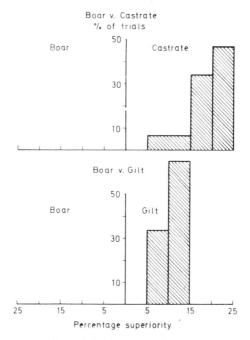

*Figure 5.5   Backfat thickness (mm)*

organs and the compartments of the alimentary tract
have been found between bulls and steers, but bulls have
a significantly heavier hide.   Bulls have greater
development of forequarter muscle which results in an
inferior distribution of valuable joints but there is
no reduction in the weight of muscle in the hindquarter
compared with steers.   There have been few experiments
made in which the carcass composition of the female has
been compared with that of bulls and steers.   One such
comparison reported by Hedrick, Thompson and Krause
(1969) suggests that the heifer carcass contains 6% less
'saleable meat' than the steer and 14% less than the
bull of the same liveweight.

## UTILISATION OF ENERGY FOR THE PRODUCTION OF LEAN MEAT

Despite the volume of literature on the comparative performance of males, females and castrates, there is little information on the relative efficiency of the complete process, that is of converting food into lean meat. As far as possible, we decided to equate lean meat with dissectible muscle (Blaxter, 1968) since this equates fairly closely with the edible portion of the carcass. The efficiency of utilisation of digestible energy for the production of lean tissue in the carcass in the different sexes has been compared and the information relating to the pig is summarised in *Figure 5.6*. It can be concluded that the overall superiority of the boar was approximately 22% over the castrate and 11% over the gilt.

The efficiency of the bull and the steer as producers of lean meat is given in *Table 5.1*. In view of the superior daily liveweight gain of the bull to the steer, the conversion of food into liveweight and carcass is approximately 9% better for the bull. As a means of producing lean meat, the bull is 17% more efficient in using dietary energy than the steer. There are few experiments which allow a comparison with the heifer. On the basis of results published by Hedrick, Thompson and Krause (1969), the heifer would appear to be 2% less efficient in producing lean tissue than the steer.

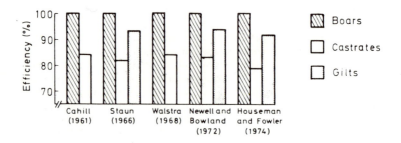

*Figure 5.6 Relative efficiencies of boars, gilts and castrates in converting digestible energy into muscle (boars = 100%)*

Table 5.1 Summary of data on feed utilisation from castration experiments with beef cattle

| | Feed consumption per unit gain (Bull = 100) | Killing out (%) (Bull = 100) | Separable lean (%) (Bull = 100) | Feed consumption per unit lean (Bull = 100) |
|---|---|---|---|---|
| Nicholls et al. (1964) | 108 | 98 | 95 | 111 |
| Bielinska et al. (1965) | 102 | 99 | 99 | 102 |
| Brannang (1966) | 108 | 103 | 92 | 121 |
| Bailey et al. (1966) | 110 | 100 | 85 | 129 |
| Spivak (1966) | 107 | 99 | 95 | 111 |
| Robertson et al. (1967) | 114 | 99 | 87 | 130 |
| Hedrick et al. (1969) | 106 | 98 | 91 | 114 |
| Hedrick et al. (1969) | 111 | 99 | 95 | 121 |
| Harte (1969) | 111 | 100 | 93 | 119 |
| Rostovcev and Cerkascenko (1966) | 110 | 102 | 96 | 116 |
| Neiman-Sørensen et al. (1967) | 120 | 100 | 92 | 131 |

The mean conversion of feed into liveweight gain for
the steer in the experiments referred to in *Table 5.1*
is 6.6. This is equivalent to a conversion of feed into
muscle tissue of 19.2. Since in the majority of these
experiments the animals have been offered a diet
containing predominantly cereals, this is equivalent to
63 Mcal digestible energy per kg muscle tissue. This
value is somewhat lower than that calculated by Bidart,
Koch and Arthand (1970) for Angus steers where 84 Mcal
digestible energy were required per kg edible product.
This would not be surprising if the Angus cattle carcasses
contained excessive amounts of subcutaneous and kidney
fat.

There are no experiments in the literature which allow
a similar calculation to be made for the efficiency of
lean-meat production for the different sexes of lamb.
According to Jacobs (1970) and Deweese *et al.* (1969),
the ram is 12-15% more efficient in converting feed into
liveweight gain than the wether and Ørskov, Fraser and
McHattie (1974) found the ram to be 13% more efficient
than the female. The results of experiments carried out
by Ørskov *et al.* (1971) showed that when slaughtered at
50 kg the ram is 9% more efficient in converting feed
into liveweight gain than the female but the difference
was reduced to 2% for the superiority of the ram in
producing carcass gain. There were no differences in
the efficiency with which feed was converted into body
protein by the ram or the female. In these experiments,
the female was found to contain 17% more ether
extractable material in its empty body than the ram at
50 kg liveweight.


## MEAT PRODUCTION FROM THE ONCE-BRED FEMALE

A method of meat production that is consistent with
high productivity is that of the once-bred female, i.e.
mating gilts or heifers early in their reproductive life,
and using their offspring for herd replacements or meat
production and then slaughtering the dam for meat soon
after parturition. In combination with early weaning,
this has the potential of being a very intensive system
of animal production.

A comparison of the relative efficiency of the once-bred
system with that of virgin counterparts is difficult
because the end point (slaughter) varies according to

the management and feeding during rearing, gestation and post-parturition. In addition, in the pregnant animal, the weight changes in maternal tissue, although often dramatic, are not always in a positive direction. Thus, in the few controlled experiments that have been carried out with the once-bred animal, there is a large range in the values for the conversion of dietary energy into carcass, which are largely a reflection of the different management conditions.

The small number of experiments that have been carried out with pigs, comparing the feed utilisation and carcass composition of pregnant and non-pregnant animals, has indicated that there may be a difference in the maternal response to pregnancy between immature gilts and mature sows. Thus, while Salmon-Legagneur (1965) reported an overall increase in subcutaneous fat and skeletal muscle during pregnancy in sows, Elsley et al. (1966) found that the difference in weight gain between pregnant and non-pregnant gilts was accountable only in the reproductive tract, mammary tissue and blood. Carcass weights were unaffected by pregnancy. In contrast, Heap and Lodge (1967) found that of the extra body weight gained by pregnant over non-pregnant animals, 48% was adipose tissue, 31% was muscle, and 21% was mammary tissue. Most of the extra fat was subcutaneous and the extra muscle was in the abdominal region.

There have been very few measurements made of the feed efficiency of gilts mated at an early age. Brooks and Cole (1973) found that feed conversion of gilts over the period from 65 kg to weaning at 5 to 6 weeks was 12% better for gilts mated early at 92 kg liveweight than for a conventional system of mating at third heat and weaning at 5-6 weeks.

Kotarbinska and Kielanowski (1973) found that the feed conversion ratio from 30 kg to slaughter at 155 kg was 3.67 for pregnant animals and 4.46 for non-pregnant animals slaughtered at 130 kg, a difference in favour of the pregnant animals of nearly 18%.

Very few results are available to allow one to assess the efficiency of the once-bred gilt as a producer of lean meat.

On the basis of the data of Hovell and MacPherson (unpublished), it was calculated that from 20 kg to slaughter at 123 kg liveweight, the digestible energy per kg of carcass produced for pregnant animals mated at the first oestrus was 19.0 Mcal, and for non-pregnant controls slaughtered at 138 kg liveweight was 18.6 Mcal. In these circumstances, the data show a similarity in the

efficiency of pregnant and non-pregnant animals, but
the non-pregnant animals were grown to a heavier weight.
Such calculations, however, do not take into account the
considerable weight loss which occurred after parturition
in the pregnant gilts (some 13 kg in one week) and also
were not adjusted for the energy required for the
products of conception.  This latter can be estimated
using data from ARC (1967) which suggests that
approximately 3.5 Mcal of digestible energy are required
per 1 kg of product, and therefore for a 15 kg 'foetal
load' about 52 Mcal are required.  The efficiency for
the pregnant animal after allowing for the products of
conception becomes 18.4 Mcal/kg carcass, a value which
is little different from that already given.

The data of Kotarbinska and Kielanowski (1973) indicated
that the conversion of digestible energy into carcass is
better for the pregnant rather than the non-pregnant
animal.  For non-pregnant animals reared from 30 kg and
killed at about 130 kg, digestible energy per kg of
carcass is 13.9 Mcal, and that for gilts reared
similarly, mated at about 90 kg and slaughtered at
151 kg, the value is  12.6 Mcal, an increase of 9.4%.
The difference between this trial and the one made by
Hovell and MacPherson was that in the Polish study the
gilts were slaughtered some 9-10 weeks after parturition,
whereas those of Hovell and MacPherson were slaughtered
about one week after birth.  Secondly, the food allowance
during pregnancy in the Polish study was much greater
than in the previous study, this being reflected in a
greater weight at slaughter for the pregnant animals.

The efficiency of the pregnant and non-pregnant gilt
for converting energy into muscle can be calculated
using results published by Kotarbinska and Kielanowski
(1973).  For maiden gilts reared from 30 kg and
slaughtered at 90 kg, 18.9 Mcal/kg of separable muscle
were required which is similar to the value given in
the comparison of gilts, castrates and boars.  However,
for gilts slaughtered at 130 kg the efficiency becomes
30.8 Mcal digestible energy per kg of muscle, and for
gilts reared along the same growth curves mated at 90 kg,
and slaughtered at 151 kg, the value is 24.4 Mcal/kg of
muscle, an improvement of some 21% over the non-pregnant
counterpart.

Breeding from heifers before they are slaughtered
allows the animals to be carried on to heavier weights
without adversely affecting the yield of lean meat.  The
breeding of heifers is not without problems and there is
a higher incidence of dystocia at first calving.  Neonatal

*Table 5.2* The liveweights and rates of gain of heifers used in calculating the efficiency of the once-bred heifer

| | Start (kg) | Conception (kg) | Post calving (kg) | Calving (kg) | Slaughter (kg) |
|---|---|---|---|---|---|
| Maiden female | 200 | | | | 410 |
| Rate of gain (kg/day) | | 0.67 | | | |
| Once-bred female | 200 | 275 | 355 | 415 | 440 |
| Rate of gain (kg/day) | | 0.67 | | 0.50 | 1.1 |

losses could be higher than with calves born to later
pregnancies owing to the lower birth weight of the
calves born to bred heifers.

Brookes and O'Byrne (1965) and Crowley (1966) have
both shown that there was little difference between the
quality of the meat from carcasses of maiden females and
bred heifers. The advantage of the bred heifer was that
carcass yield was increased by approximately 35 kg without
any increase in the proportion of fat in the carcass.
There are no experiments available which allow feed input
to be related to carcass weight for the maiden and once-
bred female. A theoretical calculation has been made
based on the following scheme to assess the relative
efficiency of the two types of female stock (*Table 5.2*).
The females were assumed to start at 200 kg liveweight
and were slaughtered at liveweights that would provide
carcasses of a similar fat content. It has been assumed
that the diet offered to the animals contained 2.5 Mcal
metabolisable energy per kg dry matter. The metabolisable
energy intakes for maintenance and the various rates of
liveweight gain were calculated using information
contained in the publication of the ARC (1965). The
metabolisable energy requirements for pregnancy in the
heifer were those suggested by Moe and Tyrrell (1972).
The maiden female was calculated to require 4415 Mcal
metabolisable energy in producing a carcass weighing
230 kg, whereas the requirement for the once-bred female
was 6750 Mcal for a carcass weighing 244 kg. These give
an efficiency of 19 Mcal/kg and 28 Mcal/kg carcass for
the maiden female and the once-bred female. Thus the
once-bred female has required an additional 2335 Mcal
but has provided 30 kg more carcass and a viable calf
containing approximately 54 Mcal.


## REFERENCES

Agricultural Research Council (1965). *The Nutrient
    Requirements of Farm Livestock. No. 2 Ruminants.*
    Agricultural Research Council, London
Agricultural Research Council (1967). *The Nutrient
    Requirements of Farm Livestock. No. 3 Pigs.*
    Agricultural Research Council, London
AIZAWA, Y. and MUELLER, G.C. (1961). *J. biol. Chem.,*
    236, 381
AMIR, S., KALI, J. and VOLCANI, R. (1967). *Growth and
    Development of Mammals,* Butterworths Scientific
    Publications, London

AUDE, D. (1927). *Rev. Franc. d'endocrin*, 5, 81
BAILEY, C.B. and HIRONKA, R. (1969). *Can. J. Anim. Sci.*, 49, 37
BAILEY, C.M., PROBERT, C.L. and BOHMAN, V.R. (1966). *J. Anim. Sci.*, 25, 132
BIDART, J.B., KOCH, R.M. and ARTHAND, V.H. (1970). *J. Anim. Sci.*, 30, 1019
BIELINSKA, K., BIELINSKI, K., BORZUTA, K., GOZDZ, H., CHRZASZCZ, T. and SLABON, W. (1965). *Roczn. Nauk roln.*, B86, 91
BLAXTER, K.L. (1968). *Proc. Sec. Wld Conf. Anim. Prod.*, p.31
BRANNANG, E. (1966). *Lantbrukshogskolans Annaler*, 32, 329
BRANNANG, E., HENNINGSSON, T., LILJEDAHL, L. and LINDHE, B. (1970). *Lantbrukshogskolans Annaler*, 36, 91
BRATZLER, L.J., SOULE, R.P., REINEKE, E.P. and PAUL, P. (1954). *J. Anim. Sci.*, 13, 171
BRODY, S. (1945). *Bioenergetics and Growth*, Reinhold Publishing Corporation, New York
BROOKES, A.J. and O'BYRNE, M. (1965). *J. R. agric. Soc.*, 126, 30
BROOKS, P.H. and COLE, D.J.A. (1973). *Anim. Prod.*, 17, 305
BUGBEE, E.P. and SIMOND, A. (1926). *Am. J. Physiol.*, 75, 542
CHAMPAGNE, J.R., CARPENTER, J.W., HENTGES, J.F., PALMER, A.Z. and KOGER, M. (1969). *J. Anim. Sci.*, 29, 887
CHARETTE, L.A. (1961). *Can. J. Anim. Sci.*, 41, 30
CLAUSEN, H. (1960). *Landøkønomisk Forsøgslaboratoriums efterårsmøde*, p.220
CROWLEY, J.P. (1966). Ph.D. thesis, University of Dublin
CURRAN, S., CROWLEY, J.P. and MCGLOUGHLIN, P. (1965). *Ir. J. agric. Res.*, 4, 93
DEWEESE, W.P., GLIMP, H.A., KEMP, J.D. and ELY, D.G. (1969). *J. Anim. Sci.*, 29, 121
DINUSSON, W.E., ANDREWS, F.N. and BEESON, W.M. (1950). *J. Anim. Sci.*, 9, 321
ELSLEY, F.W.H., ANDERSON, D.M., MCDONALD, I., MACPHERSON, R.M. and SMART, R.I. (1966). *Anim. Prod.*, 8, 391
ELSLEY, F.W.H., MCDONALD, I. and FOWLER, V.R. (1964). *Anim. Prod.*, 6, 141
FIELD, R.A. (1971). *J. Anim. Sci.*, 32, 849
FOWLER, V.R. (1967). *Anim. Prod.*, 8, 272
FOWLER, V.R. (1968). In *Growth and Development of Mammals*, Ed. G.A. Lodge and G.E. Lamming, Butterworths, London

FOWLER, V.R., TAYLOR, A.G. and LIVINGSTONE, R.M. (1969). *Proc. of Conf. on Meat from Male Animals, Meat Research Institute, Bristol,* p.51

GAUNT, R. (1954). In *Dynamics of Growth Processes,* p.183, Ed. E.J. Boell, Princeton University Press

GUILBERT, H.R. and GREGORY, P.W. (1952). *J. Anim. Sci.,* 11, 3

HART, G.H., GUILBERT, H.R. and COLE, H.H. (1940). *Calif. agric. Exp. Stn Bull.,* No. 645

HARTE, F.J. (1969). *Meat Production from Entire Male Animals,* p.153, J. and A. Churchill Ltd., London

HEAP, F.C. and LODGE, G.A. (1967). *Anim. Prod.,* 9, 237

HEDRICK, H.B., THOMPSON, G.B. and KRAUSE, G.F. (1969). *J. Anim. Sci.,* 29, 687

HOUSEMAN, R.A. (1973). *Proc. Br. Soc. Anim. Prod.,* 2, 90

JACOBS, J.A. (1970). Ph.D. thesis, University of Wyoming, Laramie, Wyoming

KEITH, T.B., DAHMEN, J.J., ORME, L.E. and BELL, T.D. (1967). *Ida. agric. Exp. Stn Bull.,* No. 488

KERCHER, C.J., THOMPSON, R.C., STRATTON, P.O., SCHOONOVER, C.O., GORMAN, J.A. and HILSTON, N.W. (1960). *Wyo. agric. Exp. Stn Mimeo Circ. 127*

KOCENOV, D.A. (1962). *Trudy Vses Nauc. Issled Inst. Zivotnovodstva,* 24, 49

KOCHANIAN, C.D. (1966). *The Physiology and Biochemistry of Muscle as a Food,* p.81, University of Wisconsin Press

KOTARBINSKA, Maria and KIELANOWSKI, J. (1973). *Anim. Prod.,* 17, 317

KROESKE, D. (1968). Report EAAP Commission on Pig Production, Dublin

LASSITER, D.B., GLIMP, H.A. and GREGORY, K.E. (1972). *J. Anim. Sci.,* 34, 1031

LERCHE, (1936). Geschlectsgeruck bie eberbastraten, *Zeitschrift fur Fleish-und Mich-hygiens,* 46, 417

LINDNER, H.R. and MANN, T. (1960). *J. Endocr.,* 21, 341

MOE, P.W. and TYRRELL, H.F. (1972). *J. Dairy Sci.,* 55, 480

NEIMANN-SØRENSEN, A., KIRSGAARD, E., AGERGAARD, E., KLAUSEN, S. and BROLUND LARSEN, J. (1967). *Landokønomisk Forsøgslaboratoriums efterårsmøde, Arbog 517*

NICHOLLS, J.R., ZIEGLER, J.H., WHITE, J.M., KEELER, E.M. and WATKINS, J.L. (1964). *J. Dairy Sci.*, 47, 179

NOTEBOOM, W.D. and GORSKI, J. (1963). *Proc. natn. Acad. Sci.*, *U.S.A.*, 50, 250

NYGAARD, L.J. and EMBRY, L.B. (1966). *S. Dak. agric. Exp. Stn Beef Cattle Day*, AS 66-13

ØRSKOV, E.R., FRASER, C. and MCHATTIE, I. (1974). *Anim. Prod.*, 18, 85

ØRSKOV, E.R., MCDONALD, I., FRASER, C. and CORSE, Elizabeth L. (1971). *J. agric. Sci.*, *Camb.*, 77, 351

PIATOWSKI, B. and JUNG, H. (1966). *Arch. Tierzucht*, 9, 301

PLIMPTON, R.F. and TEAGUE, H.S. (1972). *J. Anim. Sci.*, 35, 1166

PRESCOTT, J.H.D. and LAMMING, G.E. (1964). *J. agric. Sci.*, *Camb.*, 63, 341

PRESTON, R.L. and BURROUGHS, W. (1958). *J. Anim. Sci.*, 17, 140

PRESTON, T.R., MACDEARMID, A., AITKEN, J.N., MACLEOD, N.A. and PHILIP, E.B. (1968). *Revista Cubana de Ciencia Agricola*, 2, 183

PTAJZEK, L. (1928). *Soc. de biol. Compt. Rend.*, 99, 929

RAKO, A., MIKULEC, K., DUMANOVSKY, F., BLAGOVIC, S., MIODRAGOVIC, Z. and SARKOVIC, F. (1963). *Veterinaria Sarajevo*, 12, 307

RITZMAN, E.G., COLOVOS, N.F. and BENEDICT, F.G. (1936). *Univ. New Hampshire, Durham, Tech. Bull.*, 64

ROBERTSON, I.S., WILSON, J.C. and MORRIS, P.G.L. (1967). *Vet. Rec.*, 81, 88

ROSTOVCEV, N.F. and CERKASCENKO, I.I. (1966). *Zhivotnovodstvo, Mosk.*, 28, 33

SALMON-LEGAGNEUR, E. (1965). *Ann. Zootech.*, 14, 1

SCHUPPLI, W.P. (1911). *Jahrest Steiermärk. Landw. Landes-Lehranst*

SPIVAK, M.G. (1966). *Trudy Vses Nauc. Issled Inst. Zhivot*, 28, 109

STAUN, V.H. (1965). *Landøkonomisk Forsøgslaboratoriums efterårsmøde*, p.67

TALWAR, G.P. and SEGAL, S.J. (1963). *Proc. natn. Acad. Sci.*, 50, 226

TURTON, J.D. (1962). *Anim. Breeding Abstr.*, 30, 447

TURTON, J.D. (1969). *Meat Production from Entire Male Animals*, p.1, J. and A. Churchill Ltd., London

WALSTRA, P. and KROESKE, D. (1968). Report EAAP Commission on Pig Production, Dublin

WATSON, M.J. (1969). *Aust. J. Exp. Agric. Anim. Husb.*, 9, 164

WHETZAL, F.W., EMBRY, L.B. and DYE, L. (1966). *S. Dak. agric. Exp. Stn Beef Cattle Day,* AS 66-7

WICKENS, R. and BALL, C. (1967). *Expl Husb.,* No. 15, 64

WILKINSON, W.S., CARTER, R.C. and COPENHAVER, J.S. (1955a). *J. Anim. Sci.,* 14, 1260

WILKINSON, W.S., O'MARY, C.C., WILSON, G.D., BRAY, R.W., POPE, A.L. and CASIDA, L.E. (1955b). *J. Anim. Sci.,* 14, 866

WILSON, J.D. (1962). *Protein metabolism,* Springer-Verlag, Berlin, Gottingen and Heidelberg, 26-40

WISMER-PEDERSEN, J. (1968). *World Review Anim. Prod.,* IV, 100

# 6

# THE IMPLICATION OF DISEASE IN THE MEAT INDUSTRY

### D. R. MELROSE

*Meat and Livestock Commission,*
*Bletchley, Milton Keynes*

### J. F. GRACEY

*Veterinary and Meat Inspection Department,*
*City of Belfast*

## INTRODUCTION

There are frequent comments in the industry about
European Economic Community (EEC) export slaughterhouse
requirements and suggestions that our meat inspection
and meat hygiene requirements are either adequate or
inadequate. Thus, it is timely to have this discussion
on disease implications, which must be considered with
particular attention to the modern farm husbandry
practices, the changing pattern of meat processing and
the modern methods of meat retailing and its utilisation
in convenience foods.

Meat inspection had its beginnings with the earliest
civilisations around the Mediterranean. Ancient Egypt
proclaimed the pig unclean and the cow sacred and
banned their flesh as food for man. Moses, the first
known teacher of meat inspection was, it is recorded,
insistent that only those mammals which were cloven-
hoofed and chewed the cud were fit to eat, that they
should be slaughtered in such a way as to remove all
blood from the flesh, that meat should be destroyed if
not eaten before the third day after slaughter and that
only meat from animals 'without blemish' should be
eaten. In the late 19th century in the Pasteur era when
it was realised that certain animal diseases could be
transferred to man, and vice versa, there was considerable

concern in case tuberculosis, then common in cattle and
pigs, could be spread to humans in either cow's milk or
in meat from both these species. A Departmental
Committee in 1888 followed by a Royal Commission in 1895
were necessary to decide the extent to which carcasses
with advanced tuberculosis could be used for human
consumption and their findings formed the basis of our
meat inspection teaching and practice. The primary
function of meat inspection was then and still is, along
with meat hygiene, the protection of human health from
diseases that can be transmitted from meat animals to
man through the consumption of meat. Following the
eradication of bovine tuberculosis more emphasis has
been placed on those conditions which are less obvious
and which require more comprehensive examination, often
entailing use of laboratory aids. The changing pattern
of meat processing and marketing have also necessitated
more emphasis on prevention of meat contamination
throughout its handling. Meat and slaughterhouse
hygiene have become as important in animal health
programmes since they can provide information on those
conditions which cause economic loss but are not
manifested by outbreaks of clinical disease in the live
animal. The implications of disease on the meat
industry are interrelated but can be considered under
three headings: economic, animal health and public
health.


## ECONOMIC IMPLICATIONS

These can be summarised under the following aspects:


### COST OF MEAT INSPECTION AND MEAT HYGIENE

There could be increased costs with the more
sophisticated inspection and preventive measures
necessary to cope with the changing pattern in animal
disease and the newer procedures for meat handling which
increase the chances of contamination and therefore the
risk to human health. The meat traders associations
have questioned the justification for their having to
contribute to the inspection service on the grounds that
such a public health service should not be a charge on
a particular section of the community.

CONDEMNATIONS AND DOWNGRADING OF CARCASSES

Losses on these accounts are partly due to disease, but
injuries to the live animal either on farm or in transit
to the abattoir, faulty injection and/or castration
techniques and bad pre-slaughter handling systems can
also contribute to these losses. The *Environmental
Health Report* (1969) prepared by the Association of
Public Health Inspectors showed that 9,876 tons of meat
and offal were condemned in 1969 as unfit for human
consumption. This report covered returns from Local
Authorities representing 63% of the population of
England and Wales. In Northern Ireland it has been
shown that carcass condemnations from all causes amount
to 0.6% of the total kill of pigs, with 0.3% in cattle
and 0.2% in sheep (*Tables 6.1, 6.2* and *6.3*) figures
which are largely a reflection of the degree of
intensiveness of husbandry methods (Thornton and Gracey,
1974). Over 2% of the total production is lost in this
way and much of this loss is avoidable. The 29%
condemnation rate of bovine liver costs the industry
some £2 million annually and there is a loss of
£300,000 through the 1 million sheep liver condemnations,
in both cases the cause being fascioliasis or liver fluke
infestation. Warble fly damage to hides has been
variously estimated to cost from £1 to £3 million annually
in Great Britain. In pigs it was estimated in one report
(Norval, 1966) that 0.45 kg (1 lb) carcass weight was
lost annually on account of condemnations for every pig
slaughtered.


MEAT QUALITY

With production being increasingly geared to more
intensive systems and to leaner carcasses the quality
of the end-product has not always received sufficient
attention in assessing the merits of a system. For
example, in pigs, attention has been focussed on the
pale soft exudative (PSE) meat which appears to be
associated with the acute stress problems occurring
during transport or following exertion, leading to
collapse and very often death. More recently there
have been reports of dark meat in the carcasses from
entire bulls reared for beef. Clearly, in future much
more attention will need to be paid to meat quality.

*Table 6.1* Total carcass condemnations of pigs in Northern Ireland (1971) (total kill = 2,085,897)

| Reason | Number of carcasses | Percentage of total condemnation |
|---|---|---|
| Abscesses and pyaemia | 5408) 7383 | 42.8) 58.50 |
| Arthritis | 1975) | 15.6) |
| Fever and septicaemia | 1856 | 14.7 |
| Pneumonia and pleurisy | 861 | 6.8 |
| Imperfect bleeding | 425 | 3.4 |
| Emaciation | 315 | 2.5 |
| Peritonitis | 207 | 1.6 |
| Oedema | 97 | 0.8 |
| Bruising | 78 | 0.6 |
| Putrefaction | 58 | 0.5 |
| Tuberculosis | 32 | 0.3 |
| Uraemia | 32 | 0.3 |
| Tumours | 12 | 0.1 |
| Contamination | 10 | 0.08 |
| Nephritis and nephrosis | 6 | 0.05 |
| Pericarditis etc. | 6 | 0.05 |
| Miscellaneous | 1243 | 9.85 |
| | 12621 | |

Percentage of total kill: 0.6
Total weight of meat condemned: 732 tons

Table 6.2 Total carcass condemnations of cattle in Northern Ireland (1971) (total kill = 346,339)

| Reason | Number of carcasses | Percentage of total condemnations |
|---|---|---|
| Emaciation | 267 | 28.7) |
| Oedema | 209 | 22.4) 76.8 |
| Bruising | 132 | 14.2) |
| Fever and septicaemia | 107 | 11.5) |
| Peritonitis etc. | 45 | 4.8 |
| Abscesses and pyaemia | 28 | 3.0 |
| Insufficient bleeding | 25 | 2.7 |
| Tumours | 21 | 2.3 |
| Cysticercus bovis | 9 | 1.0 |
| Pneumonia and pleurisy | 8 | 0.9 |
| Pericarditis | 7 | 0.8 |
| Tuberculosis | 6 | 0.6 |
| Nephritis and nephrosis | 6 | 0.6 |
| Arthritis | 5 | 0.5 |
| Uraemia | 5 | 0.5 |
| Actinobacillosis | 1 | 0.1 |
| Putrefaction | 1 | 0.1 |
| Miscellaneous | 49 | 5.3 |
| | 931 | |

Percentage of total kill: 0.3
Total weight of meat condemned: 250 tons

*Table 6.3* Total carcass condemnations of sheep in Northern Ireland (1971) (total kill = 356,263)

| Reason | Number of carcasses | Percentage of total condemnations |
|---|---|---|
| Emaciation | 159 | 25.5 |
| Bruising | 150 | 24.1 |
| Oedema | 94 | 15.1 |
| Fever and septicaemia | 62 | 10.0 |
| Abscesses and pyaemia | 28 | 4.5 |
| Imperfect bleeding | 24 | 3.9 |
| Arthritis | 17 | 2.7 |
| Pneumonia and pleurisy | 15 | 2.4 |
| Peritonitis | 8 | 1.3 |
| Uraemia | 6 | 1.0 |
| Tumours | 3 | 0.5 |
| Putrefaction | 2 | 0.3 |
| Nephritis and nephrosis | 2 | 0.3 |
| Miscellaneous | 53 | 8.5 |
| | 623 | |

Percentage of total kill: 0.2
Total weight of meat condemned: 14 tons

CONSUMER REACTION AGAINST MEAT

In particular areas where there is an animal disease epidemic, consumer reaction against meat can build up owing to uninformed opinions being expressed. Hence there is a need for control over dissemination of condemnation data and disease outbreak reports.

## ANIMAL HEALTH IMPLICATIONS

The chain of meat production starts on the farm and whilst considerable attention has been paid to improving productive efficiency (including disease control) it must be accepted that the effects of disease and husbandry procedures on the end-product have not received adequate attention. However, this lack of emphasis on the end-product, i.e. the quality of meat and offal produced at slaughter, may be partly due to lack of regular feed-back of condemnation data from the factory to producer. However, this can be difficult under the present marketing system where the producer does not always sell direct to the slaughterer. In addition, the abattoir data may not be sufficiently specific to warrant costly treatment or other on-farm control measures but it should be enough to alert the producer. The farmer and his veterinarian can, however, do much to prevent disease getting into the abattoir by:

    Farm control and prevention of specific disease
        conditions.
    Prevention of spread of notifiable diseases.
    Avoiding disease spread and injuries under new
        husbandry systems.
    Control of medicinal and other products on the farm.


FARM CONTROL AND PREVENTION OF SPECIFIC
    DISEASE CONDITIONS

In addition to Public Health implications certain points relevant to the farm situation must be emphasised:

*Tuberculosis*

The progress made in the national eradication campaign
was quickly reflected in the abattoir condemnation data.
The report of Medical Officers of Health for England and
Wales for 1958 showed a 7.8% infection in cattle
slaughtered but by 1968 this incidence was only 0.4% in
a survey of selected slaughterhouses (Blamire, Crawley
and Goodhead, 1970). The position in Northern Ireland
and Scotland has also shown the same marked improvement
over this period. However, this should not result in
complacency as in some areas, e.g. the south-west of
England where there are relatively more reactors at the
on-farm test. The abattoir monitoring alone in England
and Wales revealed 32 positive cases in home-bred cattle
in 1970 and check testing of the herds of origin revealed
reactors in 16 herds. The abattoir is still an important
check-point for detecting any herd breakdowns of this
disease.

*Brucellosis*

The brucellosis control and eventual eradication
campaign is gathering momentum, there are now
6 eradication areas in Great Britain. 52.3% of the
herds are participating in the scheme and 50% of the
total bovines are in accredited herds. Northern Ireland
is brucella-accredited. By law any animal reacting to a
brucella test must only be sold for slaughter, the meat
from such animals is normally perfectly safe for human
consumption but the abattoir should be informed if
reactor animals are being sent for slaughter and proper
precautions taken by those handling carcasses and offal
from such animals.

*Salmonellosis*

With the increasing incidence of food poisoning,
possibly associated with the expanding use of convenience
foods and with the acceptance by medical authorities that
animals and animal by-products are the main reservoir of
infection (except in cases of typhoid and para-typhoid)
there is an increasing need to control or eliminate this
on the farm of origin. The intensification of stocking,

the close housing of animals, the movement of animals
between farms all provide means of spread of, for example,
salmonella organisms. The change in incidence and the
pattern of salmonella incidence investigated by
Veterinary Investigation Centres was highlighted by
Stevens, Gibson and Hughes (1967). Not only was there
a dramatic increase in the number of outbreaks in cattle
(mainly calves) from 351 in 1961 to 1728 in 1965, but
the ratio of S.typhi-murium to S.dublin isolations changed
from 1:3 in 1961 to 3:4 in 1965. This change was
attributed to the activities of calf dealers whose
premises had become infected. In the case of the calf
the main original source of infection would appear to
be the carrier adult cow but the adoption of proper
health control codes in handling calves can go a long
way to reducing the build-up and spread of infection.
In the case of pigs the prime source of infection would
appear to be from infected fish and bone meal in the
feeding-stuffs but the cubing of the ration and buying
of animal protein constituents from reliable sources
can do much to eliminate the condition. The recent
Agriculture (Miscellaneous Provisions) Act of 1972 gives
power to Government Departments to investigate fully
outbreaks caused by exotic and other strains of
salmonella which can be a major public health risk.

The human health risk has increased since contaminated
meat may spread its infecting organisms to clean meat
when processed into sausages and pies, the subsequent
cooking may not kill this organism and food poisoning
can result. The Food and Drugs Act (1955) gave
Ministers wide powers to safeguard production of safe
foods; these appear to have been effective in England
and Wales up to 1966 when the number of incidents was
2496 but thereafter the number of food poisoning
incidents increased to 4820 in 1969 (Blamire, 1972).


*Hydatidosis*

There is increasing incidence of this disease in sheep;
in 1972, data from 60 abattoirs in England and Wales
showed an incidence of 2.57% compared with 1.06% in 30
abattoirs, in 1966. It would appear that some 3% of
the sheep coming off Welsh and English farms could be
affected and there is also a 1% incidence in cattle.
This disease can be fatal in human beings (10 fatal cases

were recorded in England and Wales in 1969). The
sterilisation of all condemned meat together with
inspection of all animals slaughtered should break the
chain of infection. Since this is primarily a dog
parasite, with sheep or cattle (and occasionally humans)
being the intermediate host, the most effective point
of control must be by regular treatment of farm dogs,
especially sheep dogs, with the available tapeworm
medicines. Since the dog can harbour some six tapeworms
of importance in meat inspection there is an urgent
need for positive action against taeniasis in this
species. Prompt burial of all dead sheep and proper
disposal of offals from the on-farm slaughtered animals
are essential; the feeding of raw offal to dogs must
also be actively discouraged. The high incidence of
hydatid disease in horses in both Northern Ireland (15%
in 1970) and Great Britain (12.3% in 1963) is not fully
explained but it is considered that a different species
of the parasite may be involved in the case of the
horse.

*Liver fluke*

Considerable attention has been paid to warning farmers
of possible outbreaks of acute fascioliasis in sheep
(Ollerenshaw, 1971), but the abattoir condemnation rate
of livers remains at an unsatisfactory high level as
shown in *Table 6.4*. It has been suggested that
insufficient attention has been paid to the incidence
of chronic fascioliasis and that the loss from fluke in
cattle has been grossly underestimated. In England and
Wales, some 1 million sheep livers are condemned
annually with a possible loss of some £300,000, with a
further £1 million loss owing to unthriftiness and downgrading of carcasses. In the case of cattle there is a
29% liver condemnation rate costing the industry £2
million in liver losses alone with a further £5 million
on account of reduced weight gain and carcass quality.
This is one source of loss which should be reducable by
a link between abattoir and producer to mount an
awareness campaign by feeding back condemnation data in
order that control measures based on existing knowledge
can be applied on the farm. The fluke position in
Northern Ireland is equally as bad (*Table 6.5*) and there
are more bovine and sheep livers than carcass meat

*Table 6.4* Liver fluke condemnations in Great Britain. MAFF returns from 34 abattoirs in 1962 and 60 abattoirs in 1969 and 1972

|  | *1962* | *1969* | *1972* |
|---|---|---|---|
| **CATTLE** | | | |
| Total no. slaughtered | 677,215 | 361,989 | 379,999 |
| Condemnations | | | |
| (Part offals) | 165,109 | 127,935 | 75,768 |
| (Per cent) | 24.4 | 35.3 | 19.4 |
| **SHEEP** | | | |
| Total no. slaughtered | 2,146,691 | 1,185,231 | 1,314,437 |
| Condemnations | | | |
| (Part offals) | 128,136 | 140,695 | 75,162 |
| (Per cent) | 6.0 | 11.9 | 5.7 |

*Table 6.5* Condemnations of liver in Northern Ireland in 1971

| **CATTLE** | |
|---|---|
| Total kill | 346,339 |
| Percentage of cattle affected with liver fluke | 78.2% |
| Percentage of cattle affected with liver fluke 1955 | 67.0% |
| No. of livers condemned wholly | 146,694 |
| No. of livers condemned partially | 124,135 |
| Total weight of liver condemned | 1,118 tons |
| **SHEEP** | |
| Total kill | 356,263 |
| Percentage of sheep affected with fluke | 21.3% |
| Percentage of sheep affected with fluke 1955 | 22.9% |
| No. of livers condemned wholly | 48,583 |
| No. of livers condemned partially | 27,237 |
| Total weight of liver condemned | 43 tons |

condemned annually.  At a wholesale price of 20p per
lb the annual loss of some 1,200 tons of cattle and
sheep liver in Ulster amounts to £500,000 a staggering
loss in such a relatively small area.  The situation is
worse today than it was 20 years ago, and warrants
consideration of eradication.


*Milk spot livers in pigs*

Again, in spite of available and effective anthelmintics
there is a continuing loss as a result of ascarid
infection in pigs resulting in liver condemnations
(*Table 6.6*).  Other causes of 'milk spot', such as
toxocara worm infections, are known but more attention
to the strategic use of anthelmintics against the
ascarids together with controlled movement of pigs out
of infected pens, and proper cleansing of pens have
been suggested as a means of reducing this loss.
However, the costs of these procedures have to be
considered against the actual condemnation loss sustained.


*Warble fly*

Infestation with the parasite is notifiable in Northern
Ireland where the incidence is very low (0.11% of all
home-bred stock) but elsewhere in the UK, where control
is dependent on a voluntary preventive dressing by the
farmer, the incidence varies markedly in the different
regions.  This is shown in the results of a survey of
cattle passing through 5 livestock auction markets in
each region throughout the month of May 1973 by the Meat
and Livestock Commission (*Table 6.7*).  In the living
animal the warble fly is a major cause of 'gadding'
amongst animals at pasture with consequent risk of
injury and loss of milk and impaired growth rate in
fattening cattle.  The emergent larvae cause severe
damage to the hides seen as either 'open holes' or
'blind warbles' which are actually healed holes.  The
larvae also damage the subcutaneous and muscular tissue
of the back owing to the formation of gelatinous
material - the condition being called 'licked beef'.
The hide damage is estimated to amount to £1 million
annually but there are no precise national economic data
on the effect on milk and growth rate.

*Table 6.6*  Condemnations of pigs as a result of abscesses, pyaemia and milk spot liver (MAFF returns from selected abattoirs in Great Britain)

|  | 1962 | 1969 | 1972 |
|---|---|---|---|
| No. slaughtered | 1,334,057 | 1,495,131 | 1,643,640 |
| Abscess - Part carcass | 4067 | 8227 | 8855 |
| Per cent | 0.3 | 0.55 | 0.54 |
| Pyaemia - Whole carcass | 609 | 1593 | 1820 |
| Per cent | 0.05 | 0.11 | 0.11 |
| Milk spot livers - | | | |
| Partial | 94,922 | 92,215 | 111,811 |
| Per cent | 7.12 | 6.23 | 6.8 |

*Table 6.7*  Warble fly survey, May 1973 (MLC, unpublished data)

| Region | No. examined | % infected |
|---|---|---|
| Scotland | 9159 | 4.6 |
| Northern | 6738 | 6.3 |
| Yorks/Lancs | 8014 | 7.9 |
| East Midlands | 4965 | 25.9 |
| West Midlands | 5035 | 55.0 |
| Eastern | 5201 | 35.6 |
| Wales | 1280 | 49.9 |
| South West | 2866 | 43.0 |
| South East | 6181 | 43.0 |
| Total | 49,439 | 24.1 |

The experience in Northern Ireland where the incidence
of warbled hides has been reduced from a peak of 70%
(month of June) to 0.11% shows the value of a carefully
planned eradication scheme.


PREVENTION OF SPREAD OF NOTIFIABLE DISEASES

Meat or meat products infected with the viruses of foot-
and-mouth, swine fever and now swine vesicular disease
(SVD) will inevitably find their way into waste foods to
be fed to pigs and poultry.  The failure to sterilise
this waste food properly,  along with the risks from
contaminated food containers, has been a continual source
of trouble over the years.  This has been highlighted
over the past year with the continual reappearance of
swine vesicular disease.  The new Diseases of Animals
(Waste Foods) Order 1973, which will not be fully in
operation until after 1 July 1974, is aimed at closing
many of the loopholes in the old legislation.
The need for more stringent regulations has been
brought to the fore by the SVD epidemic but, with the
now wider international movement of meat and meat
products, the food chain has become an even more
important means of possible spread of animal disease.
This new legislation is an important aspect of our
national health control policy against notifiable
disease.  It should be noted that it is now illegal to
use waste food from aircraft as animal feed.


ANIMAL HEALTH AND MEAT HYGIENE IMPLICATIONS OF HUSBANDRY
AND OTHER FARM DEVELOPMENTS

*Housing*

The increasing cost of buildings and the need to
improve productivity have led to the use of more group
housing for all species of farm animals.  It must be
accepted that loose housing of cattle can lead to a
quicker spread of diseases such as tuberculosis,
brucellosis and salmonellosis.  In cattle, the risks of
injuries and hide damage caused by fighting have led to
de-horning or breeding of polled animals, but pneumonia
in housed cattle, sheep and pigs has also been a problem.
However, in pigs where effluent disposal and labour costs

have led to the reduction or elimination of the use of
straw, this has resulted in an increase in dirty animals
being sent for slaughter with the consequent hygiene
problems.  This system is also associated with outbreaks
of fighting and tail-biting which result in carcass
damage and also lead to condemnation.  The many
investigations into this problem have not highlighted a
single cause but temperature variation, ventilation
rates, stocking density and lack of dietary fibre are
major pre-disposing factors.  Pneumonia in housed pigs
is also quite common but unlike in cattle and sheep it
does not usually affect carcasses for meat and
condemnation losses are usually confined to the lungs.
In intensively housed lambs there is frequently a
problem of carcass taint presumably caused by the
presence of unsaturated fatty acids which may be
associated with lack of dietary fibre.

*Use of medicinal products, vaccines, etc.*

Implementation of the Swann Committee's recommendation
on the use of antibiotics and other anti-bacterial
substances should reduce the risk of the development of
antibiotic resistant strains of organisms and of
resistance transfer.  Many medicinal products approved
for animal use have a definite withdrawal period from
farm to slaughter.  This applies particularly to certain
anthelmintic preparations and to anti-bacterial agents
such as organic arsenicals.  Group feeding of animals
often means that observance of these withdrawal periods
is impracticable since animals due for slaughter are fed
alongside and not separated from younger animals.  The
presence of residues of medicinal products in tissues
can constitute a major problem for the meat hygienist
until regular screening can be carried out.  In the
meantime there is a continuing onus on the producer and
the veterinary adviser to ensure that, on the farm, the
required withdrawal periods are observed.  This aspect
is of particular importance when animals are being sent
for casualty slaughter.  There is major concern over the
increased incidence of abscesses in carcasses at
slaughter.  Norval (1966) reported from Edinburgh that
2 pigs in every 100 slaughtered had septic lesions and
a loss of 0.45 kg (1 lb) carcass weight on account of
abscesses in every pig slaughtered.  Similar data have

been reported from the Ministry of Agriculture, Fisheries
and Food (England and Wales) abattoir condemnation
returns (*Table 6.6*) and from Northern Ireland (*Tables
6.1, 6.2, 6.3* and *6.5*).  The cause is often faulty
injection techniques in the pig, the loss being
aggravated by the injections being made into the ham
which is the most expensive part of the pig.  Infected
castration wounds along with tail-biting and fighting
injuries also account for part of the pyaemia and
abscess incidence.  Similarly, in sheep, attention to
proper siting and hygiene at the time of injection will
reduce condemnation losses.  It is noteworthy that in
New Zealand it is conditional on approval of any sheep
vaccine that it is given at the neck site in order to
avoid the risk of damage in the more valuable parts of
the carcass.  The MLC have recommended either the neck
site or over the lower third of the ribs with hind leg
sites being completely ruled out.


LOSSES RESULTING FROM FAULTY HANDLING AND TRANSPORTATION

Having taken every care in producing his animals, it is
disturbing to see the results of failure by the farmer
to protect these animals against injury and consequent
carcass damage at loading or in transit to the abattoir.
Very often the unsatisfactory design of the lorry, the
state of uncleanliness, overcrowding on the lorries and
exposure to inclement weather result in animals reaching
the abattoir in an unsatisfactory state, which will down-
grade their value for meat.  Correcting these defects is
time-consuming but very much in the province of the
producer whose efforts on this should be well rewarded.
The occurrence of pale, soft exudative muscle ('watery
pork') and black cutting beef are in the main brought
about by conditions of stress (fright, excitement,
exhaustion etc.) before slaughter.  An adequate
preslaughter period of rest (12-24 h) with a maximum of
36 h is absolutely essential in the production of quality
meat of good durability.  The importance of quiet and
gentle handling of livestock cannot be overemphasised.
The mixing of different pigs and young bulls all too
frequently results in carcass meat losses.

PUBLIC HEALTH IMPLICATIONS

Certain zoonoses (i.e. diseases occurring in both man
and animals) must continue to receive very close
attention and are of particular significance in meat
hygiene.
Tuberculosis, brucellosis, salmonellosis and
hydatidosis have already been discussed. Cysticercosis,
trichinosis and leptospirosis must be also considered:

CYSTICERCOSIS

This parasitic condition, caused by the cystic stage of
the human tapeworm *Taenia saginata*, was first detected in
the UK in Belfast by McLean (1947). This was quickly
followed by reports of cases in Scotland, England and
Wales and this widespread incidence caused considerable
concern. In Great Britain some 0.18% of cattle were
found to be infected at post mortem examination (Blamire,
Crawley and Goodhand, 1970) but in Northern Ireland the
incidence in 1970 was 0.63% of the 317,273 cattle
slaughtered. The number of reported human cases in the
UK in 1970 was 168. Although this disease is not fatal
it is unpleasant and uncomfortable and the incidence is
not declining. The important factors in control are:

1. Elimination of infection in man.
2. Prevention of animals having contact with sewage
   and thereby preventing spread through sewage.
3. Detection of cysts at meat inspection.
4. Treatment of infected meat to render it safe for
   human consumption, mainly by refrigeration at
   -6.6°C for 3 weeks or -10°C for 2 weeks (this can
   cost £30 per ton and also causes deterioration in
   quality of meat).
5. Provision of adequate and efficient sewage
   purification plants.
6. Health education to raise hygiene standards e.g.
   prevention of consumption of raw beef.

TRICHINOSIS

The importance of this pig parasitic condition which
can affect humans has been highlighted by the requirement
of certain EEC countries of certification of freedom by
trichinoscopic examination. The EEC directive on intra-
Community trade (64/433) leaves it to the individual
countries to decide what examination is required. In
1972 the examination of muscle samples from 1% of pigs
slaughtered in the UK was begun. After nearly 2 years
of sampling only 2 positive cases have been recently
recorded. Individual trichinoscopic examination of
each carcass exported to certain countries is now
required. There is not a great public health risk in
this country, especially if all pig meat or pig meat
products are cooked before eating, but the possible
presence of this parasite cannot be ignored. However,
Gibson (1969) emphasised that since meat inspection
cannot be relied upon to detect all cases of *T.spiralis*
infection, the disease should be attacked on the farm by
destruction of vermin and by the boiling of swill before
it is fed to pigs.


LEPTOSPIROSIS AND OTHER CONDITIONS

Leptospirosis covers several infectious and contagious
diseases of man and animals caused by the leptospira
genus of micro-organisms. It is often a fatal condition
of man and animals, characterised by pyrexia,
haemorrhages, haemoglobinuria and jaundice. Abattoir
workers and others in regular contact with animals are
particularly at risk but proper hygienic precautions in
the abattoir will reduce this disease hazard.
    In an attempt to assess the incidence of leptospirosis
and brucellosis, as well as Q Fever and Louping Ill, in
abattoir workers, a study was carried out by Schoenell
*et al.* (1966) at Edinburgh abattoir. Antibodies to
*Brucella abortus* at a titre of 1-10 and above were
detected in 12.5% of the men while 28.1% had antibodies
to phase 2 antigen of *Coxiella burnettii* (the cause of
Q Fever). 6.2% were positive to *Leptospira canicola*
infection and 8.3% to Louping Ill antigen. More recent
work in Belfast has shown that 28.3% were positive to
*C.burnettii* (Connolly, 1968).

In addition to these infections workers in the slaughtering industry are also liable to contract bovine ringworm, contagious pustular dermatitis and anthrax.

As the eradication programme for brucellosis gets under way in Great Britain, with the concentration of the slaughter of numerous reactor animals, abattoir workers will be very prone to become infected and extra care will have to be taken. The usual meat inspection directives, including that of the EEC, of incising into udders and uteri are best disregarded in relation to brucella reactors. Instead these organs should not be handled but rejected after visual inspection only.

A knowledge of and a strict implementation of the fundamentals of hygiene are essential not only for the promotion of good health in workers but also for the production of high quality meat destined for human consumption.


## CASUALTY SLAUGHTER


No discussion on the implication of disease in the meat industry would be complete without reference to emergency slaughter, potentially the greatest source of danger to meat and men.

It was Bollinger in 1876 who pointed out that four-fifths of all cases of human food poisoning in Germany were associated with the consumption of meat from casualty animals.

As many animals suffering from illness of an acute febrile nature are slaughtered in an attempt to market the carcass, such cases add immeasurably to the responsibilities of the meat inspector. They make ante-mortem examination an absolutely essential part of meat inspection and many European countries prescribe that no case of emergency slaughter may be passed for food without a bacteriological examination of carcass and organs. In Denmark and Northern Ireland all cases of emergency slaughter must be accompanied by a veterinary certificate which in the latter country must certify:

1. That the animal will not infect other animals in the abattoir.
2. That the meat will not cause danger to health by

contaminating the premises or meat therein.
3.   The nature of any drugs administered to the animal.

The fact that such veterinary certification is not
required by legislation in Great Britain is a grave
omission. Many of the cases consigned for emergency
slaughter are not emergency slaughter cases at all, many
having been ill for days, even weeks. The fact that
some 50% of all casualty animals received at Belfast
Meat Plant are totally condemned is a measure of the
extent to which some will try to salvage animals whose
proper destiny should be an inedible by-products plant.

It is time that consideration be given to the setting
up of emergency slaughter abattoirs which would cater
solely for these particular animals.

## SUMMARY OF PROCEDURES FOR HEALTH CONTROL AND OTHER MATTERS RELATING TO MEAT PRODUCTION

The safeguarding of the country's meat supply from
disease is a team operation aimed at producing meat at
a quality and price acceptable to the consumer and with
proper attention to the elimination of public health
and animal health risks that can be associated with the
end-product. The main points of control in the process
of meat production are:

1.   The protection of the animals from the effect of
     disease on the farm and their maintenance in an
     environment which will promote positive health.
2.   The controlled administration of medicinal
     products, vaccines and other preparations on the
     farm so as to prevent any carcass residues or the
     total or partial condemnation of carcasses or
     offals.
3.   The care and protection of animals during loading
     at the farm, during transport to and unloading at
     saleyards and meat plant lairages.
4.   Ante-mortem examination to eliminate from the
     slaughter line unfit animals and, where necessary,
     to make provision for special post-mortem
     examination.

5. Post-mortem examination of the carcass and its organs immediately after slaughter, including laboratory testing. Information obtained at this stage should be fed back to the farm to help in formulating the farm health control policy and in operation of health schemes.
6. Removal of material unfit for human consumption and its efficient treatment at by-products plants located *outside* the meat plant.
7. Ensuring the maintenance of high standards of hygiene at all stages from the farm to meat processing plants, meat storage and retail premises through to the consumer.
8. Handling of waste food containing meat and meat products to prevent the spread of animal disease.
9. Strict control, including documentation and checking of animals presented for casualty slaughter.

REFERENCES

BLAMIRE, R.V. (1972). Personal communication
BLAMIRE, R.V., CRAWLEY, A.J. and GOODHAND, R.H. (1970). *Vet. Rec.*, 87, 234
CONNOLLY, J.H. (1968). *Br. Med. J.*, 1, 547
EEC Directive 64/433 of 26.6.64 on Health Problems Concerning Intra-Community Trade in Fresh Meat
GIBSON, T.E. (1969). *Vet. Rec.*, 84, 448
Joint Committee (Swann) on the Use of Antibiotics in Animal Husbandry and Veterinary Medicine. Cmnd. 4190 (1969)
MCLEAN, A. (1947). *Vet. Rec.*, 59, 517
NORVAL, J. (1966). *Vet. Rec.*, 68, 708
OLLERENSHAW, C.B. (1971). *Vet. Rec.*, 98, 152
SCHONELL, M.E., BROTHERSTON, J.G., BURNETT, R.C.S., CAMPBELL, J., COGHLAN, Joyce D., MOFFAT, Margaret A.J., NORVALL, J. and SUTHERLAND, J.A.W. (1966) *Br. Med. J.*, 2, 148
STEVENS, A.J., GIBSON, T.E. and HUGHES, L.E. (1967). *Vet. Rec.*, 80, 154
THORNTON, H. and GRACEY, J.F. (1974). *Textbook of Meat Hygiene*, 6th edn.

# II

## PREPARATION AND STORAGE

# 7

# SLAUGHTER OF MEAT ANIMALS

LIZ BUCHTER

*The Danish Meat Research Institute,
Roskilde, Denmark*

## INTRODUCTION

This chapter is intended to cover work done on the
slaughter of meat animals by the Danish Meat Research
Institute. It is thereby hoped to give some coherent
information about factors of consequence for meat
quality when slaughtering pigs and beef animals -
information which it might be useful to take into
consideration when deciding on future developments.
Slaughter processes will only be examined if they are
of consequence for meat quality. Thus, engineering and
economic problems will not be considered. Aspects of
meat quality, such as hygiene, carcass composition and
nutritive value, will not be included. The term 'meat
quality' will only cover properties which can be
perceived visually or organoleptically, and which are
known to be influenced by the immediate pre-slaughter,
slaughter and post-slaughter handling. For beef, this
is primarily meat tenderness and for pigs primarily a
complex of properties including meat colour, water-binding
capacity, and final pH. These properties are combined
to describe the PSE (pale, soft and exudative) - DFD
(dark, firm and dry) complex. The final, and perhaps
most important limitation, comes from the type of animal
used. The results are obtained on intensively raised,
fast growing, young animals. The beef animals are of

the Red Danish and Danish Friesian breeds, both of which are dual-purpose varieties, and the pigs are of Danish Landrace, which is one of the more stress susceptible breeds. Beef animals were slaughtered at 250-450 kg liveweight and pigs at 90 kg liveweight.

It can rightfully be claimed that meat quality as defined above has been of little interest to the industry. However, as animals are reared in more intensive units and as the slaughter rate increases, problems of poor meat quality will intensify. At the same time demands from the meat industry for uniform raw materials can be expected to become more important. It is in the light of these developments that the results below should be considered.

The methods and design of the experiments cannot be given in detail. Most of the results originate from experiments where the problems have been to investigate the influence of various factors on meat quality, in order to set up standardised procedures for the handling of animals from feeding and breeding experiments during their transport and slaughter. Such results have been produced under very carefully controlled conditions and the determinations of meat quality have been extensive and accurate. In experiments on commercial animals the methods have often been more primitive in order not to spoil the carcasses, but the number of animals investigated has been proportionately larger. All animals have been slaughtered in commercial abattoirs, and the handling of the animals under transport and at the abattoir has been performed by ordinary workmen under supervision.

Tenderness of beef and pork was determined both by shear force measurements with Volodkewich jaws and by taste testing performed by a trained panel. Details are given by Buchter (1972). The meat was always fully aged before the tenderness was assessed. Determination of the meat quality in pigs is more complicated. In order to get correct information of the degree and extent of PSE- and DFD-meat, it is necessary to measure the final pH value and meat colour in several muscles.

The first signs of the development of DFD-meat can usually be found by measuring the final pH value in the *quadriceps femoris* muscle in the ham and the *semispinalis capitis* muscle in the neck. The PSE-condition is most frequently determined on *longissimus dorsi*, *semimembranosus* and *biceps femoris* muscles. A programme for routine determinations of the meat quality in experimental pigs has been developed and most recently (but not up to date)

described by Barton (1971). These routine determinations
include colour measurements on fresh and cured vacuum
packed samples combined with measurements of final pH
values in several muscles. Such colour measurements have
proved to be suitable for comparing pigs of the same
breed and age. In several of the experiments the PSE-
condition has also been determined using the solubility
of the sarcoplasmic and myofibrillar proteins (S+M-
proteins) in a O.6 M KI-buffer. Blood splashing was
visually assessed in the shoulders and hams of pig
carcasses after deboning (Nielsen, 1974). In the
following examples some of the main results are given,
first for beef animals and then for pigs.

## BEEF

### TRANSPORTATION

The meat quality of young beef animals can be spoiled
by prolonging the time between the animal leaving the
farm and its slaughter. *Table 7.1* shows the results of
an investigation of the final pH values in the loin
muscle in commercial calves.

From the results it can be seen that approximately 10%
of the animals, which had passed through market places
or been kept overnight in the slaughterhouse lairage,
had developed dark cutting meat, which is found to be
inferior not only because of lower keeping quality but
also because of less desirable colour and flavour.

Our experience is that dark cutting meat can frequently
be found in young bulls and calves that have had
prolonged or rough handling prior to slaughter. On the
other hand it is very seldom found among healthy steers,
heifers, and cows irrespective of transport time and
conditions, as long as these follow the general
regulations for the humane transport of animals for
slaughter.

Blood splashing and watery meat in calves are possibly
related to the handling of the animals prior to and at
slaughter, but these problems have not been investigated
in detail by us.

All animals must, of course, be carefully handled and
transported in well-designed lorries, but as a
consequence of the above findings, the handling of young
bulls and calves must be especially considered.

Table 7.1 Frequency of dark cutting meat in the loin muscle of calves as related to trade route and holding time in the slaughterhouse lairage

| Type of animals and treatment | Trade route | Holding time on slaughterhouse | No. of animals | % calves with final pH $\geq$ 6.2 in l.dorsi |
|---|---|---|---|---|
| Commercial 250 kg liveweight calves | Farm to slaughterhouse | Slaughtered on day of arrival | 296 | 2% |
| | | Overnight stay | 34 | 9% |
| Transport and lairage conditions unknown | Farm to market to slaughterhouse | Slaughtered on day of arrival | 187 | 9% |
| | | Overnight stay | 353 | 9% |

SLAUGHTER

For beef animals no investigations have been carried out
to evaluate if or how the meat quality can be influenced
by stunning, sticking, dehiding, washing etc. A comparison of splitting the carcasses at 1/2 h post mortem and
splitting 24 h post mortem did not influence meat
tenderness. The idea of using different carcass positions
during the development of rigor mortis in order to
improve tenderness (Hostetler *et al.*, 1970) has so far
not been investigated by the Institute. We have no doubt
that the tenderness of certain muscles can be improved,
but so far we have no plans for practical application
because of the problems involved in hanging the carcass
from the obturator foramen and in the alterations of the
cuts.


CHILLING

Modern practice in beef plants tends to favour rapid
chilling of beef carcasses in order to obtain the
economic advantages of increased turn-over, better
hygiene and lower chilling losses. The possible effects
of cold shortening induced by the chilling must, however,
also be taken into consideration.
  The cold shortening effect was first investigated by
Locker and Hagyard (1963). They found that excised
pre-rigor muscles shortened rapidly and progressively,
when cooled towards 0°C. Marsh and Leet (1966) showed
the relationship between percentage shortening and
toughness, and in the same paper demonstrated how a
restrained muscle on exposure to cooling would shorten
in the areas first cooled and stretch in others, leaving
a muscle of the same total length but with alternating
zones of tough and tender meat. A suggestion was made
that such alternating zones would be found when muscles
attached to the skeleton were chilled too rapidly.
  These findings and suggestions have been fully
confirmed by us in a series of experiments where the
relationship between chilling temperature and final meat
quality was studied. An initial series of experiments
with 44 commercial cows showed clearly that chilling in
tunnels at temperatures below 0°C did induce an increased
toughness in the loin muscle. The experimental material
was then changed to 250 kg liveweight calves, as the
chilling conditions were considered to be more critical
for these animals because of the smaller meat volume and

less fat cover.  Finally, experiments were carried out
with 450 kg liveweight young bulls.
  The experiments were performed in two identical
7.5 x 8 m chilling rooms where right and left sides from
the same carcasses were exposed to different air
temperatures for the first 24 h post mortem.  The
chilling started within 3/4 h post mortem and 24 h post
mortem the sides were recombined and hung at 4°C for
further cooling and ageing.  The whole loin muscles were
cut from the sides not earlier than 2 days post mortem,
after which they were aged in plastic bags at 4°C until
7 days post mortem for calves and 9 days post mortem for
young bulls.  The results are given in *Table 7.2.*  It
can be seen that the toughness in fully aged loin muscles
from calves and young bulls increased as the air
temperature in the chilling room decreased.  Further
details are given by Buchter (1972).
  On the basis of these results it was decided that
animals from feeding and breeding experiments should be
chilled at a constant temperature of 6°C, as this
temperature is a compromise between considerations for
hygiene and considerations for meat tenderness.  As guide
lines for commercial animals, it was recommended that
calves shall not be chilled at temperatures below 6°C,
young bulls not lower than 4°C, and bigger beef animals
with more fat cover not lower than 2-3°C during the
first 24 h after slaughter.  These recommendations are,
of course, not feasible in most chilling rooms, but they
can be used, together with the rule of thumb from the
Meat Chilling Symposium (Cutting, 1972) that temperatures
in the meat should not be below 10°C for the first 10-16 h
post mortem, to design custom-built cooling programmes
for slaughterhouses.


CUTTING UP

The influence of chilling temperature on the toughness
of the meat is not as critical during the later part of
the rigor processes as during the first 12-24 h post
mortem.  It is, however, essential for the tenderness in
the loin muscle that the rigor processes are completed
before the muscles are removed from the suspension of
the skeleton.  *Table 7.3* shows the results of an
experiment in which right and left loin muscles were cut
from the skeleton at different times.  A total of 20
calves were chilled at 6°C after which the whole sides
and the muscle samples were aged at 4°C in the same room
until 7 days post mortem.

*Table 7.2*  Temperature in chilling room from 1 to 24 h post mortem as related to the tenderness in the *l.dorsi* muscle of calves and young bulls

| Temperature in chilling room | Calves | | | Young bulls | | | | |
|---|---|---|---|---|---|---|---|---|
| | Number of animals | Shear force value | | Number of animals | Shear force value | | Tenderness score | |
| | | x̄ | s.d. | | x̄ | s.d. | x̄ | s.d. |
| 10°C | 45 | 8.6 | 2.3 | – | – | – | – | – |
| 6°C | 44 | 10.5 | 4.0 | 32 | 8.1 | 2.2 | 2.7 | 1.4 |
| 4°C | 21 | 13.7 | 5.6 | 8 | 10.2 | 5.3 | 1.7 | 0.9 |
| 2°C | – | – | – | 12 | 9.6 | 4.0 | 1.8 | 2.2 |
| 0°C | – | – | – | 12 | 12.2 | 4.7 | 1.4 | 1.0 |
| Tunnel then 1°C | 7 | 24.5 | 3.7 | – | – | – | – | – |

x̄ = mean
s.d. = standard deviation

*Table 7.3* Time post mortem when muscle was cut from
the skeleton in relation to tenderness in aged loin
muscles from calves

| Time post mortem for removal of the muscles from the skeleton | Shear force value (kg) | Tenderness score (scores from -5 to +5) |
|---|---|---|
| 1 day | 21.5 | -0.44 |
| 2 days | 10.6 | +0.82 |
| 3 days | 11.0 | +0.01 |
| 4 days | 10.0 | +1.63 |
| 6 days | 11.1 | +0.77 |

It can be seen that the magnitude of the toughness,
that can be induced in the loin muscle by removing it too
early from the suspension of the skeleton, is of the same
order as the toughness that can be induced by too rapid
chilling. As a consequence of this, the industry is
advised to leave beef sides intact until at least 2 days
after slaughter.


# PIGS


## TRANSPORT, LAIRAGE AND STUNNING

All experiments concerning transport and stunning have
been carried out by a team whose members are: Patricia
Barton (group leader), N.J. Nielsen, B. Røy, C.F.
Yndgaard, and the author.
   The factors that influence the meat quality in a pig
from the time it leaves its usual environment until it
is slaughtered have an additive effect. It is, therefore,
very difficult to ascertain, how the conditions during
transport and slaughter should be improved in order to
minimise the development of PSE- and DFD-meat and the
occurrence of blood splashing. Only by changing one
factor at a time and by keeping all other factors
constant can it be hoped to obtain some details in a
large and complicated puzzle. In the following some
examples of such experiments will be given, where only
one factor in the treatment of the pigs is investigated,
while all other factors have been chosen in such a way
that they can be reproduced on different days and at
different abattoirs.

TRANSPORTATION

The first factor to be investigated was the duration of
the transport (Barton, Buchter and Røy, 1971). Progeny
testing pigs were transported 1/4 h, 1 h or 3 h in a
transport lorry of improved design (*see* below). All
pigs were electrically stunned on the floor immediately
after arrival. The results are shown in *Table 7.4*. The
results showed clearly that the shortest transport, where
loading, transport, unloading, and stunning occurred over
a short time interval, caused the highest incidence of
PSE-meat. Over longer transport periods, some of the
pigs apparently adapted themselves to the treatment.
DFD-meat (defined as final pH values $\geq$ 6.3 in the
*quadriceps* muscle) was no problem in this experiment,
possibly as a result of very careful handling and the
time of year. The rise in mean pH value in the
*quadriceps* showed, however, some signs of increasing
exhaustion over longer transport times.
 As mentioned above, the lorry used for transporting
the progeny testing pigs had been improved. The
improvements consisted of a non-slip floor and partitions
in the lorry. The effect of these installations on the
meat quality of commercial pigs was investigated
(Nielsen, 1973a) in order to convince the bacon factories
that investments in better transport vehicles could
result in better meat quality. Two containers were built
to fit the same lorry. The first container fulfilled the
standard requirements, i.e. a minimum floorspace of
0.35 $m^2$/pig, adequate natural ventilation, and shelter
from sun and rain. The second container was additionally
fitted with a non-slip aluminium floor, two transverse
partitions, and a forced ventilation system. The latter
allowed a temperature over the pigs of 10-20°C throughout
most of the year. While changing between the ordinary
and the improved container between loads, commercial pigs
were collected from several producers to give a total of
40 pigs per load. For each load the driving time was
kept within 1½-2 h. Less than 15 min after arrival, the
pigs were driven through a corridor for stunning in a
$CO_2$-tunnel. Some examples of the results are given in
*Table 7.5*. The results showed that the installation of
non-slip floor, partitions, and forced ventilation
improved the meat quality in pigs by reducing the number
of pigs with DFD-meat. No difference between the
ordinary and the improved container could be found in
$pH_1$ values in loin and ham, but an investigation of
seventy 21 lb hams (Barton, 1973) showed a significantly

Table 7.4 The influence of transport time on the meat quality in pigs

| Type of animal, general treatment, time of year | Transport time | No. of pigs | PSE % pigs with PSE in l. dorsi (colour and solubility of protein) | DFD % pigs with final pH ≥ 6.3 in quadriceps | DFD Final pH in quadriceps $\bar{x}$ | Blood splashing Visual judgement of muscle samples |
|---|---|---|---|---|---|---|
| Progeny testing pigs | 1/4 h | 123 | 40% | 2% | 5.70 | No significant difference |
| Stunned electrically on floor immediately on arrival | 1 h | 126 | 25% | 2% | 5.72 | between treatments |
| April–June | 3 h | 126 | 10% | 3% | 5.79 | |

Table 7.5 Influence of the design of the transport lorry on the meat quality in pigs

| Type of animal, general treatment, time of year | Design of container | No. of pigs | PSE | DFD | Blood splashing |
|---|---|---|---|---|---|
| | | | pH₁ measurements in l.dorsi and semimembranosus Evaluation of canned hams | % pigs with final pH ≥ 6.3 in quadriceps femoris | Visual judgement of hams and shoulders after deboning |
| Commercial pigs Transported 1½-2 h, stunned in $CO_2$-tunnel shortly after arrival January-March | Ordinary | 353 | No significant difference for pH₁ values, but better colour and consistency in canned hams from pigs transported in improved lorry | 12% | No significant difference between treatments |
| | Improved | 350 | | 7% | |

better colour and consistency in the hams produced from
pigs transported in the improved container.  It is
concluded that transport in the improved lorry caused a
decrease in both the DFD- and the PSE-condition in pigs.

STUNNING

Three types of stunning are commonly used in Denmark.
These are stunning in a $CO_2$-tunnel, electrical stunning
in a V-restrainer (Frederiksen's restrainer) and electrical
stunning on the floor.  The latter is mostly used for sows
and other animals that cannot pass through the mechanical
systems.  For electrical stunning the regulations state
that the voltage must not exceed 85 V.

In an experiment with progeny testing pigs (Barton,
Buchter and Røy, 1972) electrical stunning on the floor
was compared with stunning in a $CO_2$-tunnel or electrical
stunning in a V-restrainer.  The two latter systems were
both connected with 20 m corridors.  The pigs were
transported for 2 or 3 h in an improved lorry and stunned
within 15 min of arrival.  Examples of the results are
shown in *Table 7.6*.  The results showed that combined
efforts of passing through a corridor and being stunned
in a tunnel/restrainer resulted in a higher frequency
of pigs with DFD-meat, compared with pigs stunned
electrically on the floor.  There were very few pigs
(approximately 5%), which had PSE-meat in this experiment,
and there was no significant difference between treatments
with regard to the PSE-condition, only a slight tendency
in favour of electrical stunning on the floor.

In an attempt to answer the old question of whether
stunning in a $CO_2$-tunnel or electrical stunning in a
V-restrainer is the better method, the meat quality of
commercial pigs at a large bacon factory was investigated
before and after the factory changed from one system to
the other (Nielsen, 1973b).  The installation of the new
system ($CO_2$-tunnel), however, took longer than expected,
so that a seasonal variation was introduced.  The
experiment is, therefore, to be continued in the spring,
so that the effect of temperature etc. during transport
on meat quality can be reduced to a minimum.  However,
the difference in the number and extent of blood
splashing was found to be very different for the two
systems (*Table 7.7*).

The figures for percentage blood splashing might not
be typical for the systems.  However, experience from
other bacon factories support the general feeling that

Table 7.6 Influence of the type of stunning on the meat quality in pigs

| Type of animal, general treatment, time of year | Type of stunning | No. of pigs | PSE | DFD | |
|---|---|---|---|---|---|
| | | | % pigs with PSE in l.dorsi (colour and solubility of protein) | % pigs with final pH $\geq$ 6.3 in quadriceps femoris | Final pH in quadriceps femoris $\bar{x}$ |
| Progeny testing pigs Transported 2 or 3 h | No corridor Electrically on floor | 137 | 5% | 7% | 5.88 |
| Slaughtered within 15 min of arrival January-February | Corridor Electrically in V-restrainer or $CO_2$-tunnel | 138 | | 13% | 5.96 |

*Table 7.7* Frequency of blood splashing in meat from
pigs stunned by two systems in the same bacon factory

| *Type of animal, general treatment* | *Type of stunning* | *No. of pigs* | *% pigs with blood splashing found after deboning* | |
|---|---|---|---|---|
| | | | Shoulders | Hams |
| Commercial pigs | Electrically using 60 volts V-restrainer | 257 | 64% | 22% |
| Uncontrolled transport and time in lairage | CO$_2$-tunnel | 255 | 8% | 4% |

electrical stunning in a V-restrainer in its present form
is inferior to stunning in a CO$_2$-tunnel with regard to
the amount of blood splashing.

In *Tables 7.4, 7.5, 7.6* and *7.7* examples have been
shown of some factors which might influence meat quality
in pigs. Examples that have proved that even minor
alterations in the equipment or the handling during the
transport and slaughter of stress susceptible pigs can
improve or spoil the meat quality, both in regard to
PSE- and DFD-meat and to blood splashing. However, many
details are still required before standard solutions for
pig handling during transport and slaughter can be found.
The next series of experiments is planned to cover the
influence of holding time and conditions in the lairage.


CHILLING

Unlike beef animals, pigs should be chilled relatively
quickly. Although rapid chilling cannot prevent the
formation of PSE-meat, delayed evisceration and delayed
chilling can cause most meat to be somewhat pale and
watery.

Two types of rapid chilling methods are commonly used
in Denmark. One method, intensive batch chilling,
involves a filling phase of about 1 h, a chilling phase
of 2 to 3 h at -8 to -10$^\circ$C with an air velocity of 1.5
to 2.0 m/s, and an equalisation phase of 16 h, where the
temperature in the carcasses equilibrates to 4$^\circ$C. The
other method is the on-line chilling tunnel, where the
carcasses are chilled for 1 h at -18$^\circ$C and then for 1 h
at -7$^\circ$C before the temperature is allowed to equalise as
in batch chilling.

The reason for the introduction of rapid chilling has been low chilling losses. When the chilling tunnels were first introduced, experiments were undertaken to investigate how the meat quality was influenced. Right and left sides from a total of 42 pigs have been chilled either in a tunnel as described above or chilled more moderately at -2 to +4°C. No significant differences were found between the two chilling methods, either for the PSE-conditions, or for meat tenderness.

## CONCLUSIONS

Much of the research carried out in relation to the subjects of this paper never gets published in journals, some results are presented at meetings for meat research workers, but most of the information is only available by personal contact. Our results sometimes seem to contradict the findings of other research groups. The reason for this is, in my opinion, not because our results are not in accordance with more fundamental results and theories of the toughening of meat through cold shortening or of the formation of PSE- and DFD-meat. The explanation is more likely to be the different type of animal used.

A natural way of ending this paper would be a prophesy of future developments. According to the best of my knowledge major developments of significance for the industry can be expected to occur in the following areas:

1. The relatively slow development of the rigor processes in beef animals leaves several possibilities of influencing meat tenderness during this phase. Simple innovations are moderate cooling followed by traditional ageing or the use of different carcass positions, while other possibilities such as high temperature cooling and ageing or hot boning can only successfully be undertaken by bigger firms, which can develop and control the necessary systems.
2. In many countries slaughterhouses will either establish their own fleet of vehicles for transporting live animals or they will increase their influence over private owners of transport lorries. The new research results show that the general standard of lorries used for animal transport can easily be improved.

3. The methods used for stunning are at present
being critically questioned especially in relation
to blood splashing. New developments could include
high-frequency stunning and systems where corridors
and restrainers are avoided or modified.

## REFERENCES

BARTON, P.A. (1971). 'Some experience on the effect of
pre-slaughter treatment on the meat quality of pigs
with low stress-resistance', *Proc. 2nd Int. Symp.
Condition Meat Quality Pigs, Zeist.* Pudoc, Wagening,
p.180

BARTON, P.A., (1973). Internal report no. 01.447 II
(SVIN - TRANSPORT) of Oct. 1st

BARTON, P.A., BUCHTER, L. and RØY, B. (1971). Internal
report no. 01.396 (SVIN - TRANSPORT) of Sept. 20th

BARTON, P.A., BUCHTER, L. and RØY, B. (1972). Internal
report no. 01.415 (SVIN - TRANSPORT) of Oct. 17th

BUCHTER, L. (1972). *The Why and How of Meat Chilling*,
Meat Research Institute Symposium, Agricultural
Research Council, Bristol

CUTTING, C.L. (1972). *The Why and How of Meat Chilling*,
Meat Research Institute Symposium, Agricultural
Research Council, Bristol

HOSTETLER, R.L., LANDMANN, W.A., LINK, B.A. and FITZHUGH,
H.A. Jr. (1970). *J. Anim. Sci.*, 31, 47

LOCKER, R.H. and HAGYARD, C.J. (1963). *J. Sci. Fd. Agric.*
14, 787

MARSH, B.B. and LEET, N.G. (1966). *J. Food Sci.*, 31,
450

NIELSEN, N.J. (1973a). Internal report no. 01.447 I
(SVIN - TRANSPORT) of May 16th

NIELSEN, N.J. (1973b). Internal report no. 02.221 I
(BEDØVNING - SVIN) of Oct. 30th

NIELSEN, N.J. (1974). *Method for Evaluating Blood
Splashing in Pigs*, internal information available on
request

# 8

# CARCASS QUALITY

A. CUTHBERTSON

*Meat and Livestock Commission,*
*Bletchley, Milton Keynes*


## INTRODUCTION

The title of this chapter begs the fundamental question
of what constitutes excellence in a carcass but,
unfortunately, this question cannot be answered
succinctly and with precision because of variability in
quality requirements.

In the process of marketing meat from the farm through
to the consumer, several groups of people make judgements
about carcass quality. These are principally farmers,
wholesalers and retailers whose views about carcass
quality may not necessarily coincide. The most important
arbiter of carcass quality is the retail butcher whose
judgement probably has the greatest impact on the type
of carcass produced since he is able, for example, to
obtain information on the yield of different types of
carcasses, to modify the characteristics of carcasses
to meet his customers' requirements and also, to some
extent, to influence what the customer purchases. It
is evident, however, that the definition of carcass
quality as judged by the retail butcher becomes
modified in its movement through to the producer by
considerations such as ease of handling and processing
and trade practices. A development which may lead
increasingly to the retailer becoming less involved in
determining carcass quality is the growth of vacuum

packing of boneless trimmed cuts (even to retail level).
In that event, the meat plant operator will assume the
retailers' role as an arbiter of carcass quality.

In attempting to define quality in a carcass with
precision, it is necessary to consider the characteristics
which produce different degrees of overall carcass
excellence. Each individual butcher will have several
characteristics which he will attempt to rank in order
of importance and the items included and their order will
be different depending on the individual. At the
extreme, some will base their views on fairly precise
evidence of their customers' needs and of the factors
important in preparing meat for sale, although they may
not be aware of how to assess a carcass in order to meet
these needs. Others will have age-old beliefs of the
kind of meat their customers should eat and have little
real appreciation of the facts which are important when
cutting up a carcass. This diversity of opinion which
also exists amongst producers and wholesalers is
unfortunate, since it implies the absence of a clear
objective in evaluating a carcass and must lead to
inefficiencies in marketing meat from the producer
through to the consumer.

The object of this paper is to try to identify the
characteristics of a carcass which can be proved to be
important on the basis of objective results and to place
them in some order of importance. Methods of assessing
the chosen characteristics, both in the live animal and
in the carcass, will also be considered. The dividing
line between carcass and eating quality characteristics
is difficult to define because they are closely linked
but, as far as possible, this paper will deal with the
variables important in preparing meat up to the point
of sale and will leave consideration of the effect of
these and other characteristics on eating quality to
other contributors.

The subject of this chapter is a large one and, in the
space available, it cannot provide a comprehensive
review. It is hoped, however, that no important points
have been omitted.

## COMPONENTS OF CARCASS QUALITY

The components of carcass quality may be divided
conveniently into two, namely, the quantity of meat and
its distribution and thickness through the carcass.
These summarise the characteristics of major importance
to a retailer when buying carcasses and preparing them
for sale. In addition to these, there is the colour of
fat, particularly yellowness, which can cause problems
in marketing some beef, in particular, at certain times
of the year and from certain breeds (especially Channel
Island). Marked yellowness in the fat shows up in
contrast to carcasses with whiter fat, and it is this
contrast in colour which appears to cause carcasses with
yellow fat to be penalised by meat traders.

At the outset, it is important to define what is meant
by the term 'meat'. In the context of this chapter it
may be described as 'saleable meat', a term loosely used
to describe the meat which is sold across the counter to
the consumer. Saleable meat consists not only of the
lean meat in the carcass, but of variable quantities of
fat and bone depending on a number of factors, e.g. fat
content of carcass, fat requirements of customers, and
cutting or jointing method. The approximate allocation
of carcass tissues to the saleable components in a
typical beef carcass has been illustrated by Cuthbertson,
Harrington and Smith (1972). In contrast to beef and
pork, lambs cut traditionally produce a much higher yield
of saleable meat since this includes all the bone.
However, with the development of a market for lean heavy
lamb, bringing with it new cutting methods, less of the
bone is likely to be sold in the saleable meat.

In view of the variable nature of the saleable meat,
this chapter will deal, for the most part, with
variations in the carcass tissues and particularly in
lean meat which forms the largest component of saleable
meat. In order to simplify presentation, the factors
affecting the major carcass quality characters will be
considered in the way illustrated in *Figure 8.1*.

This approach allows the factors influencing size of
carcass to be considered first, so that factors
influencing the proportion, distribution and thickness
of the tissues can be discussed largely without having
to refer to carcass weight effects. Since beef, sheep
and pigs share somewhat similar carcass quality problems,

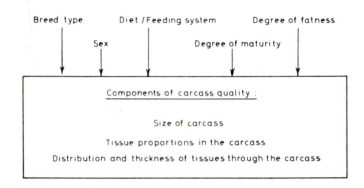

*Figure 8.1　Components of carcass quality and factors influencing them*

reference will not always be made to each of the species. The species used for particular aspects will depend, to some extent, on the evidence available.


SIZE OF CARCASS

Clearly the size of a carcass, as expressed through its weight, has a major influence on the quantity of lean meat. It also influences the size of joints and of the muscles exposed on cutting and this is of importance in relation to a meat trader's ability to use his normal cutting methods on carcasses outside his usual weight requirements. This could, for example, account in part for the producer of heavy beef carcasses tending to receive a lower value per unit weight for such carcasses in Britain.

The weight at which an animal is slaughtered is influenced by the five factors given in *Figure 8.1*. In general, breed, sex, diet/feeding system, degree of maturity and the interaction of these, influence the size of a carcass through the weight achieved at a given degree of fatness.

Breed type exerts the greatest effect on size. Results now coming forward from Clay Centre in the USA, provide some of the most comprehensive beef breed comparison information available to date, and further studies are in hand in other countries including Great Britain. The superiority of certain European beef breeds in terms of

carcass size and yield of retail cuts, is evident from
the various studies.

So far as sheep are concerned, there are very marked
breed differences as evidenced by results from the Meat
and Livestock Commission's (MLC) Sheep Recording Scheme.
Bodyweights of sheep from a wide range of breeds in the
scheme have been combined with data from experimental
flocks to estimate the mature bodyweights of most of the
common breeds in Britain. As a guide to the expected
carcass weight of lambs, grown without check and with no
store period, from different breed types, it has been
suggested that 12½% of the combined weights of the mature
parent breeds be used. On this basis, Table 8.1 sets out
the estimated carcass weights for a selection of breed
types. It must be emphasised, however, that these can
be considered only as approximate weights, since no
attempt has been made to define the optimum slaughter
weight as, for example, the weight at which the food
conversion ratio reaches the point where additional
growth uses up more feed than the additional saleable
meat produced on a value for value basis.

Table 8.1 Estimated lamb carcass weights for a
selection of breed types

| Breed | Carcass weight (kg) |
|-------|---------------------|
| Welsh Mountain | 11 |
| Southdown | 15 |
| Cheviot | 16 |
| Scottish Blackface | 17 |
| Clun Forest | 18 |
| Dorset Down and North Country Cheviot | 19 |
| Dorset Horn and Lincoln Longwool | 21 |
| Suffolk | 23 |
| Border Leicester | 25 |
| Oxford Down | 27 |

The pig breeds commonly used nowadays, at least in
Britain, are rather less disparate in their optimum
slaughter weights owing to the demise of the early
maturing pig breeds.

There are marked differences in size within breeds of
the various species and sex also has an important
influence on slaughter weight. Considerable evidence

on differences between steers and heifers has recently
become available through the operation of a beef carcass
classification scheme involving some 400 management
units in Britain. This scheme describes the weight of
carcasses, their fatness (on a scale of 5 classes where
1 = very lean to 5 = very fat), conformation (on a
scale of 5 classes where 1 = very poor to 5 = very good),
sex and age group (determined by the number of teeth
erupted). *Table 8.2* shows the average carcass weight of
steers and heifers and their average fat class together
with the distribution of weights by sex. Further MLC
data collected on the performance of semi-intensively
reared 18 month old Friesian, Hereford x Friesian and
South Devon x Friesian bulls and steers show that, at
this age, bulls had an overall advantage over steers of
12% in carcass weight and 22% in weight of lean meat.
There were also marked advantages in the cross-sectional
area of *l.dorsi* in favour of the bulls.

The diet/feeding system chosen will influence the size
of carcass at slaughter through the weight achieved at a
given level of fatness. If, for example, a high energy
diet is fed an animal will tend to reach a given level
of fatness at a lighter weight and younger age than if a
lower level of energy had been fed, since energy surplus
to the potential for lean tissue growth will be deposited
as fat.


## TISSUE PROPORTIONS IN THE CARCASS

The proportion of lean meat or fat in a carcass is of
dominant importance amongst carcasses of similar weight.
The critical importance of the leanness of joints at the
point of sale to the housewife has become evident over
the last few decades. In Britain, a study by Brayshaw,
Carpenter and Phillips (1965), and subsequent consumer
studies at the University of Newcastle-upon-Tyne, have
indicated that leanness was considered by butchers to be
the major criterion by which consumers judge quality
over the shop counter. This emphasis on leanness accords
with the trend towards the commercial production of
leaner carcasses. The trend has been particularly
evident for pigs and has been stimulated by producers
being given premiums (based on fat measurements) for
lean carcasses and, also, by the gradual recognition by
producers of the high cost of depositing fat.

Table 8.2  Weight distribution and average fat class of 511,000 steer and heifer carcasses classified between April and December, 1973

| | Number ('000s) | Average fat class | Carcass weight distribution (%) | | | | | | | Average weight (lb) |
|---|---|---|---|---|---|---|---|---|---|---|
| | | | <450 lb | 450-499 lb | 500-549 lb | 550-599 lb | 600-649 lb | 650-699 lb | >699 lb | |
| Steers | 372 | 2.8 | 2.8 | 9.3 | 15.0 | 19.9 | 19.9 | 15.5 | 17.6 | 611 |
| Heifers | 139 | 3.1 | 21.9 | 26.2 | 23.3 | 15.8 | 7.5 | 3.2 | 2.0 | 512 |

The importance of fatness to the retailer is evident from *Figure 8.2*, which shows the amount of fat which had to be trimmed from carcasses of differing total fatness in order to make the meat saleable. These results were derived from cutting tests on beef carried out by seven firms of multiple butchers with the co-operation of MLC. They show that individual firms were buying carcasses of very different total fat levels. Those buying the fatter carcasses, because of their presumed superior eating characteristics, were having to trim off about half the total fat in order to make them acceptable. Even so, there was more residual fat left in the saleable meat from the fatter carcasses. Retail cutting work carried out by MLC on pigs of different total fatness has confirmed the impact of fatness on realisation values.

Amongst animals of similar weight, breed type can have a marked influence on the percentage of lean or fat in a carcass. Late maturing breeds (i.e. breeds which reach a given level of fatness at a heavier weight than other breeds) are clearly leaner at lighter weights. When one considers carcasses of a particular weight and fatness, variations in leanness occur as a result of variations in the ratio of lean meat to bone, and breed type appears to exert a major influence on this ratio. *Table 8.3* shows results obtained by MLC for some carcasses of different breed types on different production systems. In practice, it is the level of fatness at which a given lean meat:bone ratio is achieved which is important and this has been found to vary between breeds and strains on the same feeding system. It is important to add a note of caution about drawing firm conclusions from comparisons of lean meat: bone ratios obtained by different groups of workers. This is necessary because the ratio is very sensitive to the level to which bones are cleaned before weighing.

It is generally accepted that different sexes have different lean meat potentials and that energy surplus to maintenance and bone and muscle growth requirements is deposited as fat. The superiority of bulls over steers and of steers over heifers in the weight of lean meat achieved at a given age has already been noted. Even at the same weight, bulls have a greater proportion of lean meat than steers and steers a higher proportion of lean than heifers as shown, for example, by Frood and Owen (1972, 1973) in a study comparing Friesian bulls, steers and heifers reared on high and medium planes of

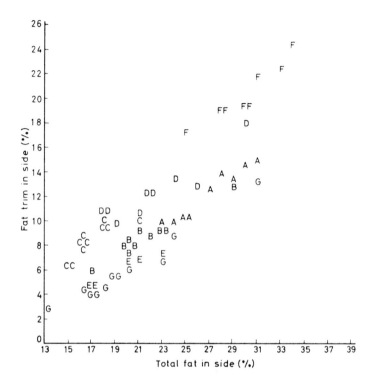

*Figure 8.2 Percentage fat trim against percentage total fat for 64 beef sides (excluding kidney knob and channel fat) trimmed by seven commercial firms (A to G)*

nutrition and serially slaughtered at 60 kg liveweight intervals from 255 to 560 kg.

Various studies have been made of differences in leanness between boars, gilts and castrates of the same weight and fed similarly. These show that boars are markedly leaner than castrates and that the advantage of boars over gilts is less marked (Fowler, Taylor and Livingstone, 1969).

The effect of dietary energy on leanness has already been mentioned and is particularly well quantified for pigs, but clear evidence on the effects of the quantity and quality of protein in the diet of pigs on carcass leanness remains to be obtained.

Table 8.3 Means and standard deviations for lean:bone ratio, fatness and weight of carcasses of different breed types and sexes from several production systems

| Breed type | Sex | Production system | n | Lean:Bone | Subcutaneous fat in side (%) | Side-weight (kg) |
|---|---|---|---|---|---|---|
| Ayrshire | Steer | Intensive | 27 | 4.02±0.23 | 9.1±1.7 | 103.1±5.0 |
| Charolais x Friesian | Steer | Intensive | 49 | 4.31±0.29 | 6.5±1.4 | 109.6±3.4 |
| Limousin x Friesian | Steer | Intensive | 26 | 4.35±0.24 | 6.7±2.3 | 108.7±15.7 |
| Friesian | Steer | Semi-intensive | 72 | 3.95±0.28 | 6.9±1.3 | 126.5±11.8 |
| Friesian | Bull | Semi-intensive | 46 | 4.40±1.0 | 4.4±1.0 | 139.7±12.2 |
| Hereford x Friesian | Steer | Semi-intensive | 97 | 3.92±0.24 | 8.3±2.2 | 110.7±11.6 |
| Simmental x Friesian | Steer | Semi-intensive | 17 | 4.26±0.25 | 6.8±1.5 | 155.8±15.8 |
| South Devon x Friesian | Steer | Semi-intensive | 10 | 4.04±0.23 | 7.0±1.6 | 132.9±11.2 |
| South Devon x Friesian | Bull | Semi-intensive | 22 | 4.50±0.38 | 5.3±1.4 | 157.8±16.8 |
| Aberdeen Angus Crosses | Steer | Intensively finished | 29 | 4.36±0.28 | 13.0±1.7 | 122.1±9.4 |
| Aberdeen Angus Crosses | Heifer | Intensively finished | 14 | 4.54±0.29 | 13.4±2.2 | 100.9±5.0 |

DISTRIBUTION AND THICKNESS OF TISSUES

The subject of meat distribution and its thickness through
the carcass is one which causes much controversy when
carcass quality is discussed between meat scientists and
those commercially active in the production and marketing
of meat.  The controversy arises through the importance
which many meat traders attach to the proportion of high
priced cuts, the distribution of lean meat, the thickness
of the exposed cut surfaces of joints and, particularly,
to the role of conformation in providing a guide to the
assessment of variations in these characteristics.

The evidence from detailed dissection studies carried
out in different parts of the world on widely different
breeds within species shows that variation in the
distribution of muscles between one part of the carcass
and another is much smaller than is often supposed to be
the case (e.g. Butterfield, 1963; Bergstrom, 1968).
This means that variations in the proportion of high
priced cuts or in the distribution of saleable meat
between high and lower valued parts largely reflect
variations in fat distribution or in cutting lines.

It is known that a butcher tries to use his skill to
make the most of a carcass by adapting his lines of
division of the cuts to the individual carcass shape.
This may, therefore, lead to rather more variation in
the yield of high priced cuts and in the proportion of
total lean falling in these cuts.  In order to test
whether carcasses of good conformation would produce a
greater proportion of high priced cuts than those of
poorer conformation, a meat trade expert was asked by
MLC to cut some 90 lamb carcasses of similar weight but
varying in conformation and fatness.  The results
revealed that, amongst carcasses of similar weight and
fatness, carcasses of good conformation did not, in fact,
produce a higher yield of the most valuable cuts.

Such variation as does exist in the yield of high
priced cuts has been shown by commercial cutting tests
to come third in importance in determining carcass
realisation value after variations in fat trim and bone
(Prescott and Smith, 1970).

It is difficult to quantify the economic importance of
the variation in the thickness of lean meat or of flesh
(lean meat + fat) on the exposed cut surfaces of joints.
Evidence obtained from MLC studies ranging over several
hundred cattle of mixed breed types indicates that there

is still a good deal of variation left in the thickness
traits even amongst carcasses of similar weight, fatness
and conformation.

Provided the thickness of flesh is obtained without
having to trim off excess fat, then thick fleshing will
be desired by the meat trade because of its implications
in preparing retail cuts which are attractive and not
wasteful. At the point of consumption, there may be
some advantages in terms of cooking loss and in ease of
carving, but there is little objective evidence available
on the subject. However, the impact of such variations
on realisation value is bound to be much less important
than, for example, variation in fat content which has
a major influence on this value.

As indicated above, variations in conformation (i.e.
the thickness of lean meat + fat in relation to the size
of the skeleton) are considered by many traders to be
almost synonymous with variations in the distribution
and thickness of meat. Some evidence on the degree of
relationship is given in *Table 8.4* which shows results
from MLC studies on beef and lamb of mixed breed types.
The correlations are all generally low, particularly at
constant weight and fatness. Nevertheless, there is a
trend for carcasses of good conformation to have larger
muscle cross-sections in the loin and to be heavier per
unit length, but the correlations are not sufficiently
good to provide a reliable guide to muscle thickness on
individual carcasses. It seems likely that, with the
growth of de-boning and fat trimming of primal or even
of retail joints at the meat plant, the possible value
of conformation to the retailer will decline.

The influence of breed type on the distribution of
lean meat appears to be little with the exception of
that which may arise through some breeds having
inherently better conformation than others and hence,
perhaps, influencing the jointing lines used by the
retailer. While the distribution of muscles may be
relatively constant between breeds, the distribution of
fat and the physical relationships of the muscles and
bones may not be so constant and, hence, introduce small
but significant differences.

The thickness of meat may be influenced by breed type
because the weight of some breeds may be distributed
over shorter, thicker bones than other breeds of
similar total weight. Some of the European beef breeds
fall into the category of high weight per unit skeletal
size.

*Table 8.4* Correlations between overall conformation scores of beef and lamb carcasses of mixed breed types and some other carcass variates

| | Beef (n=300) | | Sheep (n=435) | |
|---|---|---|---|---|
| | *Simple* | *Eliminating weight and % subcutaneous fat in side* | *Simple* | *Eliminating weight and % subcutaneous fat in side* |
| Lean in side (%) | -0.168 | 0.222 | -0.431 | 0.116 |
| Lean:bone ratio | 0.332 | 0.297 | 0.352 | 0.229 |
| High-priced cuts in side (%) | 0.079 | 0.213 | -0.157 | 0.029 |
| Lean distribution in high-priced cuts | 0.106 | 0.210 | -0.076 | 0.052 |
| Area *l.dorsi* (beef - 10th rib) (sheep - 12th rib) | 0.351 | 0.430 | 0.474 | 0.276 |
| Weight:length ratio | 0.374 | 0.439 | 0.613 | 0.273 |

Within the common commercial weight range, the effect
of sex on the distribution and thickness of the tissues
is small.  Frood and Owen (1973) in the experiment
referred to earlier, found that at the same weight, bulls
had a small but significantly greater percentage of the
high priced cuts than steers or heifers and that heifers
had less than steers.  On the other hand, there is some
evidence from an MLC study comparing Friesian bulls and
steers at the same weight, that bulls had slightly less
high priced cuts and rather less of their total lean
distributed in these cuts than steers.  Despite this
slight conflict, it appears that the development of the
crest in young bulls has more of a visual impact than an
effect on the proportion of cuts.


## ASSESSMENT OF CARCASS QUALITY

This section will consider how the tissue proportions
and the distribution and thickness of these tissues may
be assessed in the carcass and in the live animal.


### ASSESSMENT OF THE CARCASS

It will be evident from the preceding sections that
carcass assessment efforts are best directed to those
aspects of carcass quality which have the biggest impact
on realisation value.  Thus, the majority of this
section will discuss the assessment of variations in
overall fatness or leanness.

There is a wide range of methods used for assessing
variations in carcass leanness.  The method chosen will
depend on the accuracy of the answers required and on
the money available for labour, equipment and carcass
depreciation costs.  Most of the techniques use the
fatness or leanness of the carcass determined by physical
separation as the base-line, but the chemical composition
of the carcass tissues is sometimes used, particularly in
nutritional studies.  For most purposes, the base-line of
physical separation seems preferable since carcass value
judgements by consumers are made on the basis of the
physical appearance or composition of the carcass or
joint.  Further, chemical analysis does not appear to be
cheaper to operate as a base-line if sampling and
analytical work are to be carried out accurately.

The physical separation techniques used for determining the base-line range from careful separation, with scalpels and scissors, of individual muscles from bone attachments and surrounding fat, to methods involving separation by butchers' knife into fat, lean meat and bone within major joints. Other less detailed techniques (widely used in the USA, particularly for beef and pork) involve breaking carcasses down in a commercial, but to a degree standardised, manner, by jointing and then de-boning and trimming excess fat.

The technique used by the MLC (described in detail for beef and sheep by Cuthbertson, Harrington and Smith, 1972, and for pigs by Cuthbertson, 1968) to provide a base-line for its studies of, for example, breed differences, involves weighing the lean meat (including intramuscular fat), the various physically separable fat depots, and bone and trimmings within each of a number of standardised commercial joints. By adding together the results for the various joints, the composition of the whole carcass is obtained. The purpose of cutting the sides into joints before subsequent separation of the tissues is mainly to provide information to answer questions about variation in the proportion of joints from animals of the same and of different breeds, and of variation in the distribution of the tissues in these joints. This technique seems an appropriate one to use to provide a base-line for most work. It seems difficult to justify carrying out any more detailed separation, particularly that involving the separation and weighing of individual muscles since, as already observed, there is now sufficient evidence (e.g. Butterfield, 1963) to show that the variation in the distribution of the total muscle between individual muscles is quite small and much less than was believed to be the case formerly.

The MLC procedure outlined above is expensive to operate. At prices in February 1974, the loss in value as a result of the separation of a side of beef amounted to about £28. The comparable figures were £5 for lamb and £4 for pork. In addition, there is a heavy labour component. The time required to separate a side of beef (including some preliminary measurement and photographic work) is 2.5 man-days. The corresponding figures for lamb and pigs are 0.5 and 0.6 of a man-day, respectively.

Once a satisfactory procedure has been established for providing the base-line for overall leanness or fatness,

the high cost of achieving it necessitates the
consideration of methods of predicting overall values
by less expensive techniques.  A commercial type of
de-boning and fat trimming operation could be applied.
Clearly the method is difficult to standardise and, to
some extent, the variations found will reflect variations
in residual fat.  Nevertheless, the results of MLC
studies on beef, relating levels of fat trim to total %
fat in the carcass, indicate that some 80% or more of
the variation in total fat can be explained by variation
in the percentage of fat trim.  The technique has the
advantage of enabling joints at the end of the procedure
to be sold commercially with little, if any, depreciation.
Even with this technique, however, it is important, in
certain studies, to proceed beyond the de-boning and fat
trimming stage by separating the resulting saleable meat
from a proportion of the carcasses into lean and fat
(plus any bone which may be left) in order to provide a
measure of residual fatness, and to estimate the
regression relationship between percentage fat trim and
percentage total fat.

An alternative approach is to predict overall
composition by making use of the close relationship
which has been found to exist between the composition of
certain individual joints and the composition of the
whole carcass.  Results from quite extensive MLC studies
on the predictive value of alternative joints have been
given for beef and pigs by Kempster *et al.* (1974), and
Cook *et al.* (1974).  Results relating to 435 lamb
carcasses of mixed breed types are given in *Table 8.5*.
These show the value of the leanness of the best end of
neck joint as a predictor of the overall percentage of
lean in the side.

In deciding on the actual joint to choose, factors
other than predictive value have to be considered, such
as the resources available to bear the labour and
depreciation costs on the joint.  The latter can be quite
large since wholesalers are generally unwilling to be
left with a carcass minus one joint.  In this respect,
it is unfortunate that, for example, the shin and leg
joints on a beef carcass are not more valuable as
predictors.  Having selected the sample joint it is
desirable, under most circumstances, to use a sub-sampling
technique whereby one separates fully a sub-sample of the
animals on a particular treatment to estimate the
regression relationship to use in those particular

*Table 8.5* Prediction of lean in side (%) from lean in
joints (%)[1] for 435 castrated male carcasses comprising
the main British breed types (standard deviation of %
lean in side = 3.69)

| Joint | Residual standard deviation |
|-------|------------------------------|
| Leg | 2.14 |
| Chump | 1.88 |
| Loin | 1.81 |
| Best-end neck | 1.51 |
| Shoulder | 1.62 |
| Middle neck | 2.76 |
| Neck | 3.36 |
| Breast | 1.81 |

[1]Location of joints given by Cuthbertson *et al.* (1972).

circumstances.  In many cases, this will not be possible
and when selecting a predictor it is also relevant to
consider evidence on the stability of the regression
relationship across treatments.

When any form of carcass separation is precluded, it
is necessary to rely on measurements and subjective
assessments with a substantial loss in predictive
accuracy.  This is demonstrated in *Tables 8.6, 8.7* and
*8.8* where the value of selected carcass assessments is
shown for beef, lamb and pigs, respectively, using data
obtained by MLC.

A paradox of carcass assessment is that the easiest
way to predict the percentage of lean meat in a carcass
is through the assessment of fatness, particularly of
subcutaneous fat.  The relationships obtained for beef
and sheep are poorer than those obtained for pigs for
the following reasons:

1. Subcutaneous fat contributes a lower proportion of
   total fat than in the pig.
2. Subcutaneous fat is spread quite thinly and
   variably over the carcass in beef and lamb, so
   that a single fat measurement is likely to give a
   poorer guide than in pigs and the value obtained
   in beef and in lamb will be very dependent on the
   care with which the hide or skin is removed.

The specific gravity of carcasses has been used by
several workers to predict overall carcass leanness in

*Table 8.6* Residual standard deviations for the prediction of lean in beef sides (%) for sample of 390 carcasses of mixed breed type and sex

| | 10th rib | 13th rib |
|---|---|---|
| Standard deviation of percentage lean | 4.33 | |
| Subcutaneous fat score[1] | 2.87 | |
| Fat thickness over $B$[2] | 3.32 | 3.12 |
| Area *l.dorsi*[3] | 4.03 | 3.93 |
| $B$[4] | 4.26 | 4.13 |

[1] Subcutaneous fat score: a seven point scoring scale for the overall level of subcutaneous fat development, where 1 = leanest to 7 = fattest.

[2] Fat thickness over $B$ (i.e $C$): depth of subcutaneous fat over $B$.

[3] Area *l.dorsi*: cross-sectional area of exposed *l.dorsi*.

[4] $B$: the maximum depth of *l.dorsi* at right angles to $A$ (the greatest width of *l.dorsi*).

*Table 8.7* Residual standard deviations for the prediction of lean in lamb sides for sample of lambs used in *Table 8.5*

| | |
|---|---|
| Standard deviation of % lean | 3.69 |
| Subcutaneous fat score[1] | 3.01 |
| $C$ fat thickness[1] | 2.80 |
| Area *l.dorsi*[1] | 3.69 |
| $B$[1] | 3.68 |
| $B/C$ | 3.13 |

[1] Variables defined in same way as given for *Table 8.6*, but measurements taken at the 12th rib.

*Table 8.8* Residual standard deviations for the prediction of lean (%) in pig carcasses (sample: 885 carcasses comprising Large White and Landrace barrows and gilts - analysis within breed, sex and testing station)

| | |
|---|---|
| Standard deviation of % lean | 2.88 |
| *C* fat depth[1] | 2.26 |
| *K* fat depth[1] | 2.27 |
| Shoulder, mid-back, loin fat depths[1] | 2.41 |
| Area *l.dorsi*[1] | 2.49 |
| *B*[1] | 2.68 |
| *B/C* | 2.10 |

[1]See Cuthbertson and Pomeroy (1962) for definition of measurements.

beef, sheep and pigs. The indications from the literature are that in lambs it is likely to be little better than simple fat measurements and evidence from MLC work on the use of the technique for pigs shows it to be no better than simple fat measurements over the *l.dorsi* at the head of the last rib and, when combined, add little of predictive value. However, in selection work the technique could have some advantage over fat depths where the fat depths may reduce but overall fatness may not decrease *pro rata*. The development of a satisfactory air or gaseous displacement technique would seem desirable if the technique is thought to be of value and worth using more widely.

The proportion of lean meat is influenced by the proportions of fat and bone. Any prediction of leanness provided through fatness should, therefore, be capable of improvement by using some characteristic related to the lean meat to bone ratio. Unfortunately, no measurements have yet been found which predict this characteristic with sufficient accuracy. Even the cross-sectional area of the *l.dorsi* at the head of the last rib is a poor predictor of lean content on its own (Tables *8.6*, *8.7* and *8.8*) and, given fat thickness, it adds little to the accuracy of lean meat prediction.

It has already been noted that such variations as occur in the proportion of high priced cuts and in the proportion of the lean meat in them, are poorly predicted by an assessment of conformation. If, for example, there

were a difference of 3% total lean meat in the high
priced cuts occurring between two lamb carcasses of, say,
18 kg carcass weight and 55% lean (a large difference in
relation to the variation found), this would mean that
an extra 300 g of lean meat would be spread over the leg,
chump, loin and best-end neck joints at the expense of
the breast, shoulder and neck joints. It is very
difficult to believe that this could be detected by eye
even though it is obviously of economic significance.

## ASSESSMENT OF THE LIVE ANIMAL

The need to make estimates, in the living animal, of
carcass quality in terms of overall lean, fat and bone
proportions and to assess differences between animals
in the distribution of these tissues in different parts
of the carcass, exists at four levels:

1. Selection between and within breeds on central
   performance tests.
2. Selection of breeding stock on the farm.
3. Selection of animals for slaughter from, for
   example, breed comparison trials and progeny tests
   where one of the requirements may be to make
   comparisons at constant fatness.
4. Selection of commercial animals for slaughter.

There are certain important prerequisites in selecting
a suitable technique. It needs to be one which can be
applied quickly and easily under field conditions and
the equipment must be robust and transportable. In
addition, it should not cause harm to the animal or
depreciate the value of the resulting carcass.
It has been suggested by several research workers that
weight on its own is a sufficiently reliable predictor
of carcass composition. Certainly much of the variation
in leanness can be accounted for by variation in weight,
but this approach tends to assume that variations in
composition amongst animals of similar weight are
unimportant. This is far from true, as there are
considerable variations in leanness amongst animals of
similar weight and it is important to distinguish such
differences. It follows that the value of techniques
should be tested amongst animals of similar weight but,
unfortunately, many of the tests carried out in the past

have not had the results adjusted to take out the effect
of carcass weight variations.

The simplest technique is merely to handle the animal
and assess its degree of fat development and conformation.
It has already been noted that conformation, even after
allowing for fatness differences does not give a good
guide to the characteristics which have been attributed
to it over the years. So far as fatness estimation is
concerned, there are undoubtedly some who are able to
assess the degree of subcutaneous fat development with
some accuracy, but they are hard to find and in test work
it is important to achieve a common standard over time.
Further, it may sometimes be necessary to have the same
standard operating in more than one place at the same
time. Under these conditions, even with a highly skilled
judge, there are likely to be advantages in having an
objective assessment.

There is a wide range of more objective techniques
available for the *in vivo* estimation of carcass
composition and some of these have been reviewed by
Houseman (1972). In the following paragraphs, the main
types of technique will be discussed briefly.


1. *Backfat probes*

These are used, primarily on pigs, to measure the
thickness of subcutaneous fat at selected sites. A
simple metal ruler was the first probe used for this
purpose (Hazel and Kline, 1952), but Andrews and Whaley
(1955) developed the 'Leanmeter' probe which works on
the difference in electrical conductivity between fat
and lean. Good agreement has been reported between the
two probing methods, but Pearson *et al.* (1957) considered
that the simple metal ruler probe was a more reliable
tool for estimating carcass leanness.


2. *Ultrasonics*

The value of simple echo-sounding equipment for
estimating fat depths is well accepted in many parts of
the world for use on pigs, where a larger proportion of
the total fat is subcutaneous than in cattle and sheep.
Its value in estimating muscle depths and areas in pigs
is not so good owing to difficulties in identifying the

ultrasonic signals relating to particular muscle
boundaries.  Attempts to improve on the estimation of
fatness and of the underlying muscles using ultrasonics
have been made by various groups of workers in America,
Germany and Denmark.  Some of the equipment developed
works on the principle of compound scanning which, in
theory, should define the subcutaneous fat and underlying
musculature more clearly than linear scanning equipment
such as that developed in the USA and sold under the
trade name 'Scanogram'.  Results of its use in beef have,
for example, been reported by Tulloh, Truscott and Lang
(1973) in Australia.  The best equipment evolved so far
is probably that developed in Denmark in recent years.
It permits some compound scanning and a description of
the technique and results of its use have been reported
on pigs and beef by  Andersen *et al.* (1970).  The
equipment developed so far still leaves a good deal to
be desired in terms of definition of muscle boundaries
and this is likely to have some effect on the definition
of fat.  Attempts to develop improved equipment are in
hand, including development work supported by the MLC.
In the long run, it should be possible, by taking a
series of scans along and across the back of an animal,
to build up an overall picture of the composition of the
whole carcass and, at the same time, provide information
on the thickness of muscle and of the thickness and
distribution of subcutaneous fat.

Measurements of the velocity of ultrasound through the
limbs of living animals have recently been suggested as
a method of estimating lean:fat ratios at suitable sites
in the limbs (Miles and Fursey, 1974).  Such a method
has the advantage of objectivity whereas echo-sounding
techniques are subject to some errors of interpretation.

3.  *X-Rays*

This technique has been used, for example, by Dumont and
Destandau (1964) to estimate fatness, but there are
problems in taking satisfactory X-ray pictures of the
living animal and of obtaining tissue measurements from
the pictures.  However, the technique could provide
estimates of overall carcass composition and of the
distribution and thickness of the tissues.  More recently
it has been suggested that the degree of absorption of
gamma rays might be used to predict fat:lean ratios at
selected sites.

4. *Dilution techniques*

The techniques involve administering a known amount of
tracer which will become uniformly distributed throughout
a compartment in the animal body. The concentration of
the tracer at equilibrium is then measured.
Several of the dilution techniques are used to estimate
the amount of body water which is related to total fat
content. Among the tracers which have been used to
estimate body water in beef, sheep and pigs are
tritiated water, deuterium oxide, antipyrene and related
4-amino-antipyrene and N-acetyl-4-amino-antipyrene.
These techniques have given conflicting results in terms
of their ability to estimate variations in carcass
fatness. As already indicated, a crucial test of the
techniques is their ability to distinguish fatness
amongst carcasses of similar weight. In this situation
tritiated water, for example, has not proved useful
(Cuthbertson *et al.*, 1973). One of the problems of
estimating body water in the carcass from measurements
on the live animal, particularly in the case of ruminants,
is the effect of variation in the water content of the
alimentary tract. However, even if body water estimations
could give satisfactory estimates of total carcass fat,
they cannot provide information on the distribution of
fat between carcass and non-carcass parts, nor of the
distribution and thickness of the tissues in different
parts of the carcass. Similar deficiencies in
estimating carcass variations are associated with the
other dilution techniques referred to below.
One of these techniques involves measuring the uptake
of a tracer substance which dissolves uniformly in fat.
The animal is enclosed in a chamber and a known amount
of gas introduced. When equilibrium is reached, the
amount of gas absorbed can be calculated. Several gases
have been tried, including krypton (Hytten, Taylor and
Taggart, 1966) but there have been problems associated
with their application.
A number of studies have indicated a relationship
between various volume measurements of blood and its
components, and body composition. Several techniques
have been tried, including the estimation of red cell
volume by [51]Cr dilution (Doornenbaal, Asdell and
Wellington, 1962). This technique is based on the
relationship which exists between the oxygen transporting
vehicle and the oxygen consuming tissues, largely the
lean meat mass.

One other dilution technique relies on the fact that
most of the potassium in the body lies in the lean meat
mass, so that estimation of the former in the body
provides a measure of the latter. $^{42}$K has been tried as
a tracer for this purpose (Houseman, 1972).


5. *Density*

This approach involves estimating body volume and a
number of methods have been tried. These have included
air displacement and helium dilution but there have
been problems in, for example, measuring changes in
volume with sufficient precision. A different approach
has been to determine the volume of the animal from
contour lines produced by photogrammetry equipment.
This technique is an expensive one, but recently the use
of Moiré, described by Speight, Miles and Moledina
(1974), should make it possible to obtain contour lines
very much more cheaply but, perhaps, not with quite the
same accuracy. Once the contour lines have been obtained,
there still remains the task of deriving volume
measurements. As with dilution and other density
techniques, there are difficulties in drawing conclusions
on carcass composition because of the effect of non-
carcass parts, and they cannot provide information on
the thickness and distribution of the tissues.


6. *Potassium-40*

The use of potassium to assess leanness has already been
noted. A non-dilution method which is used involves
measuring the naturally occurring isotope, $^{40}$K. The
animal is shielded, as far as possible, from any
external radiation and the $^{40}$K radiation from the animal
is measured. The technique has been tried extensively in
the USA (e.g. Frahm, Walters and McLellan, 1971). However
in spite of its use, for example, in Oklahoma for the
evaluation of pig and beef breeding stock, doubts about
its value remain in relation to the high costs of the
equipment.

## 7. Electromagnetism

Recently equipment has been developed in the USA and sold under the trade name 'EMME' which estimates body composition by the difference in electrical conductivity between lean and fat. Domermuth *et al.* (1973) have reported some results of its use, but further evaluation is necessary. The technique is quick to operate and is non-destructive.

While a good deal of work has been carried out on some of the foregoing techniques, both in ruminants and non-ruminants very little work has been done comparing alternative techniques on the same sample of animals. One of the few attempts to test out several techniques on the same sample has been carried out on pigs by Houseman (1972), but the study was limited to 24 pigs varying widely in fatness. Further work of this nature needs to be implemented.

Any comparative study of techniques should examine their value, individually and in different combinations, over production characteristics such as feed conversion efficiency. In deciding which techniques should be included in any comparative study, a number of the techniques seem unsuitable. In the UK, it appears that legislation would prevent the use of certain of these techniques (e.g. backfat probes and injection of tracers) for uses other than for diagnostic purposes. Even if they were permitted for routine use, for example, in testing performance test animals, it is doubtful whether some of the techniques could be feasible for routine use under field conditions. If one takes into account the foregoing restrictions then the following techniques remain: subjective assessment, ultrasonics, density, X-rays, $^{40}$K and, perhaps, electromagnetism. Of these, the ultrasonic scanning techniques seem the most likely, in the current state of knowledge, to provide the information required. Apart from the X-ray technique, the other techniques mentioned provide no information on the distribution and thickness of the tissues which could be of importance in the future as intensive selection is applied to specific characteristics. For example, as a result of selection against fat in the subcutaneous region, it may be found that some fat becomes redistributed intermuscularly, and it is desirable that this should be detected before selection proceeds too far.

## VARIABILITY IN CARCASS QUALITY

In Britain there are marked differences between and
within regions in market requirements.  This currently
makes it impossible to obtain agreement on national
grading schemes, and accounts for the development of
carcass classification schemes which provide a common
language for describing the important carcass
characteristics in such a way that they may be used as
a basis for defining requirements in trading.  Classification schemes provide the mechanism for giving premiums
and incentive payments to producers to encourage
improvements in carcass traits and a better matching of
supply and demand.  It is evident that the existence of
such schemes should act as a stimulus and an aid to
livestock improvement schemes aimed, in part, at
improving carcass traits.

The Meat and Livestock Commission was charged with the
task of developing classification schemes in Britain and,
in this section, some of the results arising from the
operation of these will be discussed.

A classification scheme was first developed for pigs,
since more was known about variations in carcass
characteristics and methods of assessing these under
abattoir conditions than for beef and sheep.  A pig
scheme came into operation on a voluntary basis in
March 1971, and has gained wide acceptance as a basis
for defining buying grades from producers and for
specifying retailers' requirements.  At the time of
writing, some 70% of all clean pigs slaughtered
are being classified in nearly 500 management units
throughout the country.

The method of describing pig carcasses is simple and
is summarised in the Appendix to this chapter.  It
centres on two basic factors: carcass weight and back
fat thickness using the optical probe to measure fat
thickness over the *l.dorsi*.  Important features of
the scheme are that the classification information is
fed back to producers to help them in their production
and marketing plans, and that carcasses are marked
according to the classification for the use of retailers.

*Figure 8.3* shows the change in fat thickness with
weight for a sample of some 0.5 million pigs passing
through the classification scheme.  The variability in
the fat measurements is quite high, and the impact of

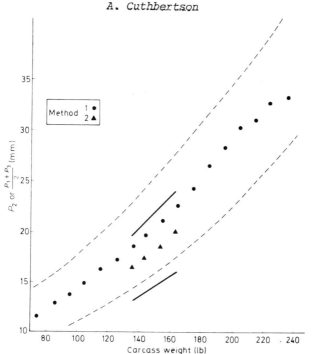

*Figure 8.3 Mean and standard deviation of fat thickness against weight for 0.5 million pigs classified in 1972-3 (definition of Methods 1 and 2 given in Appendix)*

this may be appreciated when one considers that, at any given weight, a change of 2 mm in a single fat thickness value represents a change of approximately 1% lean in the carcass.

There are very marked seasonal differences as *Figure 8.4* shows. Pigs which are grown in the summer are fatter than those grown in the winter. The reason for this is probably that during the summer months maintenance requirements are lower owing to the higher ambient temperature. *Figure 8.4* also shows evidence of a trend for pigs to be getting leaner with time, but it is not possible from the results obtained to know how much of this trend can be attributed to genetic or to environmental factors.

A classification scheme for beef (which was referred to earlier) has been developed and it came into operation

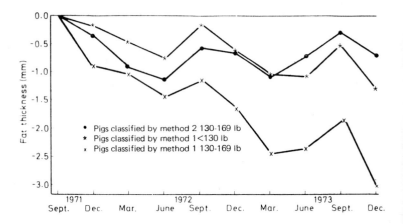

Figure 8.4   Change in the average fatness (P₂ or $\frac{P_1+P_3}{2}$ )
of classified pigs expressed as deviations from
the July-September 1971 figures (definition of methods
1 and 2 given in Appendix)

on a proving scheme basis in October 1972.  At the time
of writing, some 480 management units are participating
in the scheme and approximately 45% of all clean cattle
are being classified.  An evaluation of the proving
phase is nearly complete, and a more formal scheme is
being introduced which will include making it a
condition of participation that the classification
information is fed back to producers.  Acceptance of the
scheme as a basis for trading has been much slower for
beef than for pigs where there has been a longer history
of grade and deadweight selling.

    Although the classification for fatness involves a
subjective judgement of subcutaneous fat level it is, in
fact, linked to actual ranges of subcutaneous fat percentage in the side.  In the long run, it is hoped to be
able to find a more objective method of assessing fatness.

    *Figure 8.5* shows some results where the average fat
class has been plotted against average conformation
class for abattoirs where 100 or more carcasses were
classified in one quarter of 1973.  It illustrates the
degree of variation between abattoirs in both fatness
and conformation.

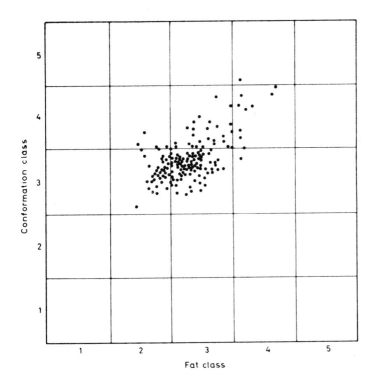

Figure 8.5 Average fat class against average
conformation class for abattoirs where 100 or more
carcasses were classified between April and June 1973

After the discussion earlier in the chapter on
conformation, it may seem somewhat illogical to
incorporate conformation in the classification scheme.
One reason for its inclusion is that the scheme is a
voluntary one and it would not have been possible to
have obtained the level of acceptance achieved had
classification for conformation not been included in the
scheme.  A further reason is that conformation in
conjunction with weight and fatness may help to
distinguish different breed types.  Given the inclusion
of conformation in the scheme, it should be possible to
assess, amongst carcasses of similar weight and fatness,
the extent to which price differentials are applied to
carcasses of different conformation.

A sheep classification scheme has been developed and, after being tested on an experimental basis, a formal scheme will be launched in 1974. In brief, carcasses are described by their weight, category (age/sex group), fatness (on a 5 point scale involving a subjective judgement of subcutaneous fat development, but linked to actual ranges of subcutaneous fat in the carcass) and conformation with 4 classes. Some results from the operation of the scheme in 16 abattoirs over a 5 month period are given in *Figure 8.6*. This illustrates the marked differences between abattoirs reflecting, to some extent, differences in demand.

In conclusion, the existence of an effective sheep classification scheme as well as schemes for beef and pigs should bring more discipline into meat trading and help traders to describe their requirements more clearly. This should lead to a better matching of supply with demand and improve the overall efficiency of marketing meat to provide the consumer with more of what is required

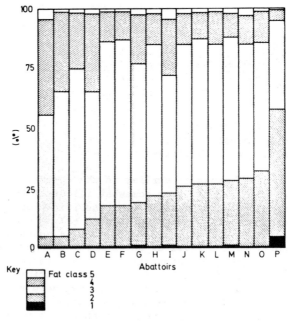

*Figure 8.6 Percentage of lamb carcasses classified falling in each fat class for each of 16 participating abattoirs (A to P)*

ACKNOWLEDGEMENT

The author would like to express his gratitude to his colleague, Dr. A.J. Kempster, for his help in the preparation of the chapter.

# APPENDIX

SUMMARY OF MLC PIG CARCASS CLASSIFICATION SCHEME

Description of carcasses is based on the following five factors:

1. Cold carcass weight.
2. Backfat thickness (*see* below).
3. Visual assessment applied to certain carcasses: scraggy, deformed, blemished, pigmented, coarse-skinned, soft fat, pale muscle, partially condemned. These are classed as 'Z'.
4. Length at the option of the participating trader.
5. Conformation at the option of the participating trader. Carcasses with poor conformation are classed as 'C'.

Backfat thickness is determined by one of the following three methods:

*Method 1* (used mainly for pork and cutter pigs)

Fat thickness is measured with an optical probe at two fixed points over the *l.dorsi:* $4\frac{1}{2}$ cm ($P_1$) and 8 cm ($P_3$) from the mid-line of the back at the head of the last rib. The measurements are added together to describe the degree of fatness.

*Method 2* (used mainly for bacon pigs)

An optical probe measurement of fat thickness is taken at $6\frac{1}{2}$ cm ($P_2$) from the mid-line of the back at the head of the last rib. This measurement, together with measurements in the mid-line at the maximum shoulder and

minimum loin fat positions, indicates the degree of
fatness.


*Method 2a*

Method 2 without the mid-line fat measurements.  Used
by traders (utilising heavy hogs, for example) who only
wish to describe fatness by a single fat thickness
measurement.


REFERENCES

ANDERSEN, B., PEDERSEN, O.K., BUSK, H., LUND, S.A. and
    JENSEN, P. (1970).  *Die Fleischwirtschaft.*, 50, 843
ANDREWS, F.N. and WHALEY, R.M. (1955).  *Rep. Ind. agric.
    Exp. Sta.*, No. 68, 27
BUTTERFIELD, R.M. (1963).  Symposium on carcase
    composition and appraisal of meat animals, C.S.I.R.O.,
    Australia, 7-1
BERGSTROM, P.L. (1968).  Symposium on methods of carcase
    evaluation, European Association of Animal Production,
    Dublin
BRAYSHAW, G.H., CARPENTER, E.M. and PHILLIPS, R.A. (1965).
    Report No. 1, Dept. of Agric. Marketing, University
    of Newcastle-upon-Tyne
COOK, G.L., CUTHBERTSON, A., SMITH, R.J. and KEMPSTER,
    A.J. (1974).  *Proc. Br. Soc. Anim. Prod.*, (in the
    press)
CUTHBERTSON, A. (1968).  Symposium on methods of carcase
    evaluation, European Association of Animal Production,
    Dublin
CUTHBERTSON, A., HARRINGTON, G. and SMITH, R.J. (1972).
    *Proc. Br. Soc. Anim. Prod.*, 113
CUTHBERTSON, A. and POMEROY, R.W. (1962).  *J. Agric. Sci.*,
    59, 207
CUTHBERTSON, A., READ, J.L., DAVIES, D.A.R. and OWEN,
    J.B. (1973).  *Proc. Br. Soc. Anim. Prod.*, 2, 83
DOMERMUTH, W.F., VEUM, T.L., ALEXANDER, M.A., HEDRICK,
    H.B. and CLARK, J.L. (1973).  *J. Anim. Sci.*, 37, 259
DOORNENBAAL, H., ASDELL, S.A. and WELLINGTON, G.H.
    (1962).  *J. Anim. Sci.*, 21, 461
DUMONT, B.L. and DESTANDAU, S. (1964).  *Ann. Zootech.*,
    13, 213

FOWLER, V.R., TAYLOR, A.G. and LIVINGSTONE, R.M. (1969).
*Meat Production from Entire Male Animals*, p. 189,
Churchill, London

FRAHM, R.R., WALTERS, L.E. and MCLELLAN, C.R. (1971).
*J. Anim. Sci.*, 32, 463

FROOD, I.J.M. and OWEN, E. (1972). *Proc. Br. Soc. Anim.
Prod.*, 128

FROOD, I.J.M. and OWEN, E. (1973). *Proc. Br. Soc. Anim.
Prod.*, 2, 78

HAZEL, L.N. and KLINE, E.A. (1952). *J. Anim. Sci.*, 11,
313

HOUSEMAN, R.A. (1972). Ph.D. thesis, University of
Edinburgh

HYTTEN, F.E., TAYLOR, K. and TAGGART, N. (1966). *Clin.
Sci.*, 31, 111

KEMPSTER, A.J., AVIS, P.R.D., CUTHBERTSON, A. and SMITH,
R.J. (1974). *Proc. Br. Soc. Anim. Prod.* (in the
press)

MILES, C.A. and FURSEY, G.A.J. (1974). *Anim. Prod.*,
18, 93

PEARSON, A.M., PRICE, J.F., HOEFER, J.A., BRATZLER, L.J.
and MAGEE, W.T. (1957). *J. Anim. Sci.*, 16, 481

PRESCOTT, J.H.D. and SMITH, P.J. (1970). Preliminary
report to the MLC on beef carcase evaluation at the
University of Newcastle-upon-Tyne, 1968-69

SPEIGHT, B.S., MILES, C.A. and MOLEDINA, K. (1974).
*Med. and biol Eng.* (in the press)

TULLOH, N.M., TRUSCOTT, T.G. and LANG, C.P. (1973).
Report submitted to the Australian Meat Bd. and
published by the School of Agriculture and Forestry,
University of Melbourne

# 9

# THE COMMERCIAL PREPARATION OF FRESH MEAT AT WHOLESALE AND RETAIL LEVELS

J. W. STROTHER

*Meat and Livestock Commission,*
*Bletchley, Milton Keynes*

## INTRODUCTION

Traditional British methods of cutting carcass meat are based on the separation of carcasses or sides of beef into primal cuts of a size and shape convenient for butchers to handle.

The vertebral column and the articulations of the fore and hind limbs provide consistent anatomical reference points, except where cutting lines run approximately parallel to the vertebral column or sternum, as in the removal of the brisket or flank from fore and hind-quarters of beef.

In the beef carcass the major exceptions to this are in the preparation of topside, silverside and thick flank from the round, and in the removal of the clod from the forequarter. In each of these cases the accepted procedure involves a certain amount of seaming between major groups of muscles and the seaming out of the femur and humerus which they enclose.

The use of skeletal rather than muscle reference points and the adoption of straight cutting lines to produce 'square cut' joints contributes to variability within the eventual retail cuts as the resulting bone-less primal cuts are aggregates of muscles severed as one cut or joint rather than seamed out individually. It is possible that evaporative and drip losses from the

severed ends of the muscles is in excess of that from
muscles seamed out in accordance with continental
practice.  However, the total surface area of muscle
exposed is likely to be less in British cutting as the
blocks of muscles are also protected to some extent by
external fat, which is either non-existent or removed
during preparation on the continent, and this may offset
losses from the larger areas of cut surface.

The point at which wholesale cutting ends and retail
cutting begins is not clearly defined.  Traditionally
the wholesaler has cut a side of beef into two quarters
between two ribs, for example the 10th and the 11th,
although regional and local variation from the 9th and
10th, to the 12th and 13th is not uncommon.

To cater for more specific demand, wholesalers also
cut quarters into large pieces such as the 'top piece',
'rump, loin and flank' or into smaller bone-in primals
shown in *Figure 9.1*.  Retailers slaughtering their own
cattle are obliged to undertake their own 'wholesale
cutting' in order to handle the carcasses more easily,
and also to produce primals which can be stored in the
refrigerator.  These primals are used in sequence to
meet the greater demand for stewing beef, braising steaks,
mince and brisket in the early part of the week and for
the higher quality and value, frying and grilling steaks
and roasting joints later in the week.

Until quite recently it has been customary for
wholesale primals to be sold bone-in to avoid loss from
evaporation, drip and the possibility of increased risk
of contamination developing when transporting boneless
primals.  The introduction of vacuum packaging of beef
has altered this situation considerably and it is
estimated that up to about 15 per cent of home killed
beef for consumption as carcass meat is now boned
centrally and delivered to the retailer as boneless
primals.  Centralised retail cutting has not made much
progress because of the technical problems of limiting
drip and colour changes in the relatively small cuts
required for sale at retail level.  Developments in
packaging or changes in the consumer attitude to the
acceptability of frozen meat may alter this situation in
time, but at present, centralised prepacking of fresh
retail cuts is costly because of the limited shelf life
of the product.  The problems of high distribution costs
caused by the need for frequent deliveries may be
compounded by the cost of returning and recycling packs
which are bacteriologically acceptable, but which may
become unsaleable as they age and darken in colour.

At the retail level, beef primal cuts are boned and cut into sub-primals, which are further sub-divided into retail cuts when required.  In the traditional small independent butcher's shop these operations may take place on a block or cutting table behind the counter, and the window display is likely to be restricted to stewing beef and cheaper cuts supplemented by pork and lamb cuts.  The more expensive cuts such as rump, sirloin and round will be displayed on the block as primals or sub-primals.  During the early part of the week when trade is mainly in cheaper cuts, rump, sirloin and topside may not be visible at all, although normally it will be produced from the refrigerator and cut on request.  On Friday and Saturday when trade is customarily brisk, a selection of the more expensive joints and steaks may appear in the window, as there will not be sufficient time for them to be cut and prepared to order for each customer.

The independent retailer who cuts meat to order in front of his customer would argue that this gives him the maximum opportunity to assess the customer's requirement, select the most suitable piece of meat and prepare it in a manner which is most likely to satisfy the customer.  Independent retailers with a high turnover and multiple butchers normally cut and prepare to retail level in a cutting room, which is not open to the customer.  The prepared cuts, joints and tray meats are displayed in a form which requires very little preparation other than cutting off a suitable piece from a length of tied, boneless roll.  In this situation presentation and standardisation within a comprehensive window display is important, as the customer is attracted by what is on show rather than by the knowledge that the butcher's skill is available to cut and prepare to order.  Cutting behind the scenes can make full use of a butcher's skill when turnover is high enough to keep him fully occupied in preparation during the quiet periods, when his presence behind the counter is less necessary, and in serving the prepared cuts when the shop is busy.

Whichever system is adopted the underlying principle of retail butchery in Britain is to bone out primal cuts and sub-divide primals into muscle groups or blocks, forming retail cuts of an economical size, which hold together in cooking or in the subsequent carving, which carve across the 'grain' of the muscle fibres to facilitate chewing, and which are acceptable by custom to the majority of customers.

Additional factors of importance in preparation include the time of cutting in relation to the time of anticipated sale, as this affects colour and drip. The ratio of visible lean to fat in a particular joint or cut is also affected by preparation and, in defining retail cuts for training or other purposes visual lean percentages are often specified, although they are difficult to verify in practice. A better alternative is to limit the amount of inter-muscular fat by specifying the original fat class of the carcass from which the cuts are to be derived, and maximum external fat thickness on joints or steaks.

Wholesale and retail cuts of lamb and pork customarily follow similar lines. The major differences are the preponderance of bone-in chops and joints which are a consequence of the smaller, initial carcass size and, in the case of boneless shoulders and legs, the fact that it is possible to prepare retail joints from whole shoulders and legs without initial separation into the primal cuts which are necessary in the larger beef carcass. Because of this similarity in approach, this chapter relates exclusively to the much more complex procedure of cutting carcass beef, in which it is believed that there is much greater opportunity for improvements to be initiated and for existing practices to be expressed in standard forms.

## BRITISH CUTTING METHODS

Although regional cutting methods show some variation, these are not generally of major importance, and they indicate the different ways in which local cutting skills facilitate the handling of carcasses of different weights and levels of finish, rather than any fundamental difference in the eventual presentation of the meat at retail level or in recommendations for its use. Perhaps the greatest difference in this respect, is the contrast between Scottish hindquarter cutting and utilisation, in which the topside is frequently sliced as round steak, and in which portions of the English rump may be utilised as a high grade mince. The variation in the quality of the separate muscles, which form the rump primal cut and which may all be included in the rump steak without fear of the consequences of the *Trades Description Act,* is such that this practice would have a good measure of support among English consumers, if their

butchers could be persuaded to adopt it.    The extent of
regional variations in primal cutting lines has been
reported by Gerrard (1971).

In an attempt to select and recommend for more widespread adoption standard procedures which lend themselves
to efficient cutting and consistently high quality in the
finished product, the Meat and Livestock Commission (MLC)
have prepared and published a technical bulletin on cutting
beef (MLC, 1974).    It contains a fully illustrated
description of the cutting and preparation of a side down
to retail level.    *Figure 9.1* shows the cutting lines of
this method superimposed on a skeletal diagram and the
following comments on the method will serve to illustrate
the use of precise anatomical points, and some of the
problems in positioning those cuts which are parallel to
the long axis of the side, and to which reference has
already been made.    *Table 9.1* indicates the yields of
primals and prepared cuts.

The fore and hindquarters are separated between the
10th and 11th ribs.    The top piece is then removed by
cutting from the 5th sacral vertebra following a straight
line which passes about one inch in front of the head of
the femur.    The leg of beef is removed by a cut through
the joint parallel to the anterior cut surface of the top
piece, and the round is then separated into three boneless primals.    The removal of the thick flank is effected
by seaming.    The aitch bone is removed and the topside
and silverside are also separated.    The femur is freed
from the silverside.    This is a traditional technique
which involves partial muscle separation, but produces a
considerable area of cut surface as indicated by the
cutting lines.    The removal of the thin flank from the
rump and loin is related more to muscle structure.    The
cut starts at the posterior cut surface of the rump and
flank, leaving *rectus abdominis* on the flank but
incorporating *obliquus abdominis internus* in the rump.
It is continued to the anterior cut surface, emerging at
a point which is determined largely by the trade view of
how much or how little 'tail' it is feasible to sell on a
sirloin steak or roasting joint.    The rump and loin are
separated by a straight cut between the 6th lumbar and
1st sacral vertebrae just clearing the anterior edge of
the ilium and approximately parallel to the anterior cut
surface of the loin.

The forequarter technique is very similar.    The shin
is removed by cutting through the elbow joint.    The clod
and neck are taken off by cutting along the humerus
through the articulation with the scapula continuing in a

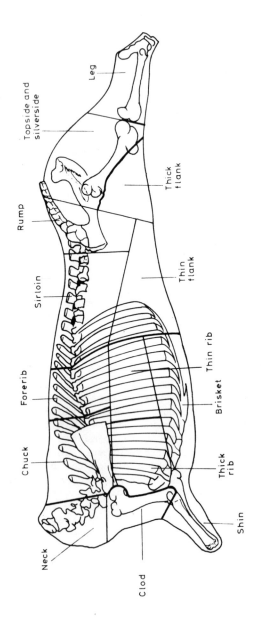

*Figure 9.1  An English method of cutting beef*

*Table 9.1* An English method of cutting beef (average side weight 136.4 kg MLC Fat Class 3)

| | Bone-in cuts | | Boneless (trimmed) cuts | |
|---|---|---|---|---|
| | Average weight (kg) | % of the side | Average weight (kg) | % of the side |
| **HINDQUARTER** | | | | |
| Leg | 5.7 | 4.2 | 2.8 | 2.1 |
| [1]Thick flank | 8.6 | 6.3 | 6.1 | 4.5 |
| [1]Topside | 9.1 | 6.7 | 7.3 | 5.3 |
| [1]Silverside | 11.5 | 8.4 | 7.0 | 5.1 |
| [1]Rump | 9.4 | 6.9 | 5.0 | 3.7 |
| [1]Sirloin | 11.8 | 8.6 | 7.2 | 5.3 |
| Thin flank | 7.5 | 5.5 | 4.6 | 3.4 |
| Cod-udder fat | 2.1 | 1.5 | – | – |
| [1]Fillet | – | – | 1.8 | 1.3 |
| Lean trim | – | – | 4.7 | 3.4 |
| Usable meat | – | – | 46.5 | 34.1 |
| Kidney | 5.3 | 3.9 | 0.8 | 0.6 |
| Kidney knob & channel fat (KKCF) | | | 4.5 | 3.3 |
| Fat trim | – | – | 7.1 | 5.2 |
| Bone & waste | – | – | 11.8 | 8.7 |
| Losses | – | – | 0.2 | 0.1 |
| Sub-total hindquarter | 70.9 | 52.0 | 70.9 | 52.0 |
| **FOREQUARTER** | | | | |
| Shin | 4.0 | 2.9 | 1.8 | 1.3 |
| Clod | 7.2 | 5.3 | 3.3 | 2.4 |
| Neck | 5.0 | 3.7 | 2.7 | 2.0 |
| Brisket | 12.0 | 8.8 | 5.7 | 4.2 |
| Thick rib | 9.0 | 6.6 | 5.8 | 4.3 |
| Thin rib | 2.9 | 2.1 | 1.8 | 1.3 |
| [1]Fore rib | 6.7 | 4.9 | 4.9 | 3.6 |
| Chuck & blade | 18.7 | 13.7 | 13.5 | 9.9 |
| Lean trim | – | – | 5.9 | 4.3 |
| Usable meat | – | – | 45.4 | 33.3 |
| Fat trim | – | – | 6.7 | 4.9 |
| Bone & waste | – | – | 13.0 | 9.5 |
| Losses | – | – | 0.4 | 0.3 |
| Sub-total forequarter | 65.5 | 48.0 | 65.5 | 48.0 |
| **SIDE TOTALS** | | | | |
| Usable meat | – | – | 92.0 | 67.5 |
| Kidney | – | – | 0.8 | 0.6 |
| KKCF | – | – | 4.5 | 3.3 |
| Fat trim | – | – | 13.7 | 10.0 |
| Bone & waste | – | – | 24.8 | 18.2 |
| Losses | – | – | 0.6 | 0.4 |
| Total side | 136.4 | 100.0 | 136.4 | 100.0 |
| [1]%High value cuts | | 41.8 | | 28.8 |

straight line through the 6th cervical vertebra.  The
clod and neck are separated along muscle seams.  The
anterior removal point of the brisket is determined by
the removal of the shin, but the depth of brisket
required determines the position of the line along which
it is removed, and also the depth of the thick and thin
rib joints.  These two joints are also affected by the
length of rib tops that it is considered feasible to
include in the bone-in fore rib or boneless fore rib roll.
*Figure 13.1* illustrates a standard four-bone fore rib
and a six-bone chuck which includes the blade, although an
alternative which incorporates a six-bone fore rib is
also recommended.


## AMERICAN CUTTING


As a basis for considering alternatives to existing
British cutting methods, the MLC has examined methods
used in other countries, including the USA and France.
Perhaps the most unexpected feature is the marked degree
of similarity in the appearance of the primal cuts when
inserted on an outline carcass diagram, *Figure 9.2*, the
yields of which are given in *Table 9.2*.

The example illustrated consists of a 12 rib forequarter with a single wing rib on the hindquarter.  When
the top piece is similar to the English cut of the same
name it is called the Chicago Round.  Alternatively the
cutting line may include a portion of the English rump
with the thick flank (the Diamond Round), or leave the
thick flank attached to the English rump (the New York
Round).  It should be noted that the term 'sirloin' in
American nomenclature relates to what in the English
cutting method is termed 'rump' and the American
expression 'short loin' is used to describe the sirloin.
The breakdown of the remainder of the hindquarter and
the forequarter into primals is very similar to the
English method, although it is simplified to facilitate
use of the band saw by the adoption of cutting lines
which pass through bones instead of following the
contours of the joints.  However, sub-primal and retail
cuts which are produced differ quite markedly from
modern English retail cuts (*Figure 9.3*).  They reflect
the relative values of labour and meat, the use of the
band saw to reduce labour to the minimum, greater
consumer acceptance of bone-in cuts of all descriptions
and a pattern of consumption based much more on grilling

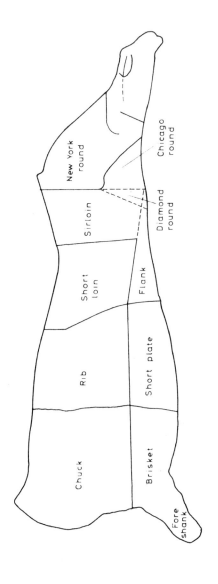

*Figure 9.2  An American method of cutting beef*

*Table 9.2* An American method of cutting beef calculated on a side weight including KKCF 136.4 kg

| | Primal | | Retail | |
|---|---|---|---|---|
| | Average weight (kg) | % of the side | Average weight (kg) | % of the side |
| **HIGHER VALUE CUTS** | | | | |
| Chuck (5 ribs) | 35.4 | 26.0 | - | - |
| Boston cut | - | - | 2.2 | 1.6 |
| [1]Arm - pot roast | - | - | 4.8 | 3.5 |
| [1]Cross rib - pot roast | - | - | 2.3 | 1.7 |
| [1]Blade - pot roast | - | - | 12.8 | 9.4 |
| Boneless chuck (steaks & ground beef) | - | - | 7.0 | 5.1 |
| Rib (7 ribs) | 12.3 | 9.0 | - | - |
| [1]Ribs - short cut | - | - | 8.7 | 6.4 |
| Cap meat | - | - | 0.1 | 0.1 |
| [1]Short ribs | - | - | 1.0 | 0.7 |
| Stewing beef | - | - | 0.8 | 0.6 |
| Short loin (1 rib) | 10.9 | 8.0 | - | - |
| [1]Porterhouse steaks | - | - | 4.1 | 3.0 |
| [1]T-bone steaks | - | - | 2.0 | 1.5 |
| [1]Club steaks | - | - | 1.1 | 0.8 |
| Grinding beef | - | - | 0.1 | <0.1 |
| Sirloin (rump) | 12.3 | 9.0 | - | - |
| [1]Wedge bone steaks | - | - | 1.2 | 0.9 |
| [1]Round bone steaks | - | - | 1.4 | 1.0 |
| [1]Flat bone steaks | - | - | 4.0 | 2.9 |
| [1]Pin bone steaks | - | - | 2.6 | 1.9 |
| Round | 31.4 | 23.0 | - | - |
| Tip (thick flank) | - | - | 3.6 | 2.6 |
| Top (topside) | - | - | 9.4 | 6.9 |
| Outside (silverside) | - | - | 9.1 | 6.7 |
| Boneless round (steaks & ground beef) | - | - | 3.8 | 2.8 |
| **LOWER VALUE CUTS** | | | | |
| [1]Fore shank | 5.5 | 4.0 | 4.1 | 3.0 |
| Brisket | 6.8 | 5.0 | 4.5 | 3.3 |
| Short plate | 11.0 | 8.0 | 4.1 | 3.0 |
| Flank | 6.8 | 5.0 | 4.1 | 3.0 |
| Kidney & suet | 4.0 | 3.0 | 4.1 | 3.0 |
| Fat trim | - | - | 20.4 | 15.0 |
| Bone & waste | - | - | 13.0 | 9.5 |
| TOTAL SIDE | 136.4 | 100.0 | 136.4 | 100.0 |

[1]Bone-in cuts

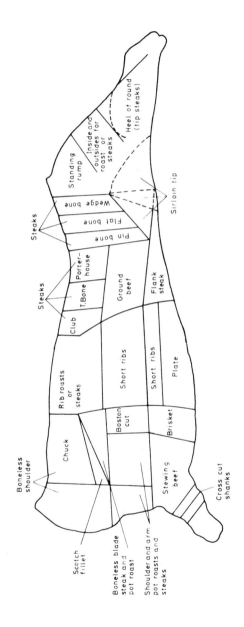

*Figure 9.3   An American retail breakdown*

(broiling), restaurant meals and convenience foods than in Britain. Another important factor is the relatively high value of grinding beef for hamburger or ground beef production and the negligible importance of the roasting joint in American consumption.

To a foreigner perhaps the most attractive feature of American retail cutting is the great range and variety of steaks which are produced for grilling. This is presumably closely related to the production and slaughter of relatively immature beef from beef breeds fed on a continuously high plane of nutrition. Perhaps the least attractive feature is the production of bone-in steaks or pot roasts from that part of the forequarter which we should term the chuck and the clod (American chuck and arm steaks and pot roasts). These cuts are not easy to carve when cooked on the bone and tend to disintegrate completely when prepared and cooked as boneless steaks.

A certain amount of boning out is undertaken after band sawing, especially in the preparation of the rolled brisket and plate, the boneless arm joint, and to a lesser extent in the preparation of very high quality boneless or mini-bone quality cuts from the fore rib.

Where labour is available American muscle boning techniques include removal of the *supraspinatus* ('Scotch fillet') and *infraspinatus* ('boneless blade steak') and the production of an 'inside chuck roll' from the muscles underlying the scapula. Another boneless cut is the 'shoulder clod pot roast' formed from the lateral and long heads of the *triceps brachii* muscles.

In terms of its application to British cutting, muscle boning of the chuck, with the object of producing second grade frying steak and a chuck roll for pot roasting, has considerable possibilities. The MLC has also tested individual hindquarter muscles as frying or grilling steaks comparable with the American top, bottom and eye of round steaks cut from the top side and silverside. If the complete muscles are sliced to form steaks with little waste, then a greater return can be obtained by producing steaks from a hindquarter than from cutting in the form of the traditional English joints. Variation between and within muscles in terms of tenderness and juiciness has been noted and it is doubtful if utilisation of the more suitable muscles for steaks and preparing the remainder as joints would yield an improved commercial return because of the wastage in muscle separation which necessitates a considerable amount of trim.

## FRENCH CUTTING

On initial inspection, the major difference between this
method (*Figure 9.4*) and the English and American methods
is the separation of the fore and hindquarters into the
5 rib forequarter and the 8 rib or 'pistola cut'
hindquarter.  The second major feature concerns the
removal of the 'Raquette' or shoulder which is taken off
the forequarter rather like an English shoulder of lamb.
Primals are then cut from the two quarters along lines
which are very like those used in English cutting,
although the application of additional time and skill
is indicated by the removal of the shin and leg during
which the *extensor carpi radialis* and *semimembranosus*
are each removed at their point of insertion, instead
of cutting through them at the joint as in standard
English practice.
   The real difference and skill, however, lies in preparing
retail joints by boning out and muscle separation of the
boned or partially boned cuts (yields are given in
*Table 9.3*).  In the case of the Raquette for example,
five distinct muscle groups are separated and two
individual muscles, *supraspinatus* and *infraspinatus*, are
used for grills.  The extent to which the French trade
apparently separate and utilise for different purposes
some very small muscles is remarkable, and one wonders
if there is a distinctive market and differential price
for each, or if the separation is simply to ensure the
removal of every vestige of intermuscular fat and
connective tissue.
   The French method ensures that variation between
muscles will not detract from the quality of the cooked
product.  The extent to which older cattle, including
cow beef, have been used in the production of carcass
meat for domestic consumption and the French preference
for very rare beef may each have contributed to the
development of French butchery, but the cost in man
hours is immense.  In contrast to the one and a half
hours' work needed to fully prepare an American side to
the retail level and up to three and a half hours for
the English side, the French butcher may take up to
seven hours to prepare a full side.  This includes the
time spent on fatting up certain joints with pork fat
arranged to form a decorative effect over the lean meat.

*Table 9.3* A French method of cutting beef calculated on a side weight including KKCF 167 kg

| | Bone in cuts | | Retail (untrimmed) cuts | |
|---|---|---|---|---|
| | Average weight (kg) | % of the side | Average weight (kg) | % of the side |
| HINDQUARTER – 8 RIBS | | | | |
| ALOYAU – Milieu de train de côtes | (37.6) | (22.5) | | |
| Hanche (rump) | 10.9 | 6.5 | | |
| [1]Rumsteck | | | 6.6 | 3.9 |
| [1]Aiguillette (rump flap) | | | 6.6 | 3.9 |
| [1]Bavette d'aloyau (sirloin flap) | 2.2 | 1.3 | 2.2 | 1.3 |
| Filet | 4.6 | 2.8 | | |
| [1]Tête de filet | | | 1.8 | 1.1 |
| [1]Milieu de filet | | | 2.0 | 1.2 |
| [1]Queue de filet | | | 0.8 | 0.5 |
| Coquille – shell loin + 2 ribs | 7.5 | 4.5 | | |
| [1]Faux filet – coquille boned and rolled | | | 5.6 | 3.3 |
| Milieu de train de côtes – 6 bone fore rib | 12.5 | 7.5 | | |
| [1]Milieu de train – désossé – boneless fore ribs | | | 9.7 | 5.8 |
| Sub-total | 37.7 | 22.6 | 30.8 | 18.4 |
| CUISSE | (44.8) | (26.8) | | |
| Gîte – leg of beef | 4.4 | 2.6 | 4.4 | 2.6 |
| [1]Tende de tranche – topside | 14.7 | 8.8 | 13.3 | 8.0 |
| [1]Araignée – small muscle on medial surface of aitch bone | 0.5 | 0.3 | 0.5 | 0.3 |
| Tranche grasse – thick flank | 8.0 | 4.8 | | |
| [1]Plat – *quad. femoris* | | | 3.2 | 1.9 |
| [1]Rond ) other | | | 2.3 | 1.4 |
| [1]Mouvant ) separate | | | 1.7 | 1.0 |
| Nourrice) muscles | | | 0.9 | 0.5 |
| Semelle – silverside | 13.0 | 7.8 | | |
| [1]Gîte a la noix – *biceps femoris* | | | 7.0 | 4.2 |

*Table 9.3 (continued)*

| | Bone in cuts | | Retail (untrimmed) cuts | |
|---|---|---|---|---|
| | Average weight (kg) | % of the side | Average weight (kg) | % of the side |
| **CUISSE** (*continued*) | | | | |
| Nerveux - flexor group; shield | | | 3.3 | 2.0 |
| [1]Rond de gîte - *semi-tendinosus* | | | 2.8 | 1.7 |
| [1]Bavette de flanchet- goose-skirt | 1.0 | 0.6 | 1.0 | 0.6 |
| Sub-total | 41.6 | 24.9 | 40.4 | 24.2 |
| Queue | 1.6 | 1.0 | 1.6 | 1.0 |
| [1]Onglet | 1.2 | 0.7 | 1.2 | 0.7 |
| Sub-total hindquarter | 82.1 | 49.2 | 74.0 | 44.3 |
| **FOREQUARTER** | | | | |
| Collier - basses - côtes | (21.9) | (13.1) | | |
| Basses côtes | 10.4 | 6.2 | | |
| [1]Pièce parée | | | 0.4 | 0.2 |
| Derrière de paleron | | | 0.8 | 0.5 |
| [1]Filet mignon | | | 0.4 | 0.2 |
| [1]Basses côtes | | | 8.8 | 5.3 |
| Collier | 11.5 | 6.9 | | |
| Salière | | | 0.9 | 0.5 |
| Veine maigre | | | 7.2 | 4.3 |
| Veine grasse | | | 3.5 | 2.1 |
| Raquette - shoulder | 24.0 | 14.4 | | |
| Macreuse - pot au feu | | | 8.0 | 4.8 |
| [1]Macreuse - bif teck | | | 4.0 | 2.4 |
| [1]Jumeau - bif teck | | | 1.9 | 1.1 |
| Jumeau - pot au feu | | | 0.8 | 0.5 |
| Gîte | | | 4.0 | 2.4 |
| [1]Hampe - skirt | 0.7 | 0.4 | 0.7 | 0.4 |
| Caparaçon | (29.5) | (17.7) | | |
| Panneau | 14.2 | 8.5 | | |
| Bavette | | | 3.0 | 1.8 |
| Plat de côtes couvert | | | 8.0 | 4.8 |
| Plat de côtes découvert | | | 3.2 | 1.9 |
| Pis | 15.3 | 9.2 | | |
| Flanchet | | | 2.7 | 1.6 |
| Tendron | | | 4.6 | 2.8 |
| Milieu de poitrine | | | 4.0 | 2.4 |
| Gros bout de poitrine | | | 4.1 | 2.5 |
| Sub-total forequarter | 76.1 | 45.5 | 71.0 | 42.5 |
| Hindquarter | 82.1 | 49.2 | 74.0 | 44.3 |
| Forequarter | 76.1 | 45.5 | 71.0 | 42.5 |
| Bone trimmed | 8.8 | 5.3 | 22.0 | 13.2 |
| Total side weight | 167.0 | 100.0 | 167.0 | 100.0 |

[1]Roasting and grilling

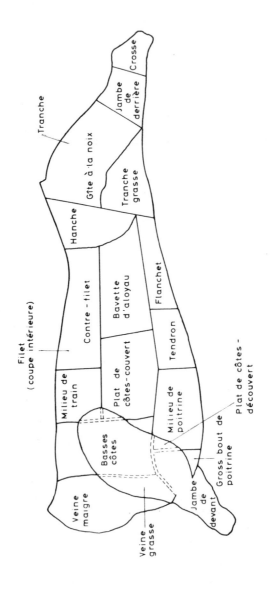

Figure 9.4  A French method of cutting beef

## WORK OF THE MLC ON NEW CUTTING METHODS

Although the French cutting method is of great interest
to the retail trade it is clear that it is related to
culinary practices which are so vastly different from
those in Britain that the end product would not
generally find favour in British butchers' shops, even
if the labour were freely available to undertake the
preparation.  Interest in the French method is therefore
particularly related to the recommendations for using
certain muscles which it might be possible to remove for
frying or grilling without the need to greatly modify
the traditional British techniques.  It is also important
to avoid increasing the labour content of retail
preparation unless this is absolutely necessary and
justifiable in terms of the increased value of the
product.  The work of the MLC has been concentrated on
forequarter cutting methods on the grounds that it is
in the forequarter that the greatest scope for upgrading
stewing or braising meat into roasting or grilling meat
exists.  Two methods have been devised, one of which is
aimed at the retailer and his normal retail trade, and
the other at the retailer supplying the lower end of the
catering market with boneless rolled joints principally
for pot roasting and with some steaks and stewing beef.
Sides ranging from 130 to 175 kg have been cut to these
specifications and typical yields of usable meat are
shown in *Table 9.4*.

CUTTING METHOD

1.  Remove the shin leaving the *extensor carpi radialis,*
    attached to the forequarter.  *Figure 9.6* (3)
2.  Remove the brisket.  Bone and roll.  *Figure 9.6* (8)
3.  Remove a five bone roasting rib and thin rib.
    *Figure 9.6* (4)
    (a) *Retailing Method*
        (i)   Remove the vertebrae and rib bones.
        (ii)  Cut through the eye muscle (*longissimus
              dorsi*), to the medial surface of the
              *latissimus dorsi* and the surface of the
              scapula.
        (iii) Seam out and remove when there is sufficient
              flap to form a rolled joint.  Roll and tie.
              *Figure 9.6* (cutting line Cl-C2)

Table 9.4 An MLC method of forequarter cutting (forequarter weight 80.06 kg)

| Joint description | Cooking method | Retailing | | Catering | |
|---|---|---|---|---|---|
| | | Weight of usable meat (kg) | Joint as % of quarter weight | Weight of usable meat (kg) | Joint as % of quarter weight |
| Shin | S | 1.95 | 2.4 | 2.63 | 3.3 |
| Extensor carpi radialis | B | 0.68 | 0.8 | | |
| Brisket | PR | 6.53 | 8.2 | 6.53 | 8.2 |
| Extended roasting rib | R | 7.72 | 9.6 | 7.10 | 8.9 |
| Extended thin rib | | | | 4.52 | 5.6 |
| Supraspinatus | F/G | 1.39 | 1.7 | 1.39 | 1.7 |
| Infraspinatus | F/G | 2.15 | 2.7 | | |
| Brachialis | B | 1.16 | 1.4 | 6.67 | 8.3 |
| Triceps brachii | F/G | 3.37 | 4.2 | | |
| Subscapularis | B | 1.33 | 1.7 | 1.33 | 1.7 |
| Neck | S | 4.69 | 5.9 | 4.69 | 5.9 |
| Chuck roll | B | 3.28 | 4.1 | 4.64 | 5.8 |
| Chuck rib roll | B | 3.25 | 4.1 | 4.84 | 6.0 |
| Thin rib roll | B | 2.94 | 3.7 | 1.47 | 1.8 |
| Braising pieces | B | 2.57 | 3.2 | 2.88 | 3.6 |
| Stewing pieces | S | 4.61 | 5.8 | 6.31 | 7.9 |
| Mince pieces | M | 7.38 | 9.2 | 8.91 | 11.1 |
| Fat trim | | 8.91 | 11.1 | 16.15 | 20.2 |
| Bone & waste trim | | 16.15 | 20.2 | | |
| Usable meat | | 55.00 | 68.7 | 55.00 | 68.7 |
| TOTAL | | 80.06 | | 80.06 | |

KEY: F/G Frying/grilling; B Braising; PR Pot roasting; S Stewing; M Mince; R Roasting.

    (iv)   Roll the remaining section, removing the
           tip of the scapula and attached muscles.
           If at this stage the joint is too thick the
           *latissimus dorsi* may be removed. Roll and
           tie. *Figure 9.6* (cutting line B1-B2)

  (b)  *Catering Method*
    (i)   Mark a point on the rib immediately above
         the inner tip of the scapula.
    (ii)  Mark a second point along the 10th rib at a
         distance equal to the length of the eye
         muscle (*longissimus dorsi*).
   (iii)  Inscribe a line on the external surface
         after measuring 4" along the ribs from these
         two points.
    (iv)  Peel back the external fat along this line
         until the *latissimus dorsi* is completely
         exposed. Peel back this muscle until the
         tip of the scapula and the muscles which are
         immediately attached can be removed.
         Separate the extended roasting rib from the
         thin rib by cutting between the points which
         were first marked on the ribs. Bone and roll
         both joints.

4.  Peel back the *trapezius* exposing the *supraspinatus*
    (blade steak). Remove the latter muscle.
5.  Remove the *extensor carpi radialis*. *Figure 9.5* (2)
6.  Remove the humerus.
7.  Inscribe a line for the removal of the neck. *Figure
    9.6* (4)
8.  Seam out the scapula and attached *subscapularis/
    triceps brachii/infraspinatus*. Leave the fat attached
    to the forequarter-portion. Separate into individual
    muscles and slice. *Figure 9.5* (1)
9.  Remove the neck.
10. Seam out the thick fat layer and attached muscles of
    the clod. *Figure 9.6* (7)
11. (a)  *Retailing*
      Peel back and remove the *trapezius* from the chuck
      and chuck rib tops. Remove all bones. Cut into
      two chuck rolls and a chuck rib boll. Roll and
      tie. *Figure 9.6* (5)
   (b)  *Catering*
      Peel back and remove the *trapezius*. The chuck
      roll can then be removed by a cut from the tip
      of the eye muscle (*longissimus dorsi*), parallel
      to the natural edge of the forequarter. Bone
      and roll. Bone and roll the remaining portion
      termed the chuck rib roll.

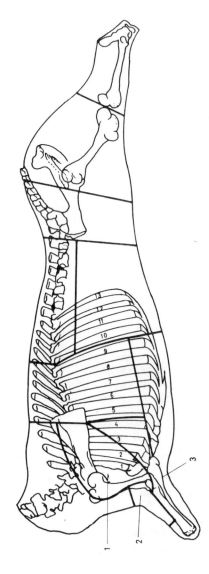

1. Frying muscle block
2. Braising muscle
3. Shin

Figure 9.5 Meat and Livestock Commission forequarter cutting method (note this side is quartered between the 9th and 10th ribs)

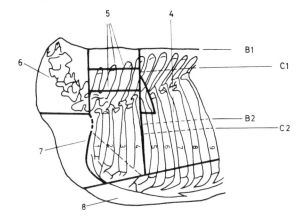

4. Rib roasting joints
5. Braising joints
6. Neck
7. Clod & attached muscles
8. Brisket

Figure 9.6   Meat and Livestock Commission forequarter cutting method

The methods are summarised and illustrated in *Figures 9.5* and *9.6*. Features include the adoption of muscle separation techniques not dissimilar to those used in the removal of the French Raquette and in seaming out individual shoulder muscles in both French and American methods. The chuck rolls also resemble the American 'Inside Chuck Rolls' although the splitting of the *longissimus dorsi* and seaming between the thin rib muscles was developed by the MLC as a means of producing rolled joints of acceptable size from the larger forequarter.

The work of the MLC is not yet complete although the retail and catering methods have each been demonstrated to and used by retail meat traders. On the completion of further tests it is intended to publish a detailed and illustrated description of the two methods.

REFERENCES

GERRARD, F. (1971). *Meat Technology,* 4th edn.
    Leonard Hill, London
MEAT AND LIVESTOCK COMMISSION (Oct. 1974). *Beef Cutting,*
    Meat and Livestock Commission, Milton Keynes

The following organisations are not specifically referred to in the text of this chapter but they are useful sources of information on American and French cutting methods:
NATIONAL LIVESTOCK AND MEAT BOARD, 36 South Wabash
    Avenue, Chicago, Illinois 60603
ECOLE PROFESSIONELLE DE LA BOUCHERIE DE PARIS ET DE LA
    RÉGION PARISIENNE, 37 Boulevard Souet, Paris 12e

# 10

## PROCESSED MEAT

*The Wall's Meat Company Limited,
Willesden, London*

### INTRODUCTION

There is a surprising difference of view regarding the
precise meaning of the term 'processed meat'. In many
warmer countries the realities of practical experience
have divided meat and meat products into two clearly
distinguishable categories: highly perishable products
for immediate consumption and processed products which
have been subjected to procedures such as heavy smoking,
salting or drying to confer considerable shelf stability.
As a consequence there is, in many parts of Europe, a
concept of inevitable association between 'processing'
and 'preservation' and this view has been accentuated by
an emphasis on public health aspects in those countries
where the veterinarian plays a dominant part in Meat
Science and Technology.

In Britain, however, the position is somewhat different.
With our equable climate and isolated island position
we have developed considerable expertise in making and
distributing on a national scale a variety of products
which are not fully preserved. The modern processor,
moreover, is finding opportunities for creative
development in applying newer technology, such as control
of temperature in production and distribution, and the
housewife is increasingly transferring at least some of
her domestic food-preparation chores into the factory.
The British concept of 'processed meats' is therefore
much wider than 'preserved meats'.

A useful definition which appears in the British food
regulations, for example the Meat (Treatment) Regulations,
1964, will be used in the present discussion:
> '*Processed*' in relation to meat, includes curing by
> smoking and any treatment or process resulting in a
> substantial change in the natural state of the meat
> but does not include boning, paring, grinding, cutting,
> cleaning or trimming.

In effect this usually means that the end product
contains significant quantities of non-meat ingredients
or has undergone extensive physical change.


## NATIONAL CHARACTERISTICS IN MEAT PRODUCTS


At first sight the changes which can be produced in meat
by processing seem to be limited but closer examination
shows that over the centuries an extensive technology
has developed reflecting the versatility of the basic
raw material.  In fact the technology is broad enough to
permit a great deal of specialisation associated to a
large extent with national identity.  Each of the
continents has its own characteristic practices and even
within Europe and within the British Isles we have many
local differences.

Some of the differences are quite basic and may, for
example, relate to consumption of parts of the carcass
which are delicacies for some communities and abhorrent
to others.  The reasons may often be historical and no
longer relevant but they undoubtedly lead to rejection
of products which contain nutritive protein and essential
fatty acids, such as arachidonic acid.  With modern
knowledge and techniques some of these distinctions are
becoming illogical but a full discussion of this topic
is perhaps more concerned with meat by-products than
processed meats.

There are, however, widespread similarities of practice
and these are being recognised increasingly as attempts
are made to harmonise food legislation and facilitate
international trade.  We are learning to understand the
subtleties of technology and terminology in different
countries and the complexity of the position with
products such as Salami where a name may be used in many
different countries, without translation, to denote
different local styles of product.  Corned beef is such
a case and is currently attracting attention as a first
example of the difficulties of agreeing to a world wide

Codex Alimentarius Standard for even an apparently
unsophisticated well-known product. Other products with
differing local names may be superficially similar while
differing in essential character. In these cases one of
the greatest areas of difficulty is in translation.
'Wurst' in German does not have precisely the same meaning
as 'Sausage' in English.

The published literature on the technology of this wide
family of products is limited. Most of the general texts
refer primarily to Continental and North American
practice (Grau, 1969; Price and Schweigert, 1971),
comparable British publications (Lawrie, 1966; Gerrard,
1969, 1971) tending to concentrate on broad principles
or on trade practice. The present discussion is
therefore aimed at summarising newer technological trends
in Britain and discussing their relationship with the
products of other countries.


## THE CHARACTER OF BRITISH PROCESSED MEATS

One distinguishing feature of British meat-eating
practice is that we have no built-in reservations about
processing meat in close association with less expensive
foods such as flour and potatoes. This has resulted in
a range of widely used products of varying meat content
and correspondingly varying price level. Our tradition
includes not only the Roast Beef of Olde England but
also the Cornish Pasty and Irish Stew. Meat is, therefore,
an integral component of the normal diet rather than an
extra luxury to be kept separate and dispensed in
appropriate amounts according to the family budget.

We are fortunate in Britain in having, among the basic
statistics available to us, a series of Annual Reports
by the National Food Survey Committee on Household Food
Consumption and Expenditure. *Table 10.1* shows figures
taken from the report for 1970 and 1971 which are
typical of recent years.

It will be seen that the average domestic consumption
of processed meats and meat products was just over 1 lb
(454 g) per person per week, slightly less than
carcass meat *per se*. Together these products provided
a fifth of the protein and energy in the diet and a
corresponding or even greater amount of iron and various
B vitamins. (Poultry meat has been excluded from the
category of 'meat' for the purposes of this discussion.)

*Table 10.1* Household consumption of meat and meat products[1] contribution to the nutrient content of the total diet

| | *Ounces per person per week* | *Gramme equivalent* |
|---|---|---|
| CARCASS MEAT | | |
| Beef and veal | 7.94 | 226 |
| Mutton and lamb | 5.39 | 153 |
| Pork | 3.03 | 86 |
| Total | 16.36 | 465 |
| OFFALS | | |
| Bones | 0.16 | 5 |
| Liver | 0.80 | 23 |
| Other offals | 0.49 | 14 |
| Total | 1.45 | 42 |
| PROCESSED MEATS AND MEAT PRODUCTS | | |
| Bacon and ham, uncooked | 5.11 | 145 |
| Bacon and ham, cooked, including canned | 0.92 | 26 |
| Corned meat | 0.39 | 11 |
| Other cooked meat not purchased in cans | 0.68 | 19 |
| Other canned meat | 1.85 | 53 |
| Sausages, uncooked, pork | 2.34 | 66 |
| Sausages, uncooked, beef | 1.32 | 38 |
| Meat pies and sausage rolls, ready to eat | 0.71 | 20 |
| Quick frozen meat and meat products | 0.55 | 16 |
| Meat pies (not ready to eat), pasties, puddings, paste, spreads, liver sausage, cooked sausage, rissoles, haslett, black puddings, faggots, haggis, hogs pudding, polony, scotch eggs | 2.20 | 63 |
| Total | 16.07 | 457 |

[1]Total from products not itemised separately.

in England, Scotland and Wales in 1971 and percentage
(after National Food Survey Committee, 1973)

Contribution to total dietary content (%)

| Protein | Energy | Iron | Vitamin $B_1$ | Vitamin $B_2$ | Nicotinic acid equivalent |
|---|---|---|---|---|---|
| 6.8 | 3.0 | 9.2 | 0.9 | 3.5 | 8.5 |
| 3.8 | 2.3 | 2.9 | 1.4 | 2.7 | 5.3 |
| 1.8 | 1.7 | 0.8 | 5.2 | 1.2 | 2.8 |
| 12.4 | 7.0 | 12.9 | 7.5 | 7.4 | 16.6 |
| 0.8 | 0.2 | 3.3 | 0.6 | 5.5 | 1.9 |
| 2.8 | 3.2 | 1.5 | 4.8 | 1.6 | 2.7 |
| 2.1 | 1.9 | 1.3 | - | 0.7 | 2.0 |
| 6.5[1] | 3.6[1] | 7.6[1] | 5.5[1] | 4.3[1] | 7.2[1] |
| 12.2 | 8.9 | 13.7 | 10.9 | 12.1 | 13.8 |

Bacon and uncooked cured products were the largest category of processed meats, amounting to some 32% of the total, but sausages were a close second with 29%. Over 80% of the households purchased bacon during the week of the survey and over 70% purchased sausages and these products are an important component of the national diet contributing together some 6% of the total protein and energy value.

## BASIC PROCESSING PROCEDURES

Probably the single most important feature of meat for the processor, apart from its basic water-binding properties, is its response to treatment with sodium chloride. At high concentration this is a powerful preservative and at low concentration a desirable flavour adjunct but its greatest value is in the intermediate range of concentration, say 2% to 8% in the aqueous phase of the product, where it acts as a 'protein conditioner'. Various changes are produced which are only partly understood but their effects are readily apparent in a swelling of the tissue and solubilisation of the protein (Hamm, 1960, 1973; Sherman, 1961a, 1961b, 1961c, 1962). Worthwhile control over microbial growth can also be achieved with this range of salt concentration and exploitation of these effects forms the basis for a wide range of products with useful keeping qualities and attractive eating quality different from those of the original meat.

It is, however, not a simple matter to introduce salt in the appropriate concentration into the complex myofibrillar system. If meat is merely immersed in strong salt solution the immediate effect is for water to move outward from the tissue rather than the much slower movement of salt inwards. This causes shrinking and desiccation which can produce irreversible damage. 15% salt concentration has been indicated as a threshold for such damage (Kopp, 1971), this being possibly related to effects on the connective tissue; most workers have found optimum water binding at salt concentrations of 10% or less depending on the conditions of test. On the other hand, if the salt concentration of the brine is too low a considerable quantity is needed to introduce an adequate amount of salt into the finished product since raw lean meat contains up to 80% of water. Addition of large amounts of water may not always be compatible with other

requirements, including apparent wetness of the product
and legal restrictions on in-going meat content of
processed meat, and the processor usually has to aim at
a compromise. Greater uniformity and control of swelling
particularly with intermediate brine levels can be
achieved by the addition of polyphosphates at levels in
the final product up to about 0.5%. Again there is very
little theoretical explanation although the effect is
presumably parallel to the biological action of ATP on
the muscle proteins and involves chelation of heavy metals
such as calcium and magnesium which are integral to the
muscular tissue. Pyrophosphate is said to be the most
effective of the polyphosphates but this has limited
solubility in brine and most commercial polyphosphates
also contain the more soluble triphosphate often with a
proportion of high molecular weight material.

Various attempts have been made to quantify the usefulness of the meat protein system for manufacturing
products. One which has gained wide acceptance in the
USA and Europe is based on observations that dilute
solutions of meat proteins are powerful emulsifiers of
vegetable oil (Swift, Lockett and Fryar, 1961). The
quantity of oil emulsified under standardised conditions
is used as an index of meat processing quality, often in
conjunction with a further figure for the total amount
of protein which can be extracted from the meat (Saffle,
1968; Sulzbacher, 1973).

Probably more important for British meat technology is
the ability of meat protein extracts to form strong gels
on heating (Sherman 1961a, 1961c; Trautmann, 1966;
Samejima *et al.*, 1969). This varies with the quality of
the meat and particularly with pH of the tissue which may
lie anywhere between 5.3 and 7.0. However, pH also
affects the swelling of the tissue, the rate of diffusion
of salt and the rate of growth of micro-organisms in the
finished product. Detailed evaluation of product
performance is therefore usually directed to examining
practical systems rather than to attempting to develop
all-embracing principles.

The foregoing comments refer mainly to the lean
component of meat. The fat component is also an
important contributor to eating quality particularly in
modifying a dry leathery texture which can develop in
all-lean products. The 'fat' in such cases is, of course,
fatty tissue in which globules of glyceride fat are
enclosed in a cellular structure comprising 5 to 20% of
the total weight. The fat is normally not released on
heating unless the cells are damaged and control of fat

cell damage is therefore an important aspect of processing many products. In applications where true fat-in-water emulsions are formed the emulsification often depends on the age and condition of the fat tissue since lipase action can cause significant soap formation (Sherman 1961b).

## BASIC MEAT PROCESSING EQUIPMENT

A number of items of equipment are in common use in the industry. For reproducible introduction of brine into whole meat pieces it is becoming common practice to use complex multi-needle injection machines. The needles may be arranged in a special pattern for given cuts of meat (Vahlun, 1973) and may be flexible for use with bone-in meat.

A variety of devices is also in use for reducing the size of meat pieces and removing lean from bone. Some are merely alternatives to the sharp knife and skill of the butcher but others have a highly specialised action. The old fashioned mincer or grinder is one such machine which not only shears the meat but also works it violently in the process. This can have a valuable tenderising as well as cutting action and it is unfortunate that its use in Britain is so often associated with inferior meat. In other countries it is normal practice also to grind meat of the highest quality.

In many types of product the extreme cell damage produced by the grinder is a disadvantage. For these applications the high speed cutter is widely used. This usually takes the form of a bowl chopper in which a series of sharp blades revolves at high speed in a contoured bowl. Rotation of the bowl carries the meat through the path of the knives and a variety of effects can be produced by varying the rates of rotation and shape of the knives (Thamm, 1966). One incidental implication of using this machine is that it works on defined batches of meat and imposes batch technology on the user. For continuous processing a modified type of cutter is needed.

Most of these machines aim at a clean cutting effect with a minimum of ancillary tissue damage and this also applies to the larger scale machines used for slicing bacon and cooked meats. The ideal slicer must produce slices which are uniform in thickness and attractive in appearance without break up or smearing of the surface;

it must be capable of operating at high speed without
breakdown and of being dismantled easily for cleaning.
These are demanding requirements and there is continual
search for improved machines.

Among other basic items of equipment of the meat
processor are various types of smoke house, although
these are much less common in Britain than on the
Continent, and a variety of devices for cooking meat in
water or moist air.


## UNCOOKED COMMINUTED PRODUCTS


The classic example of a comminuted uncooked product is
the British fresh sausage. This differs from sausages of
other countries in that it usually contains added cereal
together with sulphite, or metabisulphite, as a
preservative. It is consumed in substantial quantities
throughout the country as we have already seen, and its
public health record is second to none in the field of
meat products. The official food poisoning statistics
showed that in the 1960s there was an average of just
over one outbreak of food poisoning a year attributable
to sausages as compared with a nationwide consumption in
the region of $10^{10}$ sausages per year.

One essential requirement of the sausage is that it must
be able to withstand its final exposure to that most
British of domestic utensils - the frying pan - as
compared with Continental sausages which tend to be eaten
cold or after heating in hot water or steam. Whereas the
latter type of sausage ideally consists of a matrix of
precoagulated protein with entrapped fat the preferred
British sausage is a coarse mixture of pieces of lean
meat swollen with dilute brine, chopped fat tissue,
carbohydrate binder such as dried yeastless rusk, and an
aqueous protein-containing mobile phase suitably seasoned.
This is all contained in a collagen casing.

The performance of this system on heating gives little
evidence of the presence of a classical meat emulsion.
The overall response is a succession of gel formations
and denaturation of the various carbohydrates and proteins
present, often at quite well defined temperatures together
with release of meat juices and molten fat. Superimposed
on this is a pattern of physical effects caused by
shrinkage of connective tissue and development of pressure
due to escaping gases including steam.

One method of studying this system is by conventional histological techniques but this can be singularly unrewarding. Possibly the most useful approach is to measure the losses of aqueous and fatty constituents during a standardised frying procedure. A typical example can be seen in *Figure 10.1* which shows that the main effect in the early stages of cooking is the escape of fat. This reaches a maximum where the product begins to 'cook' and appreciable amounts of water begin to boil off.

This type of testing can be used, for example, to check the effect of temperature of the meats in the chopper. Low temperatures (*c*. 0°C) give maximum retention of water during cooking while higher temperatures (*c*. 15°C) give maximum fat retention. The former effect is clearly related to hydration of the muscle proteins and the latter to the rheology of the chopping operation. The overall effect, however, is that the processor must seek a compromise temperature suited to his own particular product usually by experimentation on a production scale.

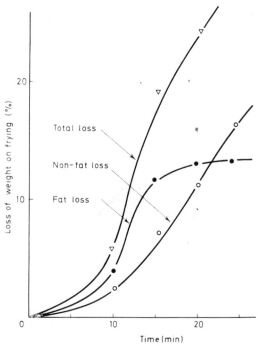

*Figure 10.1   Loss of weight on frying of British fresh pork sausages*

A major source of variation in traditional sausage is variation in the size and thickness of the natural casing and its tendency to shrink. Many of these difficulties can now be avoided by using regenerated collagen for the manufacture of edible artificial casings either in the form of pre-formed casing or by co-extrusion of meat and a surface coating of collagen (*Figure 10.2*).

*Figure 10.2  Continuous production of British fresh sausage by co-extrusion of sausage meat and collagen casing.  (Photograph by courtesy of The Wall's Meat Company Ltd.)*

Keeping properties of the British sausage have been investigated by various workers (Dyett and Shelley, 1965, 1966; Dowdell and Board, 1968, 1971) and, as with many traditional perishable foods, a characteristic harmless microbial flora has been found to develop.  The product recognisably becomes progressively 'mature', 'acid' and 'sour' often with increasing consumer acceptance during the earlier stages.  Loss of attractive appearance occurs well before noticeable effects on flavour and this is a further safeguard for the consumer. Such built-in warning signals combined with a specific activity of the sulphite preservative against pathogens are completely consistent with the public health record of the product.

## COOKED COMMINUTED PRODUCTS

A feature of Continental meat processing is the wide range
of highly processed comminuted products some of which are
cooked and others essentially uncooked but matured under
conditions which encourage drying and controlled microbial
growth.

In such products part, at least, of the meat can
usually be comminuted finely in the presence of salt to
develop maximum binding properties and it seems probable
that some of the muscle protein is involved in fat
emulsification. Even this is being questioned however
(van den Oord and Visser, 1973). The dependence of
quality on rheology in the chopper has been studied
extensively and a great deal of background information
is available (Hamm and Rede, 1973). Controlled cooking,
usually in hot water or moist air, is a feature of many
of the processes, often in conjunction with smoking.

The main British products of this type are in loaf or
canned form for slicing, such as Chopped Pork and Ham or
Pork Luncheon Meat, the latter usually containing a
proportion of added cereals. Similar material is also
used as the filling of cold-eating pork pies. Continental
types of sausage are increasing in popularity and we have
our traditional range of British 'small goods' such as
Liver Sausage and Black Pudding.

## UNCOOKED WHOLE-MEAT PRODUCTS

The main uncooked meat product, other than sausage, which
is distributed in large amounts is bacon. Much of this
is in the form of back bacon and is sold either unsmoked
or after relatively light treatment with cold smoke.
'Bacon' in other countries tends to be belly bacon heavil'
treated with hot smoke and to have a number of different
characteristics.

The traditional Wiltshire bacon curing process involved
injection of brine and immersion in a curing 'pickle'
followed by long maturation resulting in a product with
complex chemical and microbiological characteristics
(Gardner, 1973). Such lengthy processes are becoming
increasingly unattractive in view of the capital
requirement for equipment and material-in-process and
unwillingness on the part of the consumer to pay the

economic cost. Many variations of the basic process
have therefore been developed (Jones and Strang, 1973;
Vahlun, 1973). In some cases a shortened form of the
Wiltshire process is in use but in others the time
required to introduce the curing ingredients has been
reduced drastically (Barrett *et al.*, 1965). One of the
main differences between these types of bacon is in the
microbial load which, in the case of some of the more
rapid cures, is initially low. This type of bacon is
therefore often used for the growing proportion of the
commodity which is sold in a sliced pre-packaged form,
usually in vacuum packs. Variation of processing arises
from a number of factors, particularly localised variation
in deposition of curing ingredients and of natural pH of
the meat (Vetterlein and Hannan, 1966), where wide
differences can be found between even the differing
muscles in a single rasher. In some types of bacon
polyphosphate has been introduced to assist the evenness
of cure and to control frying characteristics; such bacon
is often subjected to a high temperature smoking procedure
in excess of 130°F, which in some respects partly pre-
cooks it.

The subtleties of these various types of bacon process
are, however, overshadowed by the implications of the
normal presence of nitrite in the product (Walters, 1973).
This acts not only as a preservative against spoilage
organisms and pathogens but also produces the characteristic flavour and colour of cured meat, no other substance
being known to produce this wide range of effects.
However, very small amounts of N-nitrosopyrollidine (of
the order of parts per thousand million) have been shown
to form in bacon on frying at the temperature of a very
hot frying pan (Crosby *et al.*, 1972) and a long-term
programme of research is in progress to establish the
acceptability of low levels of such nitrosamines in foods
and the possibilities of controlling their formation by
modification of technique. Addition of ascorbate is one
measure which is being tested for ability to impede
nitrosamine formation without affecting the preservative
action of the nitrite particularly towards *Clostridium
botulinum* (Roberts, 1973). Some care is needed in using
ascorbate in the presence of lipid since this is not only
a labile reducing agent but also seems to act as a
pro-oxidant in some cases (Sato, Hegarty and Herring,
1973; World Health Organization, 1973).

A common constituent of bacon is nitrate which was
present in the traditional Wiltshire product in
appreciable amounts often in excess of 1000 ppm. This

has currently been reduced to an upper limit of 500 ppm
and suggestions are being made that it should be
eliminated altogether.  In bacon however it has a valuable
role of being available for conversion to nitrite and of
controlling spoilage after the product has been sliced and
exposed to air, and since this part of the handling is
usually carried out by the consumer relatively limited
information is available.  It can be shown moreover that
even if the product is initially free of nitrate appreciab
amounts can be formed from nitrite during the early stages
of storage with subsequent reconversion at the end of
shelf life.


## COOKED WHOLE PRODUCTS


Considerable expertise has developed in recent years in
the processing of products such as cooked ham.  The
traditional product is an expensive, rather dry, bone-in
product sliced by hand in the grocer's shop.  In some
cases (e.g. York ham) it receives no cooking and is merely
matured for a long period in dry conditions.  Research on
absorption and retention of brine in cooked products has
led in recent years to a realisation that this can be
greatly facilitated by subjecting the injected meat to a
working operation, usually in a tumbler or paddle mixer.
It is by no means clear why this should have a major
effect on the hydration characteristics of the meat
except that there is an obvious parallel in traditional
beating of steak to tenderise it.  The simplest
explanation is that working produces a thick creamy
protein exudate which gels readily on heating and both
sticks pieces of meat together and seals channels for
escape of juices released by cooking.  Examination of the
cell structure under the microscope however suggests that
working also causes considerable breakdown or weakening
of internal membranes which could modify the texture and
affect permeability characteristics in respect of water
and ions.  Whatever the reason the result can be a much
more succulent ham partly because of the retention of
juices which would otherwise be lost in cooking and
partly because of the more controlled action of the
larger-volume weaker brines.  Eating quality can now be
controlled to a degree not previously considered possible
and it is increasingly common practice to dispense with
the layer of jacket fat which was previously retained in
an attempt to counteract dryness.  Comparative testing

has shown optimum consumer acceptance of this newer type
of ham at retained injection levels of 25-30% followed
by thorough cooking and it is interesting that the
resulting end product has a similar water content on a
fat-free basis to the original raw meat (*c*. 78%). There
is at least an implication therefore that our natural
preference is for meat which is as juicy as the raw meat
eaten by our forefathers. Juiciness and succulence are
the hall marks of high quality in meat products and the
ham system seems to provide opportunity for systematic
study.

It has been claimed that addition of extra water to ham
is nutritionally to the considerable disadvantage of the
consumer (Ingram and Rhodes, 1971), but the classical
tables of McCance and Widdowson (1960) show ham as
containing about 16% protein and at least 40% fat. This
compares with a virtually identical protein content in
modern ham but a fat content of 15% or less. This is
well in line with modern nutritional thinking.


## PORTION CONTROL


The techniques used with ham to produce coherent loaves
of meat suitable for trouble free slicing are also
applicable to uncured products where the usual objective
is to produce meat which can be sliced uniformly and
reproducibly to give a known portion of meat per slice.
This is particularly desirable for modern catering
practice and a range of products both cooked and uncooked
is now made in this way often using a regenerated
cellulose casing to form and shape the product.

Similar techniques (Shaw, 1974) can also be used for
making chunks of processed meat suitable for incorporation
in prepared dishes but with superior or more reproducible
eating quality. One of the advantages offered is the
ability to break down large pieces of sinew, one of the
commonest causes of consumer criticism.

A practical problem with this new technology is the
difficulty of naming the end products in such a way that
the consumer will appreciate their character without
being deterred by the thought of extensive manipulation
by processes which she can only partly comprehend. In
some cases such as Vienna Steak there is established
precedent but in others it is difficult to decide on
sensible informative labelling practice. It would be
regrettable if imaginative new developments in food

technology were to be stifled by difficulties of
labelling.


## PRE-RIGOR PROCESSING

A test of water-binding properties on pre-rigor meat
shows substantially higher values than for post-rigor
meat and there is great interest in utilising this effect
(Mandigo and Henrickson, 1966). The addition of salt
before the onset of rigor, usually by comminuting in the
presence of salt, prevents the loss of water-binding
properties (Johnson and Henrickson, 1970; Hamm and Rede,
1972), and such procedures are widely practised in those
countries such as Germany which restrict the use of
polyphosphates in meat products. It is not easy to see
all the implications of pre-rigor processing and a number
of programmes of research are in progress. One incidental
advantage in an energy hungry world is that it is at least
as logical to raise the temperature of fresh slaughtered
meat directly from $40^{\circ}C$ to a cooking temperature of $65^{\circ}C$
or more, rather than cool it mechanically to $5^{\circ}C$ and then
re-heat by a lengthy and inefficient conventional
procedure.


## METHODS OF THE FUTURE

The definition of processing at the beginning of this
discussion laid emphasis on change in the product rather
than preservative effect. Irradiation procedures which
have been a near-technical possibility for many years
would aim at preservation with negligible change in the
product. Are these then 'processes'? In some countries
irradiation even has the status of an additive. For the
moment however the imponderables of unprovable freedom
from toxicity and problems associated with off-flavours
continue to discourage firm commitment to the method.
The application of these procedures therefore remains
one of the potential new processes of the indefinite
future.

REFERENCES

BARRETT, J., GALBRAITH, C., HOLMES, A.W. and HERSCHDOERFER, S.M. (1965). *Proc. 1st Int. Congr. Fd Sci. Tech. 1962*, 4, 747, Gordon and Breach, New York

CROSBY, N.T., FOREMAN, J.K., PALFRAMM, J.F. and SAWYER, R. (1972). *Nature, Lond.*, 238, 342

DOWDELL, M.J. and BOARD, R.G. (1968). *J. appl. Bact.*, 31, 378

DOWDELL, M.J. and BOARD, R.G. (1971). *J. appl. Bact.*, 34, 317

DYETT, E.J. and SHELLEY, D. (1965). *Proc. 1st Int. Congr. Fd Sci. Tech. 1962*, 2, 393, Gordon and Breach, New York

DYETT, E.J. and SHELLEY, D. (1966). *J. appl. Bact.*, 29, 439

GARDNER, G.A. (1973). *Proc. Inst. Fd Sci. Technol.*, 6, 130

GERRARD, F. (1969). *Sausage and Small Goods Production*, 5th edn., Leonard Hill, London

GERRARD, F. (1971). *Meat Technology*, 4th edn., Leonard Hill, London

GRAU, R. (1969). *Fleisch und Fleischwaren*, 2nd edn., Parey, Berlin

HAMM, R. (1960). *Adv. Fd Res.*, 10, 355

HAMM, R. (1973). *Fleischwirtschaft.*, 53, 73

HAMM, R. and REDE, R. (1972). *Fleischwirtschaft.*, 52, 331

HAMM, R. and REDE, R. (1973). *Fleischwirtschaft.*, 53, 1279

INGRAM, M. and RHODES, D.N. (1971). *Br. Nutr. Foundation Information Bull.*, 6, 39

JOHNSON, R.G. and HENRICKSON, R.L. (1970). *J. Fd Sci.*, 35, 268

JONES, A.N. and STRANG, T. (1973). *Proc. Inst. Fd Sci. Technol.*, 6, 168

KOPP, J. (1971). *Fleischwirtschaft.*, 51, 1647

LAWRIE, R.A. (1966). *Meat Science*, Pergamon, Oxford

MCCANCE, R.A. and WIDDOWSON, E.M. (1960). *The Composition of Foods*, HMSO, London

MANDIGO, R.W. and HENRICKSON, R.L. (1966). *Fd Technol., Champaign*, 20, 186

National Food Survey Committee (1973). *Household Food Consumption and Expenditure: 1970 and 1971*. HMSO, London

PRICE, J.F. and SCHWEIGERT, B.S. (eds.) (1971). *The Science of Meat and Meat Products*, 2nd edn., Freeman, San Francisco

ROBERTS, T.A. (1973). *Proc. Inst. Fd Sci. Technol.*, 6, 126

SAFFLE, R.L. (1968). *Adv. Fd Res.*, 16, 105

SAMEJIMA, K., HASHIMOTO, Y., YASUI, T. and FUKAZAWA, T. (1969). *J. Fd Sci.*, 34, 242

SATO, K., HEGARTY, G.R. and HERRING, H.K. (1973). *J. Fd Sci.*, 38, 398

SHAW, R. (1974). In the press

SHERMAN, P. (1961a). *Fd Technol., Champaign*, 15, 78

SHERMAN, P. (1961b). *Fd Technol., Champaign*, 15, 87

SHERMAN, P. (1961c). *Fd Technol., Champaign*, 15, 90

SHERMAN, P. (1962). *Fd Technol., Champaign*, 16, 91

SULZBACHER, W.L. (1973). *J. Sci. Fd Agric.*, 24, 589

SWIFT, C.E., LOCKETT, C. and FRYAR, A.J. (1961). *Fd Technol., Champaign*, 15, 468

THAMM, S. (1966). *Fleischwirtschaft.*, 46, 25

TRAUTMANN, J.C. (1966). *J. Fd Sci.*, 31, 409

VAHLUN, S. (1973). *Proc. Inst. Fd Sci. Technol.*, 6, 152

VAN DEN OORD, A.H.A. and VISSER, P.R. (1973). *Fleischwirtschaft.*, 53, 1427

VETTERLEIN, R. and HANNAN, R.S. (1966). Paper read at 12th European Meeting of Meat Research Workers, Norway

WALTERS, C.L. (1973). *Proc. Inst. Fd Sci. Technol.*, 6, 106

World Health Organization (1973). *Food Additives Series*, No.3, World Health Organization, Geneva

# 11

# MICROBIAL PROBLEMS IN HANDLING AND STORAGE OF FRESH MEATS

D. A. A. MOSSEL, K. E. DIJKMANN and J. M. A. SNIJDERS
*Institute for the Science of Food of Animal Origin,*
*University of Utrecht, The Netherlands*

## INTRODUCTION

### GENERAL PRINCIPLES OF MICROBIAL CONTROL

Microbial problems in the handling and storage of fresh meats have two distinct aspects: (i) those concerned with the protection of the consumer against food-borne diseases of microbial origin; and (ii) avoiding the deterioration of the commodities as a result of microbial attack. This has a strictly microbiological justification in that the ecological and physiological attributes of the groups of organisms that present public health problems are quite distinct from those of the spoilage agents. Although measures effective against the first group of micro-organisms may fail to control microbial deterioration, inhibition of spoilage agents will however avoid outgrowth of pathogens. The motivation behind the study of microbial problems related to fresh meats is to control their microbiological condition in both respects. The only system that will assure quality is the preventative one, wherein reliable manufacturing, handling and distribution codes are laid down and their application secured by instrumental control and human supervision.

In such an approach it is essential to distinguish clearly between microbial contamination and proliferation (Mossel, 1971). It was originally assumed (Hauser, 1886) that the muscular tissue of healthy animals at slaughter was sterile, but later this was questioned (Reith, 1926). Recent investigations suggest that *in vivo* contamination is indeed sporadic, both in time and in location. When it occurs it is mostly limited to mesophilic Gram-positive organisms (Nottingham, 1960; Vanderzant and Nickelson, 1969). The older conception that the Gram-negative bacteria, and particularly the psychrotrophic ones, stem from the environment, is thus supported by evidence from recently developed techniques for the examination of deeper layers of animal tissues.

Except in rare cases, the mere contamination of a carcass with micro-organisms does not present an immediate health risk. The harm lies in the subsequent proliferation of the initial contaminant, because this will lead to numbers of viable organisms that endanger the health of the consumer, or lead to deterioration phenomena. Because it is usually difficult, and occasionally impossible, to control contamination and since the natural antimicrobial properties of fresh muscle are very limited (Ingram, 1949; Buttiaux and Catsaras, 1964; Konowalchuk and Speirs, 1973) the strategy of assuring microbial quality rests on the prevention of the multiplication of most microbial contaminants. This does not eliminate the pressing need to minimise contamination because of the ever present risk of spreading pathogens. This could cause contamination of other, initially sound, food items.


## CONTAMINATION AND ITS CONTROL

### ORGANISMS OF HEALTH SIGNIFICANCE

In most countries overtly diseased animals will not present a health problem to the consumer because such animals show clinical syndromes during ante and/or post mortem examination and are rejected by the veterinary inspectors. A real menace is presented by some bacteria, helminths and protozoa which provoke a 'healthy carrier state'. These can only be partly monitored by macroscopic, microscopic or immunological procedures (Ruitenberg, Kampelmacher and Berkvens, 1967; Kampelmacher, 1969; Work, 1971). Little control is feasible without costly eradication campaigns. The trade and particularly the consumer should be fully informed about this since it discourages the consumption of raw

meats and dishes made with fresh meat.  Prevention is
thus fully in the hands of the consumer.

In principle, the situation is more promising for some
bacteria that may be carried by healthy slaughter animals,
particularly organisms of the Salmonella group.  In most
countries these are the main causes of food-borne diseases
in humans.  It has been shown that the following measures
will lead to a considerable reduction of Salmonella
carrier rate in pigs and calves, the most susceptible
animal types (Edel *et al.*, 1973):  (i) use of suitably
decontaminated feedstuffs only; (ii) modern, sanitary
modes of animal husbandry, including control of the water
supply on farms and adequate protection from insects and
rodents; (iii) holding and transport of slaughter animals
under conditions suitably designed by veterinary experts;
(iv) carefully laid-out slaughter lines, where every step
has been evaluated for the possibilities of minimising
cross contamination risks; and (v) breaking of the
Salmonella cycle by suitable sewage treatment and
monitoring.  It seems that the purpose of attaining
virtually Salmonella-free fresh meats can be attained in
this way.  Where this is so far impossible for economic
or technological reasons, or deliberately neglected,
public health officials will be forced to apply one of
the following approaches:  (i) decontamination of fresh
meats by a dose of 0.10-0.20 Mrad of ionising radiation
(Mossel, Krol and Moerman, 1972); or (ii) intensively
motivating the consumer to refrain from the consumption
of raw meats and to avoid recontamination of cooked, fried,
roasted or grilled meats from cutting boards, knives etc.
previously contaminated by fresh meats (Mossel,
Kampelmacher and Noorle Jansen, 1966; Kampelmacher *et al.*,
1971).

A most pertinent question is how the success or failure
of attempts aimed at improvement should be monitored.  It
must be stressed that a suitable 'Presence or Absence'
(P-A) test for Salmonellae should always be used.  This
sort of examination is most useful for an evaluation of
the sanitary quality of meats (Hobbs and Wilson, 1959;
van Schothorst and Kampelmacher, 1967; Rey, Kraft and
Rust, 1971; Carpenter, Elliot and Reynolds, 1973).
A difficulty is that negative results may be erroneous
owing to very unequal distribution of the pathogens.
Since Salmonellae are transferred to pathogen-free meats
by contamination from infected lymph nodes or intestinal
contents, it is obvious that a quantitative examination
for *E.coli* may yield useful additional information.

Although this is common practice when monitoring fresh seafoods (Thomas and Jones, 1971) it has, so far, only been done occasionally for fresh meats (Pantaléon and Baudeau, 1958; Gulistani, 1972). It has also been suggested that the entire Enterobacteriaceae group be used for this purpose (Hechelmann *et al.*, 1973). This criterion should be handled with care, however, as many types of Enterobacteriaceae, particularly the psychrotrophic species of various genera (Mossel and Zwart, 1960) may stem from quite different ecological niches and hence obscure the sanitary picture. Once the proportion between the Enterobacteriaceae group and *E.coli* and Salmonellae has been found consistent (Drion and Mossel, 1975), in many instances a count of Enterobacteriaceae may be helpful.

*Staph. aureus* is occasionally encountered on fresh meats. However, it does not seem to thrive very well under normal storage conditions (McCoy and Faber, 1966) and the formation of enterotoxin is inhibited by the proliferation of the non-fermentative Gram-negative bacteria which predominate on fresh meats (Collins-Thompson, Aris and Hurst, 1973).

SPOILAGE ASSOCIATION

Fresh meats are initially contaminated from (i) ubiquitous niches (soil, dust, water and manure); (ii) specific foci (spoiling foods and inadequately cleaned equipment); and (iii) the *in vivo* niches referred to earlier. This non-specific primary contamination does not usually contribute to the ultimate microbial deterioration: selection occurs. Thus fresh meats mostly spoil during chill storage as a result of slime formation by Gram-negative, rod-shaped bacteria with a non-fermentative glycolytic pathway. Such a specific group of micro-organisms predominating the spoilage microflora of a food is called the food's spoilage association (Westerdijk, 1949). Spoilage associations are determined by the various intrinsic parameters of the food and its customary mode of storage and distribution (Mossel, 1971). In the case of fresh meats the decisive parameters are the following: (i) the structural condition of the meat ('native', partly or severely comminuted, or even further structurally damaged); (ii) its pH, normally in the range 5.6-6.2 (Ingram and Dainty, 1971); (iii) the water activity at the meat surface (depending on the relative humidity of the

environment and accessibility to interior moisture);
(iv) the chilling temperature; and (v) the degree of
access of oxygen.

The approximate taxonomy of the spoilage association
of meats has been known for many years (Haines, 1933;
Ingram, 1949) but only recently have sufficient criteria
been evaluated to permit accurate grouping (Shewan, 1971;
Davidson, Dowdell and Board, 1973). Some problems,
particularly with regard to yellow pigmented rods (Halls
and Board, 1973) do still exist and apparently cannot be
resolved easily.  In *Table 11.1,* an approach has been
made to incorporate the most recent taxonomic findings in
a rather simple determinative key.  The occurrence of the
Gram-positive main groups in this table results from the
role played by some of these types in the spoilage of
meats, stored in the absence of oxygen or under deliberately
increased carbon dioxide pressure (*vide infra*).  The main
spoilage types on chilled fresh meats are members of the
*Pseudomonas/Acinetobacter/Moraxella/Flavobacter/*
*Alcaligenes* group, as defined in *Table 11.1.*

This group of psychrotrophic Gram-negative, non-
fermentative rods is almost ubiquitous, being the
predominating microflora of surface waters (Poffé *et al.*,
1973) and from there spreading to claws and hides of
animals.  Hence it is virtually impossible to completely
avoid contamination of freshly slaughtered carcasses with
such bacteria.  Systematic efforts should be made, however,
to limit their numbers, because the time taken for slime
to develop on raw meats under given conditions of chilled
storage is directly related to the initial numbers of
colony-forming units of the members of this spoilage
association per unit area (Mossel and Ingram, 1955;
Farrell and Barnes, 1964).  Niches of such organisms
which should be avoided particularly are wooden and other
surfaces in chill rooms, residues of meats stored in
chill rooms and the transportation vehicles used therein
(Empey and Scott, 1939; Buttiaux and Catsaras, 1964).

Table 11.1  Preliminary grouping of bacteria

| Group | Cytochemistry (Gram stain) | Morphology | Biochemistry[1] | Example |
|-------|----------------------------|------------|-----------------|---------|
| 1 | + | rods, no spores | catalase + | Corynebacterium |
| 2 | + | ditto | catalase θ | Lactobacillus |
| 3 | + | rods, with spores | catalase + | Bacillus |
| 4 | + | ditto | catalase θ <br> glucose 5-1[2] | Clostridium |
| 5 | + | coccus | catalase + <br> glucose type 1-2 <br> glucose type 3 | Micrococcus <br> Staphylococcus |
| 6 | + | ditto | catalase θ <br> glucose type 3 <br> glucose type 4 | Streptococcus <br> Leuconostoc |
| 7 | θ | rods[3], no spores | oxidase θ <br> glucose type 1-2 <br> motile <br> non-motile <br> glucose type 3-4 <br> lactose + <br> citrate θ <br> citrate + <br> motile | Xanthomonas <br> Acinetobacter <br><br> Escherichia <br> Enterobacter |

|  |  |  | non-motile | Klebsiella |
|  |  |  | lactose θ |  |
|  |  |  | urea θ | Salmonella |
|  |  |  | urea + | Proteus |
| 8 | θ | rods[3], no spores | oxidase + |  |
|  |  |  | glucose type 1 | Alcaligenes/Comamonas[4] |
|  |  |  | motile | Moraxells/Flavobacter[1] |
|  |  |  | non-motile |  |
|  |  |  | glucose type 2[5] | Pseudomonas |
|  |  |  | glucose type 3 | Beneckea/Vibrio[4] |
|  |  |  | glucose type 4[6] | Aeromonas |

[1]Reactions for approximately 90% of strains.

[2]Glucose type 1 = no attack; 2 = oxidative dissimilation; 3 = fermentative anaerogenic dissimilation; 4 = aerogenic fermentative dissimilation; 5 = strict anaerobic dissimilation.

[3]Gram-negative cocci do exist and are even of great significance as causes of infective diseases; however they do not play a role in foods.

[4]Respectively, peritrichous and polar flagellae.

[5]Glucose type 1 types of Pseudomonas also exist.

[6]Anaerogenic Aeromonas types are encountered occasionally.

## MICROBIAL PROLIFERATION AND POSSIBILITIES
## FOR ITS LIMITATION

Pathogenic bacteria are generally mesophiles. Hence no
growth will occur below 5°C (Angelotti, Foter and Lewis,
1961). Two classic exceptions are *Listeria monocytogenes*
that grows slowly and *Cl.botulinum,* type E (Schmidt,
Lechowich and Folinazzo, 1961). This latter organism
will not be a risk with chilled meats, however, since
their psychrotrophic spoilage association swiftly outgrows
the clostridia. Obvious organoleptic defects will occur
long before a public health risk develops (Schmidt,
Lechowich and Nank, 1962).

Chilled storage is not a fully effective weapon against
the spoilage association of fresh meats, because the
minimum temperature of these organisms lies somewhere
around -7°C (Haines, 1934). However, a reduced temperature
will retard the growth of slime-forming organisms,
temperatures below 2°C being effective in delaying
deterioration (Hess, Marthaler and Ruck, 1962).

It is a matter of technological judgement whether this
lower temperature range will be the means chosen to
prolong freshness, or whether other extrinsic parameters
will be employed. Reduction of the relative humidity of
the air in chilled stores is effective because the
spoilage association of psychrotrophic Gram-negative rods
is sensitive,to $a_w$ reduction (Mossel, 1971). When this
approach is chosen, two limitations will prevail.
Firstly, storage at reduced relative humidity of meats
which are no longer protected by their fascia will lead
to loss of weight. Secondly, meat thus stored will spoil
eventually when psychrotrophic organisms, such as bacteria
of the genera Micrococcus, Microbacterium, Streptococcus
and Lactobacillus (Barlow and Kitchell, 1966; Gardner,
1971) and some moulds, which are all rather insensitive
to reduced $a_w$, predominate. A summary of the spoilage
association which may develop in this instance is given
in *Table 11.2.*

The oxygen partial pressure $(p_{O_2})$ value in the gaseous
environment of stored meats is also listed in *Table 11.2.*
This is justified since a further increase in storage
life of chilled meats is often sought by reducing $p_{O_2}$
either by vacuum packaging (Shank and Lundquist, 1963;
Pierson, Collins-Thompson and Ordal, 1970; Roth and Clark,
1972) or by storage under a certain partial pressure of
carbon dioxide (Moran, Smith and Tomkins, 1932; Ogilvy
and Ayres, 1953; Adams and Huffman, 1972). When this

Table 11.2 A review of the more $a_w$ tolerant, psychrotrophic food spoilage organisms with reference to their susceptibility to $a_w$, pH and $pO_2$ values.

| Taxonomic genus | Limit of $a_w$ value for growth | Growth at pH <4.5 | Sensitivity to decreased $pO_2$ |
|---|---|---|---|
| **BACTERIA** | | | |
| Micrococcus | 0.95-0.91 | − | + |
| Group D - streptococci | 0.93 | − | − |
| Lactobacilli | 0.91 | + | − |
| Corynebacterium, Arthrobacter, | 0.98-0.95 | − | +− |
| Listeria, Microbacterium | | | − |
| Some Clostridium and Bacillus species | c. 0.95 | − | −+ |
| **YEASTS** | | | |
| Candida, Debaryomyces, Monilia, | 0.91-0.87 | + | − |
| Torulopsis | | | |
| Osmophilic Saccharomyces species | c. 0.70 | + | − |
| Halotolerant Debaryomyces species | c. 0.80 | ... | ... |
| **MOULDS** | | | |
| Alternaria, Aspergillus, Botrytis, | 0.88-0.80 | + | +− |
| Cladosporium, Fusarium, Margarinomyces, | | | |
| Mucor, Penicillium, Rhizopus, | | | |
| Sporotrichum, Thamnidium, and a few | | | |
| others | | | |

− = no growth, or insensitive; $\overset{-}{+}$ = mostly insensitive; + = growth or sensitive; $\overset{+}{-}$ = variable response; ... = insufficient data available.

approach is followed the association shifts in the sense
that Micrococcus species (and a few yeasts and moulds
which tolerate $a_w$ values of the order 0.92) will now also
be inhibited. A carefully devised combination of reduced
initial load of spoilers, lowered temperature and $a_w$, and
increased $p_{CO_2}$, has enabled shipments of unfrozen meat
from Australia and New Zealand to Great Britain to be made
(Ingram, 1949). This shows how far conventional methods
can be effective in overcoming microbiological problems
related to fresh meat storage and transportation.

The time-honoured practice of treating meat surfaces
with vinegar (Biemuller, Carpenter and Reynolds, 1973)
or lactic acid (Erickson and Fabian, 1942) in instances
where the natural protection barrier has been impaired,
may still be useful.

Nevertheless attempts have also been made to apply
newer techniques in this area. Treatment with $\gamma$-rays
($\sim 0.1$ Mrad), immediately before shipping, has been the
most intensively studied innovation. There is no doubt
this considerably increases the storage life of carcasses
under refrigeration (Rhodes, Roberts and Shepherd, 1967;
Tiwari and Maxcy, 1972). In addition it successfully
eliminates Salmonellae from meats (Mossel et al.,1972)
and thus solves one of the most pressing needs in Meat
Microbiology. However, to maintain first class quality
of meats, the dose cannot be much increased above 0.1
Mrad. This was empirically established by hedonic rating
(Mossel, Krol and Moerman, 1972) and confirmed by
molecular investigations (Satterlee, Wilhelm and Barnhart
1971). As in meat storage in artificial atmospheres
there remains a need for concomitant hygiene. Legislator
are apparently awaiting the outcome of the systematic
investigations which are being carried out under the
auspices of the United Nations (Hickman, 1973) before
authorising this preservative procedure.

Amongst the other non-conventional procedures, the
economically most important one is the preservation of
minced meats by the addition of sodium sulphite. There
is evidence that sulphite at high concentrations is a
strong bactericidal agent (Dijkmann, 1966), while at a
level of $\sim 0.03\%$ it acts as a most effective preservati
for fresh meats (Christian, 1963; Dowdell and Board,
1971). In considering the acceptability of the use of
chemical additives in food, an important point of
consideration is whether they are essential or not. It
has been demonstrated that optimal plant sanitation and
vacuum packaging can also be most effective in improving
the keeping quality of minced meats.

# MONITORING THE MICROBIOLOGICAL CONDITION
## OF MEATS

The groups of organisms which must be kept under control
and, hence, followed by analytical methods, have been
reviewed in the previous sections. Here we recommend
the methodology which has been found useful in daily
practice.

## DRAWING SAMPLES FROM CONSIGNMENTS

The first and most important problem to be resolved is
that of adequate investigation of a consignment.
     Enormous intra carcass, inter carcass and inter bag
spreads are generally encountered. However, sufficient
statistical guidance is available to assist the quality
assurance bacteriologist in attempts to base his
acceptance schemes on sound investigational or routine
sampling plans (Ingram *et al.*, 1974).

## SAMPLING OF A CHILLED CARCASS

The topography of sampling of carcasses has been the
object of many studies (e.g. Buttiaux and Catsaras, 1964;
Carpenter, Elliot and Reynolds, 1973; Catsaras, Gulistani
and Mossel, 1974). It is a most pertinent aspect of
microbiological monitoring of meats. The distribution of
organisms over carcasses is so variable that strict
procedures cannot be given.
     The most productive mode of releasing the organisms
from sampling sites has also been studied extensively.
In the case of carcasses there is no doubt that the
dissection technique is the most reliable one (Bartels
and Klemm, 1962; Kampelmacher *et al.*, 1971; Patterson,
1972). But, because it is rather elaborate and slightly
damages the carcasses, simpler alternatives have been
suggested. Swabbing methods are quite reliable
(Manheimer and Ybanez, 1917; Mossel and Büchli, 1964;
Patterson, 1971), although the organisms they remove may
not exceed 10% of the actual numbers present and these
recovery percentages may vary widely (Gisske and Klemm,
1963; Snijders, 1974). It is well established that
impression techniques may be very misleading, because
here the recovery greatly depends on factors like: (i)
penetration of the microflora (Elmossalami and Wassef,

1971); (ii) degree of association of microbial cells in chains or clusters (Greene, Vesley and Keenan, 1962; Mossell, Kampelmacher and Noorle Jansen, 1966; Maunz and Kanz, 1969); (iii) experimental details (e.g. humidity of the surface under examination and that of the agar used, intensity and homogeneity of impressing, etc.); and (iv) substrate linked parameters (e.g. the state of the surface - fresh or scalded).

## SAMPLING OF BAGS OF FROZEN BONELESS MEATS

Frozen boneless meats are an important item in the international meat trade. Once a sufficient number of bags per consignment has been taken to account for variations between bags, the best method of releasing organisms from the bags chosen poses a problem. It has been established that defrosting (at not too high an environmental temperature) and subsequent hand pressing of the bags to release drip is a reliable and reproducible mode of collecting micro-organisms (Mossel, Krol and Moerman, 1972). Obviously the time required for sufficient defrosting depends on the size of the bags.

## MICROBIOLOGICAL-ANALYTICAL METHODOLOGY

There is often a need for dilution and plating to be done outside a fully equipped laboratory. In these instances rapid dilution and plating, based on loop (Kitchell, Ingram and Hudson, 1973) or micropipette (Donnelly et al., 1970) techniques, are most useful and sufficiently reliable (Snijders, 1974). For the detection or enumeration of specific groups of organisms we recommend the following methods:

### 1. *Salmonellae*

In many instances a resuscitation treatment (Mossel and Ratto, 1970) using buffered peptone water is required. This is followed by enrichment in tetrathionate brilliant green broth at 43°C (Edel and Kampelmacher, 1973) and isolation on brilliant green-lactose-phenol red agar. For identification β-galactosidase, urease, lysine decarboxylase and agglutination patterns (Catsaras, Seynave and Sery, 1972) can be recommended.

## 2. *E.coli*

A most useful direct enumeration method is culturing
press fluids or meat dilutions in MacConkey's agar at
44°C (Clegg and Sherwood, 1947). For the purpose of
achieving sufficient anaerobiosis and rapid, accurate
temperature control the use of plastic pouches (Mossel
and Vega, 1973) or Miller-Prickett tubes (1939) is
recommended.

## 3. *Enterobacteriaceae*

The currently employed medium is violet red-bile-glucose
agar (Mossel, Mengerink and Scholts, 1962) in poured
plates, covered with 10-20 ml sterile violet red-bile-
glucose agar (Holtzapffel and Mossel, 1968). In order
not to miss psychrotrophic Enterobacteriaceae (Mossel
and Zwart, 1960) incubation at 30°C is necessary. This
temperature does not inhibit pathogenic types (Taylor
and Schelhart, 1973). A justified objection to this
incubation temperature is that a few Gram-negative rods,
not belonging to the Enterobacteriaceae group, may
interfere. This is met by assessing the mode of attack
of isolates on glucose (Mossel and Martin, 1961),
preferably by stabbing into tubes with violet red-bile-
glucose agar (Mossel and Ratto, 1970).

A much less generally recognised source of false
positives is the frequent occurrence of Aeromonas species
on chilled meats. We have observed that these may form
some 30% of the colonies obtained by current techniques
for the enumeration of coli-aerogenes bacteria. Their
elimination requires an oxidase test (Steel, 1961). We
have found a double layer tube most useful for the
combined purpose of testing for mode of attack on glucose,
oxygen requirement, motility and oxidase formation.
(The top layer consists of peptone salt motility agar,
the bottom one of violet red-bile-glucose agar).

Which procedure is followed in any particular instance
has to be decided on the basis of the required accuracy
of counts and the benefit/effort relationship. The
number of samples to be examined and the maximum
acceptable time delay in monitoring work, often restrict
the examination to gross counting of typical colonies
obtained at 37°C in layered plates.

4.   *The psychrotrophic Gram-negative non-fermentative
     rods*

During the initial development of the bacterial
population on freshly slaughtered animals enormous shifts
in the original microflora occur.  Hence in studies of
this sort the use of a medium selective for Gram-negative
rods is appropriate.  The medium for this purpose
contains 1 µg/ml crystal violet (Olson, 1961; Mossel,
1965).

For quality monitoring of meats found in commerce,
where mostly the specific Gram-negative association is
predominant, a simple plate count on agar suffices.

A point of dispute may be the temperature of incubation
in both instances.  Ideally, plates should be incubated
at 1°C (Barnes and Impey, 1968) but this requires rather
long and somewhat impracticable incubation times.  We
have established that a temperature of *c.* 17°C in
combination with a 2 days incubation period may be an
acceptable compromise (Mossel and Ratto, 1973).  Although
this temperature is not inhibitory to the most
psychrophilic psychrotrophes (Griffiths and Haight, 1973),
it retards the growth of mesophiles (Board, 1965)
sufficiently for routine monitoring.  In ecological
research work, an adequate selection of colonies obtained
may always be further checked for identity with the Gram-
negative psychrotrophic group.


5.   *Gram-positive spoilers*

A medium of the type recommended by Beerens and Tahon
Castel (1968) has proved useful for the tentative
enumeration of Gram-positive spoiling bacteria, with
incubation at *c.* 17°C (Gray and Jackson, 1973).

The preparation of this medium requires blood.  In its
absence phenylethanol agar (Lilley and Brewer, 1953) may
be used.


6.   *Moulds and yeasts*

We have shown earlier that oxytetracycline agar is the
most suitable medium for the examination of dried foods
for moulds and yeasts, but that it may not be entirely
appropriate for the examination of fresh meats because
it does not fully suppress non-fermentative rods (Mossel
*et al.*, 1970; Dijkmann, 1974).  The addition of 50 µg/ml

of gentamicin to oxytetracycline medium is most helpful
in eliminating most of the interference by bacteria
particularly when spread plates are used instead of
poured plates.


## AN ATTEMPT AT MICROBIOLOGICAL STANDARDS

It seems certainly appropriate to deal with the
possibility of having microbiological standards for fresh
meats.   Indeed in such a chapter as this it cannot be
omitted, since the use of standards seems to be one of
the most difficult problems in the field of Meat
Microbiology.


### PRINCIPLES

The introduction of microbiological standards for foods
has been frequently opposed.   We have demonstrated
repeatedly, however, that no scientifically justified
objections can be raised against the use of standards,
provided the following five basic principles are
accepted:   (i) Specifications serve to *verify* good
manufacturing and distribution practices, but are useless
to attain good quality.   (ii) All criteria for which
numerical limits are suggested must be ecologically
sound and absolutely required for adequate quality
monitoring.   (iii) Standards should be derived from the
results of surveys of products taken off production lines
which have been previously examined for good manufacturing
practices.   (iv) The numerical values thus obtained should
include tolerances, acknowledging the usual coefficients
of variation encountered in distribution of organisms in
or on foods and in carrying out microbial enumerations.
(v) An essential part of a microbiological standard is
the very exact procedure to be used in assessing the
numerical values specified (Mossel, 1973).

When drafted and used in this way, standards are very
helpful and cause no difficulty in daily practice
(Mossel, 1974).   Microbiological examination of foods
without standards to gauge the results obtained is
pointless.   In principle no standards should be needed
when proper preventative measures are taken to reduce
quality fluctuations and hence substandard quality.   But
how can Public Health Authorities of importing countries
establish the faithful adherence to such good manufacturing

and distribution practices without adequate spot checks
on the final product?

## WHOLESOMENESS OF CARCASSES

The public health monitoring of meat by veterinarians
varies widely between countries.  On the European
continent every animal is examined by a veterinarian
both while on the hoof and after slaughtering.  When
neither pathology nor abnormal physico-chemical values
are observed, the carcass is passed.  Where doubt exists
bacteriological examination is carried out.  A rather
pressing strategic problem here is a more modern mode of
bacteriological monitoring of muscle samples and organs
at the stage of veterinary examination of carcasses for
wholesomeness.

The methods used so far are mostly still based on von
Ostertag's classical approach (1913).  It has been
suggested that the task of the public health
veterinarian might be greatly facilitated by (i) a
reliable mode of sampling of organs and tissues, and (ii)
a quantitative-diagnostic approach to the monitoring
proper.

With reference to the latter, there is no doubt that
a sole 'presence or absence' approach is fraught with the
risk of false positives as a result of contamination from
the environment (Nickerson, Proctor and Goldblith, 1956)
unless the tests are systematically carried out in laminar
airflow cabinets (Heldman, Hedrick and Hall, 1968).  To
overcome this serious difficulty we have suggested the
use of solid media only (Mossel and Visser, 1960).  For
the purpose of harvesting any sparse foci of *in vivo*
infection we have carried out some model experiments.
These have shown that the following procedure is
effective and relatively easy:

   (i)   Examination of the material within one hour after
        drawing by dissection (Habisreitinger, 1973).
  (ii)   Surface sterilisation by the 'deep frying'
        technique of Névot (1947).
(iii)   Aseptic dissection of a fragment of about 10 g of
        tissue.
 (iv)   Aseptic preparation of a 1:10 triturate with
        peptone saline (Pantaléon *et al.*, 1970).
  (v)   Plating 0.2 ml aliquots of such dilutions on to
        the following media, all contained in approximately
        15 cm diameter plates:

(a)  Plain blood-agar for the total aerobic
     microflora.
(b)  Plain blood-agar to be incubated
     anaerobically for the pathogenic anaerobes.
(c)  Plain blood-agar + 2 μg/ml crystal violet
     (Packer, 1943) for the Gram-negative
     pathogens.
(d)  Plain blood-agar + 40 μg/ml nalidixic acid
     (Beerens and Castel, 1968) for the Gram-
     positives.

Even when the working conditions are not fully aseptic,
no more than one or two colonies per plate will be
derived from the environment (Mossel and Drion, 1954);
hence plates containing more colonies point in the
direction of primary infection. Valid diagnostic
indications from the universal plate, will gain support
or be invalidated by the selective counts.


ACCEPTANCE CRITERIA FOR FRESH MEATS

In the last five years we have carried out the type of
survey required for the setting of standards for two
types of wholesale fresh meat, viz. for frozen boneless
meats (Mossel, Krol and Moerman, 1972) and for carcasses
(Catsaras, Gulistani and Mossel, 1974). The suggestions
derived from these surveys are summarised in the upper
part of *Table 11.3*.
In addition we have made a smaller scale survey on
consumer size cut-up meats, such as steaks and cutlets.
Where the slicing had been done by a sanitary procedure
increases in counts did not generally exceed one log.
cycle; cf. middle part of *Table 11.3*.


COMMINUTED FRESH MEATS

The main types of comminuted meat found in the trade are:
(i) pure ground muscle, such as hamburgers, beefburgers,
filet americain and tartar steak; and (ii) continental-
style minced meat, containing in addition the less
valuable parts of the carcass.
There seems to be no reason at all to exclude
comminuted meat products from attempts to apply standards.
As a matter of fact, most of the outbreaks of food-borne
salmonellosis in countries where minced meat products are
an important item of the menu are caused by these foods.

Table 11.3 Results of surveys on the bacteriological condition of fresh meats of good quality. Data expressed as log. 95 percentile, unless otherwise indicated.

| Type of meat | Counts expressed per | Salmonella | E.coli | Entero-bacteriaceae | Staph. aureus | Total aerobic count |
|---|---|---|---|---|---|---|
| WHOLESALE MEATS | | | | | | |
| Carcasses | 100 cm$^2$ | <0 | <70%+ve/10 cm$^2$ <br> <20%+ve/ 1 cm$^2$ | 3 | ** | 5 |
| Frozen boneless | 1 ml hand-pressed drip | <-1 | ** | 4 | ** | 6.5 |
| CONSUMER-SIZE CUTS | 100 cm$^2$ | <1 | <50%+ve/ 1 cm$^2$ | 4 | ** | 6 |
| MINCED MEATS | 1 g | 0 | ** | 5-6* | 2-3* | 7-8* |

** = no survey carried out.

* = strongly dependent on type of meat, particularly beef versus pork.

The problem of setting a standard for Salmonellae in these products has been discussed repeatedly (e.g. Hobbs and Wilson, 1959; Mossel, 1973). Yet, the five basic principles mentioned allow a balanced approach. If at present, under the best possible conditions, it appears possible only to require the absence of Salmonellae from say, one gramme aliquots – as contrasted to the 25-100 g aliquots customarily examined in the monitoring of dried foods (Mossel, Shennan and Vega, 1973) – then there is no harm in temporarily using this very relaxed standard, provided the one gramme aliquots are taken from a larger representative sample. However, it should be made a point of principle that such standards will only be used pending essential improvement of the sanitary quality of fresh meats by the vertical integration approach (Mossel, 1973).

As for additional criteria – mainly Enterobacteriaceae, *Staph. aureus* and total count as indices of sanitation and proper refrigeration – a survey made in the Netherlands has revealed that the log 95 percentile values of these counts encountered in minced meat found in commerce, are relatively high, viz. of the order of 6.5, 4.5 and 8.5, respectively (van der Meijs, 1970). Comminuted muscle made and stored under the best possible conditions appeared to show corresponding values of *c.* 5.5, 2.5 and 7.5 (cf. *Table 11.3*). Beef products are significantly better than pork. Obviously it would be unrealistic to expect that, almost overnight, the meat industry would be in a position to mimic what has been achieved in experimental pilot plants. However there is ample room for improvement and this should gradually be attained by the meat industry. This will certainly not occur without pressure – recently Law, Yang and Mullins (1971) had to conclude that the microbiological quality of ground beef in the USA had not at all improved since the mid-fifties!

The considerations just presented show a final, implicit advantage of the introduction of microbiological standards for foods: it leads to continuous reduction of the index counts and hence to better consumer's protection (Wodicka, 1973).

## ACKNOWLEDGEMENT

The authors are greatly indebted to Professor J.G. van Logtestijn D.V.M. for a critical review of their manuscript.

REFERENCES

ADAMS, J.R. and HUFFMAN, D.L. (1972). *J. Fd Sci.*, 37, 869

ANGELOTTI, R., FOTER, M.J. and LEWIS, K.H. (1961). *Amer. J. Public Health*, 51, 76

BARLOW, J. and KITCHELL, A.G. (1966). *J. appl. Bacteriol.*, 29, 185

BARNES, E.M. and IMPEY, C.S. (1968). *J. appl. Bacteriol.*, 31, 97

BARTELS, H. and KLEMM, G. (1962). *Fleischwirtschaft.*, 14, 645

BEERENS, H. and TAHON CASTEL, M.M. (1968). *Ann. Inst. Pasteur*, 111, 90

BIEMULLER, G.W., CARPENTER, J.A. and REYNOLDS, A.E. (1973). *J. Fd Sci.*, 38, 261

BOARD, R.G. (1965). *J. appl. Bacteriol.*, 28, 437

BUTTIAUX, R. and CATSARAS, M. (1964). *Ann. Inst. Pasteur Lille*, 15, 165

CARPENTER, J.A., ELLIOT, J.G. and REYNOLDS, A.E. (1973). *Appl. Microbiol.*, 25, 731

CATSARAS, M., SEYNAVE, R. and SERY, C. (1972). *Bull. Acad. Vet. France*, 45, 379

CATSARAS, M., GULISTANI, A.W. and MOSSEL, D.A.A. (1974). *Rec. Méd. Vét.*, 150, 287

CHAPMAN, G.H. and BERENS, C. (1935). *J. Bacteriol.*, 29, 437

CHRISTIAN, J.H.B. (1963). *Food Preservation Quart.*, 23, 30

CLEGG, L.F.L. and SHERWOOD, H.P. (1947). *J. Hygiene*, 45, 504

COLLINS-THOMPSON, D.L., ARIS, B. and HURST, A. (1973). *Can. J. Microbiol.*, 19, 1197

DAVIDSON, C.M., DOWDELL, M.J. and BOARD, R.G. (1973). *J. Fd Sci.*, 38, 303

DIJKMANN, K.E. (1966). *Tijdschr. Diergeneesk.*, 91, 1449

DIJKMANN, K.E. (1974). To be published

DONNELLY, C.B., LESLIE, J.E., MESSER, J.W., GREEN, M.T., PEELER, J.T. and READ, R.B. (1970). *J. Dairy Sci.*, 53, 1187

DOWDELL, M.J. and BOARD, R.G. (1971). *J. appl. Bacteriol.* 34, 317

DRION, E.F. and MOSSEL, D.A.A.(1975). *J. Hyg. Camb.* (in preparation)

EDEL, W., GUINEE, P.A.M., SCHOTHORST, M. VAN and KAMPELMACHER, E.H. (1973). *Can. Inst. Fd Sci., Technol. J.*, 6, 64

EDEL, W. and KAMPELMACHER, E.H. (1973). *Bull. World Health Organ.*, <u>48</u>, 167

ELMOSSALAMI, E. and WASSEF, N. (1971). *Zentrabl. Vet. Med.*, <u>18</u>, 329

EMPEY, W.A. and SCOTT, W.J. (1939). *Commonwealth Sci. Ind. Research Organ. Bull.*, No.126

ERICKSON, F.J. and FABIAN, F.W. (1942). *Fd Res.*, <u>7</u>, 68

FARRELL, A.J. and BARNES, E.M. (1964). *Brit. Poultry Sci.*, <u>5</u>, 89

FERON, V.J. and WENSVOORT, P. (1972). *Pathol. Europ.*, <u>7</u>, 103

GARDNER, G.A. (1971). *J. Fd Technol.*, <u>6</u>, 225

GISSKE, W. and KLEMM, G. (1963). *Fleischwirtschaft.*, <u>15</u>, 288

GRAY, R.J.H. and JACKSON, H. (1973). *Antonie van Leeuwenhoek*, <u>39</u>, 497

GREENE, V.W., VESLEY, D. and KEENAN, K.M. (1962). *J. Bacteriol.*, <u>84</u>, 188

GRIFFITHS, R.P. and HAIGHT, R.D. (1973). *Can. J. Microbiol.*, <u>19</u>, 557

GULISTANI, A.W. (1972). *Arch. Lebensm. Hyg.*, <u>23</u>, 172

HABISREITINGER, K. (1973). *Schlacht- u. Viehhof Ztg.*, <u>73</u>, 457

HAINES, R.B. (1933). *J. Hygiene*, <u>33</u>, 175

HAINES, R.B. (1934). *J. Hygiene*, <u>34</u>, 277

HALLS, N.A. and BOARD, R.G. (1973). *J. appl. Bacteriol.*, <u>36</u>, 465

HAUSER, G. (1886). *Arch. exper. Pathol.*, <u>20</u>, 162

HECHELMANN, H., ROSSMANITH, E., PERIC, M. and LEISTNER, L. (1973). *Fleischwirtschaft.*, <u>53</u>, 107

HELDMAN, D.R., HEDRICK, T.I. and HALL, C.W. (1968). *J. Dairy Sci.*, <u>51</u>, 1356

HESS, E., MARTHALER, A. and RUCK, C. (1962). *Fleischwirtschaft.*, <u>14</u>, 497

HICKMAN, J.R. (1973). In *Internat. Atomic Energy Agency, Vienna, Publ. STI/317*, 659

HOBBS, B.C. and WILSON, J.G. (1959). *Monthly Bull. Min. Health London*, <u>18</u>, 198

HOLTZAPFFEL, D. and MOSSEL, D.A.A. (1968). *J. Fd Technol.*, <u>3</u>, 223

INGRAM, M. (1949). *J. Royal Sanit. Inst. (Lond.)*, <u>69</u>, 39

INGRAM, M. and DAINTY, R.H. (1971). *J. appl. Bacteriol.*, <u>34</u>, 21

INGRAM, M., BRAY, D.F., CLARK, D.S., DOLMAN, C.E., ELLIOTT, R.P. and THATCHER, F.S., (Eds.) (1974). *Microorganisms in Foods. II. Sampling for Microbiological Analysis*, University of Toronto Press, Toronto

KAMPELMACHER, E.H. (1969). *Proc. V Symp. World Assoc. Vet. Fd Hyg.*, 15

KAMPELMACHER, E.H., MOSSEL, D.A.A., SCHOTHORST, M. van and NOORLE JANSEN, L.M. van (1971). *Alimenta*, 10, 70

KITCHELL, A.G., INGRAM, G.C. and HUDSON, W.R. (1973). In *Sampling - microbiological monitoring of environments*, Ed. R.G. Board and D.W. Lovelock, Academic Press, London, p.43

KONOWALCHUK, J. and SPEIRS, J.I. (1973). *Can. J. Microbiol.*, 19, 177

LAW, H.M., YANG, S.P. and MULLINS, A.M. (1971). *J. Amer. Diet. Assoc.*, 58, 230

LILLEY, B.D. and BREWER, J.H. (1953). *J. Amer. Pharmac. Assoc. Sci. Ed.*, 42, 6

MCCOY, D.W. and FABER, J.E. (1966). *Appl. Microbiol.*, 14, 372

MAN, J.C. DE, ROGOSA, M. and SHARPE, M.E. (1960). *J. appl. Bacteriol.*, 23, 130

MANHEIMER, W.A. and YBANEZ, T. (1917). *Amer. J. Public Health*, 7, 614

MAUNZ, U. and KANZ, E. (1969). *Gesundheitswes. Desinfekt.*, 61, 129

MEIJS, C.C.J.M. VAN DER (1970). *Tijdschr. Diergeneesk.*, 95, 1180

MILLER, N.J., GARRETT, O.W. and PRICKETT, P.S. (1939). *Fd Res.*, 4, 447

MORAN, T., SMITH, E.C. and TOMKINS, R.G. (1932). *J. Soc. Chem. Ind.*, 51, 114

MOSSEL, D.A.A. (1965). *Fette. Seifen. Antrichm.*, 67, 903

MOSSEL, D.A.A. (1971). *J. appl. Bacteriol.*, 34, 95

MOSSEL, D.A.A. (1973). *Brit. Fd Manufact. Inds. Research Assoc. Techn. Circular*, No. 526

MOSSEL, D.A.A. (1973). *Komp. Allmänt Veterinärmöte Sver. Vet. Förb.*, I, 23

MOSSEL, D.A.A. (1974). *Caveant artifex et emptor*, Inaugural Address, University of Utrecht

MOSSEL, D.A.A. and BÜCHLI, K. (1964). *Lab. Practice*, 13, 1184

MOSSEL, D.A.A. and DRION, E.F. (1954). *Netherl. Milk Dairy J.*, 8, 106

MOSSEL, D.A.A. and INGRAM, M. (1955). *J. appl. Bacteriol* 18, 232

MOSSEL, D.A.A., KAMPELMACHER, E.H. and NOORLE JANSEN, L.M. VAN (1966). *Zentralbl. Bakteriol. Parasitenk. Abt. I, Orig.*, 201, 91

MOSSEL, D.A.A., KLEYNEN-SEMMELING, A.M.C., VINCENTIE, H.M BEERENS, H. and CATSARAS, M. (1970). *J. appl. Bacteriol.*, 33, 454

MOSSEL, D.A.A., KROL, B. and MOERMAN, P.C. (1972).
*Alimenta,* 11, 51

MOSSEL, D.A.A. and KRUGERS DAGNEAUZ, E.L. (1959).
*Antonie van Leeuwenhoek,* 25, 230

MOSSEL, D.A.A. and MARTIN, G. (1961). *Ann. Inst.
Pasteur Lille,* 12, 225

MOSSEL, D.A.A., MENGERINK, W.H.J. and SCHOLTS, H.H.
(1962). *J. Bacteriol.,* 84, 381

MOSSEL, D.A.A. and RATTO, M.A. (1970). *Appl. Microbiol.,*
20, 273

MOSSEL, D.A.A. and RATTO, M.A. (1973). *J. Fd Technol.,*
8, 97

MOSSEL, D.A.A., SHENNAN, J.L. and VEGA, C. (1973).
*J. Sci. Fd Agric.,* 24, 499

MOSSEL, D.A.A. and VEGA, C.L. (1973). *Health Labor. Sci.,*
10, 303

MOSSEL, D.A.A. and VISSER, M. (1960). *Ann. Inst. Pasteur
Lille,* 11, 193

MOSSEL, D.A.A. and ZWART, H. (1960). *J. appl. Bacteriol.,*
23, 185

NEVOT, A. (1947). *Controle bactériologique pratique des
denrées alimentaires d'origine animale,* Flammarion,
Paris, p.205

NICKERSON, J.T.R., PROCTOR, B.E. and GOLDBLITH, S.A.
(1956). *Fd Technol.,* 10, 305

NOTTINGHAM, P.M. (1960). *J. Sci. Fd Agric.,* 11, 436

OGILVY, W.S. and AYRES, J.C. (1953). *Fd Res.,* 18, 121

OLSON, H.C. (1961). *J. Dairy Sci.,* 44, 970

OSTERTAG, R. VON (1913). *Handbuch der Fleischbeschau.*
Vol. II, 555, Enke, Stuttgart

PACKER, R.A. (1943). *J. Bacteriol.,* 46, 343

PANTALÉON, J. and BAUDEAU, H. (1958). *Bull. Soc. Sci.
Hygiène Aliment.,* 46, 137

PANTALÉON, J., BILLON, J., GILLES, G., LE TURDU, P.,
ROSSET, R., CUMONT, F., GANDON, Y. and RICHOU-BAC, L.
(1970). *Hygiène des denrées animales et d'origine
animale. Techniques de laboratoire,* Ministère de
l'Agriculture, Paris

PATTERSON, J.T. (1971). *J. Fd Technol.,* 6, 63

PATTERSON, J.T. (1972). *J. appl. Bacteriol.,* 35, 569

PIERSON, M.D., COLLINS-THOMPSON, D.L. and ORDAL, Z.J.
(1970). *Fd Technol.,* 24, 1171

POFFE, R., BEVERS, J., DUMON, P. and VERACHTERT, H.
(1973). *$H_2O$,* 6, 497

REITH, A.F. (1926). *J. Bacteriol.,* 12, 367

REY, C.R., KRAFT, A.A. and RUST, R.E. (1971). *J. Fd Sci.,*
36, 955

RHODES, D.N., ROBERTS, T.A. and SHEPHERD, H.J. (1967).
  *J. Sci. Fd Agric.*, 18, 576
ROTH, L.A. and CLARK, D.S. (1972). *Can. J. Microbiol.*,
  18, 1761
RUITENBERG, E.J., KAMPELMACHER, E.H. and BERKVENS, J.
  (1967). *Fleischwirtschaft.*, 47, 1217
SATTERLEE, L.D., WILHELM, M.S. and BARNHART, H.M. (1971).
  *J. Fd Sci.*, 36, 549
SCHMIDT, C.F., LECHOWICH, R.V. and FOLINAZZO, J.F. (1961).
  *J. Fd Sci.*, 26, 626
SCHMIDT, C.F., LECHOWICH, R.V. and NANK, W.K. (1962).
  *J. Fd Sci.*, 27, 85
SCHOTHORST, M. VAN and KAMPELMACHER, E.H. (1967).
  *J. Hygiene*, 65, 321
SHANK, J.L. and LUNDQUIST, B.R. (1963). *Fd Technol.*, 17,
  1163
SHEWAN, J.M. (1971). *J. appl. Bacteriol.*, 34, 299
SNIJDERS, J.M.A. (1974). *Netherl. J. Vet. Sci.*, 99, 551
STEEL, K.J. (1961). *J. gen. Microbiol.*, 25, 297
TAYLOR, W.I. and SCHELHART, D. (1973). *Appl. Microbiol.*,
  25, 940
THOMAS, K.L. and JONES, A.M. (1971). *J. appl. Bacteriol.*,
  34, 717
TIWARI, N.P. and MAXCY, R.B. (1972). *J. Fd Sci.*, 37,
  901
VANDERZANT, C. and NICKELSON, R. (1969). *J. Milk Fd
  Technol.*, 32, 357
WESTERDIJK, J. (1949). *Antonie van Leeuwenhoek*, 15, 187
WODICKA, V.O. (1973). *Fd Technol.*, 27, No. 10, 52
WORK, K. (1971). *Acta Pathol. Microbiol. Scand.*, B,
  Suppl. no. 221

# III

## COMPOSITION OF MEAT

# 12

# MEAT COMPONENTS AND THEIR VARIABILITY

### R. A. LAWRIE

*Department of Applied Biochemistry and Nutrition,
University of Nottingham*

## INTRODUCTION

Meat signifies the edible flesh of those animals which
are acceptable for consumption by man. The terms
'edible' and 'acceptable', however, have different
interpretations according to the common practice and
aesthetic beliefs of the country, religion or group
concerned. Such a definition would include organ meat,
such as liver and tongue, but would exclude inedible
offal, such as hooves and horns.

According to the scientific view, however, meat is the
post mortem aspect of the three hundred or so anatomically distinct muscles of the body, together with the
connective tissue in which the muscle fibres are
deposited and such *inter*muscular fat as cannot be
trimmed off without breaking up the muscle as a whole.
*Intra*muscular fat is automatically included, being
physically inseparable. The material thus defined
represents by far the major edible portion of the carcass.
Its composition reflects - and can be understood with
reference to - the biological influences which affect
the nature of muscle.

The second definition will be used in the present
context; derivative aspects of composition (e.g.
differences between butcher's joints) will be excluded.

## TYPICAL COMPOSITION

The chemical composition of a typical mammalian muscle, after *rigor mortis* but before post-mortem degradative changes, is presented in *Table 12.1*.  It will be evident that three quarters of the muscle consists of water and that about one fifth is protein.  The remainder comprises lipid, carbohydrate, miscellaneous soluble non-protein substances (both organic and inorganic) and traces of vitamins.

In pre-rigor muscle most of the water is immobilised by the physical configuration of the proteins.  As glycogen is converted to lactic acid post mortem, however, the water-holding capacity of proteins falls and an increasing proportion of the water becomes free, tending to form visible exudation.  The water content of muscle thus tends to decrease on storage owing to exudation; less perceptibly, it decreases because of evaporation. About 5% of the total water remains directly bound to hydrophilic groups in proteins (Hamm, 1960), indeed the total water of muscle is probably present in a number of different physicochemical associations which can be separately assessed.

Proteins form by far the most important solid constituent.  Of the muscle proteins, about two thirds consist of the contractile elements of the myofibrils, the most important members being myosin and actin.  In muscle after *rigor mortis* they are combined as actomyosin; this complex persists on subsequent storage.  Myofibrillar proteins are insoluble in pure water but they will dissolve in a solution containing greater than 3.5% salt.

The second major group comprises the water-soluble proteins of the fluid sarcoplasm.  These include the muscle pigment myoglobin and the enzymes of the glycolytic pathway.  On storage after *rigor mortis* they are gradually proteolysed to peptides and amino acids.

The third group of muscle proteins comprises those of the connective and supporting tissue, and of insoluble (mainly mitochondrial) enzymes.  The major constituent, collagen, is insoluble in water and salt.  Although it undergoes subtle changes subsequent to *rigor mortis*, it is not normally broken down to its constituent amino acids.

The intramuscular lipids include triglycerides, phospho-lipids, fatty acids and other derivative fat-soluble substances, the total content being typically

2.5%. On storage, lipids tend to be hydrolysed and unsaturated fatty acids to be oxidised to more or less volatile products.

Before *rigor mortis* most of the 1% of carbohydrate is present as glycogen whereas after *rigor mortis* most is present as lactic acid. The soluble nitrogenous constituents tend to increase after *rigor mortis*. Thus hypoxanthine accumulates as the end product of purine nucleotide breakdown. The amount of free amino acids and of volatile amines is enhanced by concomitant proteolysis and there is a redistribution of ions between the aqueous phase and proteins.

Notwithstanding the normality of the constituents listed in *Table 12.1*, therefore, these may be expected to undergo some changes as the time after *rigor mortis* increases. It should be emphasised, however, that these occur in sterile muscle. Where there is microbial spoilage, various compositional changes may be anticipated which will be more or less marked, according to the organism concerned and such factors as the time and temperature of storage.

The most noteworthy deviations from the composition given in *Table 12.1* arise, however, from the influence of such factors as the species, breed, sex and age of the animal, its diet and the particular muscle being studied. Consideration will now be given to these effects.


# VARIABILITY


## 1. WATER

For a given muscle from mature animals there is little difference in water content between rabbit, sheep, pig, ox and whale.

In general, breed, sex and plane of nutrition affect the water content only in so far as the latter bears a reciprocal relation with the percentage of intramuscular fat. Thus, for example, whereas the water content of *l.dorsi* from a steer could be 74%, that from a comparable bull might be about 77%. Implantation of hexoestrol in steers raises the water content of the muscles as the percentage of fat is diminished (Lawrie, 1960). Again, whereas the moisture content of *l.dorsi* from pigs on a high plane of nutrition for 6 months was found to average 72%, the average was 74% from corresponding pigs on a low plane of nutrition (McMeekan, 1940). The water content

*Table 12.1* Chemical composition of typical adult mammalian muscle after *rigor mortis* but before degradative changes post mortem (after Lawrie, 1974; Scopes, 1970; Maruyama and Ebashi, 1970)

| Components | % Wet weight | |
|---|---|---|
| 1. WATER | | 75.0 |
| 2. PROTEIN | | 19.0 |
| (a) myofibrillar | | 11.5 |
| myosin[1] (H and L meromyosins, and | | |
| proteins associated with them) | 6.5 | |
| actin[1] | 2.5 | |
| tropomyosin | 1.5 | |
| troponins A, B and T | 0.4 | |
| $\alpha$ and $\beta$ actinins | 0.4 | |
| M. protein etc. | 0.2 | |
| (b) Sarcoplasmic | | 5.5 |
| glyceraldehyde phosphate | | |
| dehydrogenase | 1.2 | |
| aldolase | 0.6 | |
| creatine kinase | 0.5 | |
| other glycolytic enzymes | 2.2 | |
| myoglobin | 0.2 | |
| haemoglobin and other extracellular | | |
| proteins | 0.4 | |
| other unspecific proteins | 0.4 | |
| (c) Connective tissue and organelle | | 2.0 |
| collagen | 1.0 | |
| elastin | 0.05 | |
| mitochondrial etc. | 0.95 | |
| 3. LIPID | | |
| neutral lipid, phospholipids, | | |
| fatty acids | | |
| fat-soluble substances | | 2.5 |
| 4. CARBOHYDRATE | | |
| lactic acid | 0.90 | |
| glucose-6-phosphate | 0.15 | |
| glycogen | 0.10 | |
| glucose, traces of other | | |
| glycotic intermediates | 0.05 | |

*Table 12.1 (continued)*

| Components | | % Wet weight |
|---|---|---|
| 5. MISCELLANEOUS SOLUBLE NON-PROTEIN SUBSTANCES | | 2.3 |
| (a) Nitrogenous | | 1.65 |
| creatine | 0.55 | |
| inosine monophosphate | 0.30 | |
| di- and tri-phosphopyridine nucleotides | 0.10 | |
| amino acids | 0.35 | |
| carnosine, anserine | 0.35 | |
| (b) Inorganic | | 0.65 |
| total soluble phosphorous | 0.20 | |
| potassium | 0.35 | |
| sodium | 0.05 | |
| magnesium | 0.02 | |
| calcium, zinc, trace metals | 0.03 | |
| 6. VITAMINS | | |
| Various fat- and water-soluble vitamins, quantitatively minute | | |

[1]Actin and myosin are combined as actomyosin in post-rigor muscle.

of muscles from pigs which had been severely under-
nourished for a year has been found to rise to 83%
(Dickerson and Widdowson, 1960). Under genetic
influences, however, the reciprocal relationship between
intramuscular fat and water may be altered. Thus the
water content of the muscles of doppelender cattle is
about 10% less than those of normal animals despite a
concomitantly lower content of intramuscular fat (Lawrie,
Pomeroy and Williams, 1964).

With increasing animal age the water content of muscles
decreases, both on a whole-tissue basis and on a fat free
basis, falling from about 78% in the *l.dorsi* of a 12 day-
old calf to 74% in that of a 3 year-old steer. After
maturity, however, water *on a fat free basis* changes
relatively little.

Characteristic variations in water content between the
same eight muscles from pigs and cattle are shown in
*Table 12.2*. It is clear however that the relationship
between the muscles, using this parameter, is not identical
in each species.

*Table 12.2* Characteristic water content of muscles from
pigs and cattle (after Lawrie *et al.*, 1963; 1964)

| Muscle | % *moisture* | |
|---|---|---|
| | *Pigs* | *Cattle* |
| *l.dorsi* (lumbar) | 76.3 | 76.5 |
| *l.dorsi* (thoracic) | 76.9 | 77.1 |
| *psoas major* | 78.0 | 77.3 |
| *rectus femoris* | 78.5 | 78.1 |
| *triceps* (lateral head) | 78.7 | 77.2 |
| *superficial digital flexor* | 78.9 | 78.7 |
| *extensor carpi radialis* | 79.0 | 74.8 |

## 2. PROTEIN

Although species has some effect in determining the
total nitrogen content of the musculature (characteristic
values for pork and beef being 3.45% and 3.55%
respectively, on a fat free basis; Anon, 1961, 1963) this
parameter includes non-protein constituents. More striking
differences are shown by the individual proteins. Thus,
the relative proportions of the sarcoplasmic proteins
differ in corresponding muscles of the domestic species,

as shown by electrophoresis (Giles, 1962). These
differences reflect not only differing concentrations of
the glycolytic enzymes responsible for converting
glycogen to lactic acid, but also different contents of
the muscle pigment, myoglobin. Typical contents of
myoglobin in *psoas* muscles of rabbit, sheep, pig, ox and
horse are 0.02%, 0.35%, 0.45%, 0.60% and 0.70% (Lawrie,
1953). In sperm whale and seal myoglobin concentration
may be as high as 8% on a wet weight basis (Sharp and
Marsh, 1953). Excepting the marine mammals, these
differences in myoglobin concentration are reflected by
corresponding differences in both concentration and
activity of the insoluble mitochondrial enzymes effecting
respiration. Thus the relative activity of cytochrome
oxidase in the *psoas* muscle is four times as much in pig
as in rabbit and half as much again in horse as in pig
(Lawrie, 1953).

In respect of the water-insoluble proteins of the
myofibrils there are species differences in the
electrophoretic patterns of those proteins overall and
between their components, such as the tropomyosins and the
low MW constituents associated with the myosin molecules
(Parsons *et al.*, 1969; Champion, Parsons and Lawrie,
1970). The relative concentrations of $N^\epsilon$-methyl-lysines
in the proteins concerned may be partly responsible for
interspecies differences in mobility.

Species also affects the nature of the connective tissue
in muscle. Corresponding muscles of pigs, for example,
have more collagen than those of cattle (Lawrie, Pomeroy
and Cuthbertson, 1963; Lawrie *et al.*, 1964), and
considering the relative tenderness of pork, clearly its
nature must also be different between the two species.

Differences in protein due to breed are much less clearly
established. Among cattle, Brahmin-type animals are said
to have proportionately more connective tissue protein
than those of European breeds and Aberdeen Angus less than
larger breeds such as Friesian, but this factor does not
appear to have been studied systematically. Among pigs
myoglobin concentration is higher in the *l.dorsi* muscles
of Large White than Landrace animals (Lawrie and Gatherum,
1962), whereas myofibrillar protein is lower in Large
White than in Landrace or Welsh pigs (Lawrie and Gatherum,
1964). The concentration of certain enzymes of the
glycolytic pathway are higher in the muscles of those
porcine breeds which show a marked tendency to produce
'pale, soft exudative' flesh than in those yielding
normal flesh (Briskey, 1964). Again, the concentration
of salt-soluble and heat-labile collagen is greater in

muscles from the former breeds (McClain *et al.*, 1969).

To date no consistent effect of sex on either the quantity or quality of muscle proteins appears to have been
established. On the other hand, age is associated with
distinct compositional differences. The concentrations
of both myofibrillar and sarcoplasmic proteins (and of
total nitrogen) increase with age. Thus, in *l.dorsi*
muscles of cattle, adult values for these parameters are
attained at about 10 months of age. Connective tissue
protein, however, although representing 32% of the total
nitrogen at birth, constitutes only 14% of the total
nitrogen in the adult animal (Lawrie, 1961). Its nature
changes, however, the collagen becoming more insoluble with
age (Gross, 1958) there being a greater degree of cross
bonding between the polypeptide chains in older collagen.

In cattle, myoglobin swiftly increases up to 24 months
of age (0 to *c.* 0.5%; thereafter the concentration rises
by only *c.* 0.4% over the subsequent 8 years (Lawrie,
1961). A similar two-phase increment in myoglobin with
age is found in other species such as pig and horse. The
activity and concentration of the enzymes concerned with
respiration increases concomitantly with the myoglobin
(Lawrie, 1953). Typical values are shown for the *psoas*
muscles of the horse in *Table 12.3*.

The differentiation of the musculature into 300 or so
anatomically distinct units signifies more or less subtle
differences in their function and mode of action.
Muscles can be conveniently classified as 'red' and
'white' (Needham, 1926). The former tend to develop
power slowly but continuously by a highly developed
aerobic metabolism: the latter develop power swiftly
but intermittently by anaerobic mechanisms. But the
distinction is by no means clear and there is a continuous
spectrum of properties between muscles. This is
reflected in their protein components.

The overt 'redness' of a muscle is largely determined
by its content of myoglobin, but the relative proportions
of other proteins is also determined by the predominant
energy-yielding mechanism of the various muscles. Thus
the enzymes of the glycolytic pathway are found at much
higher concentration among the sarcoplasmic proteins of
white than of red muscles, whereas those of the oxidative
pathway are present in much higher concentration in red
than in white muscles (Lawrie, 1953). Again, quite apart
from their higher relative concentration in red than in
white muscle, the myofibrillar proteins have different
enzymic activity in the two types of muscle. Thus the
adenosine-triphosphatase activity of myosin is

*Table 12.3* Comparative rates of increase of myoglobin and cytochrome oxidase in *psoas* muscles of horses in relation to age.  Rates of increase expressed as percentages per year of average adult value for these factors (after Lawrie, 1953)

| Age (years) | Myoglobin concentration | Cytochrome oxidase activity |
|---|---|---|
| 0–2 | 42.8 | 37.9 |
| 2–12 | 2.1 | 0.6 |

intrinsically greater in white muscle (Seidel *et al.*, 1964).  Moreover the sarcoplasmic reticulum, by which much of the myofibrillar activity is regulated, is also much more highly developed in white muscle (Sreter, 1964). Again there are differences in the low molecular weight components of myosin between red and white muscles (Parsons *et al.*, 1969; Perrie and Perry, 1970; Lowey and Risby, 1971).  Training and exercise tend to change the metabolism pattern from a white to a red type.

Differences between muscles which are even more subtle than those signified by their relative redness, are also reflected by their complement of enzyme proteins (*Table 12.4*), in the overall pattern of myofibrillar proteins (Champion, Parsons and Lawrie, 1970), and in the constitution of specific myofibrillar components such as myosin (Parsons *et al.*, 1969).  Further aspects of subtle differentiation are exemplified in *Table 12.5* by the variability between certain porcine muscles in total nitrogen (roughly proportionate to total protein) and in hydroxyproline (roughly equivalent to connective tissue protein).  Moreover the proportion of the total connective tissue due to elastin and collagen varies. Bendall (1967) found that whereas elastin represented 5% of the total in bovine *l.dorsi,* it constituted 40% in *semitendinosus.*

A particularly striking indication of how much variability in protein concentration can be found out, with the influences so far considered, recently arose in analysing the *psoas* muscles of closely inbred cattle. The hydroxyproline content was 450 µg/g (i.e. normal) in one animal of the group and 2850 µg/g  in another.

*Table 12.4* Activity of certain enzymes in various beef muscles (after Berman, 1961)

| Muscle | Lactic[1] dehydrogenase | Glutamic[2] dehydrogenase |
|--------|-------------------------|---------------------------|
| l.dorsi | 3.5 | 660 |
| Semimembranosus | 3.7 | 730 |
| Serratus ventralis | 1.1 | 960 |
| Rectus abdominis | 2.6 | 675 |
| Semitendinosus | 3.7 | 810 |
| Trapezius | 2.1 | 835 |

[1]Moles lactate oxidised per kg wet weight per 3 min.

[2]Micromoles glutamate oxidised per kg wet weight per h.

*Table 12.5* Characteristic values for total nitrogen and hydroxyproline in muscles from pigs (after Lawrie, et al., 1963; 1964)

| Muscle | Total nitrogen (% fat free) | Hydroxyproline (µg/g wet wt.) |
|--------|------------------------------|-------------------------------|
| l.dorsi (lumbar) | 3.77 | 670 |
| l.dorsi (thoracic) | 3.69 | 527 |
| psoas major | 3.58 | 426 |
| rectus femoris | 3.41 | 795 |
| triceps (lateral head) | 3.46 | 1680 |
| superficial digital flexor | 3.35 | 1890 |
| sartorius | 3.41 | 850 |
| extensor carpi radialis | 3.36 | 2470 |

## 3.  LIPIDS

Within the three common domestic species, the percentages
of intramuscular total lipid vary widely as a result of
different feeding regimes so that it is difficult to find
a valid basis for comparison.  Nevertheless, cattle tend
to have a higher percentage of intramuscular lipid than
sheep and the latter tend to have more than pigs (Callow,
1948).  The percentage of intramuscular lipid in non-
domesticated species, such as the whale and various
antelope, tends to be relatively low (Sharp and Marsh,
1953; Wismer-Pedersen, 1969).  A more characteristic
effect of species on intramuscular lipid is in respect
of its iodine number.  Whereas that of *l.dorsi* in the
blue whale is 120, typical values for this muscle in pigs,
cattle and sheep are 60, 57 and 54, respectively (Callow,
1958; Lawrie, 1961; Lawrie and Gatherum, 1964).  The
generally lower values in ruminants reflect hydrogenation
of ingested fat by rumen micro-organisms.  The fat of the
horse, which deposits unsaturated fatty acids from herbage
largely unchanged, has an iodine number of *c.* 75-80 (Dahl,
1958).  The influence of the proportion of unsaturated
fatty acids in determining differences in iodine number
can be seen from *Table 12.6*; the relatively high
concentration of linoleic acid in porcine lipids is
evident.

*Table 12.6*  Typical fatty-acid composition of fats from
cattle, sheep and pigs (after Dugan, 1957)

| Fatty acid | % fatty acid in fat | | |
|---|---|---|---|
|  | Cattle | Sheep | Pigs |
| SATURATED |  |  |  |
| Palmitic | 29 | 25 | 28 |
| Stearic | 20 | 25 | 13 |
| UNSATURATED |  |  |  |
| Oleic | 42 | 39 | 46 |
| Linoleic | 2 | 4 | 10 |
| Arachidonic | trace | 1.5 | 2 |

Breed also influences the percentage, and iodine number, of intramuscular lipid. Thus, in steers of a beef breed (Hereford), a typical value for the percentage of intramuscular lipid in *l.dorsi* (lumbar) would be *c.* 7.1% and that in a dairy breed (Friesian) *c.* 6.4%, on similar planes of nutrition (Callow, 1962). The differential in favour of beef-type animals increases with age (Lawrie, 1961). In West Highland cattle intramuscular lipid may reach 17% of the wet weight and the iodine number may be as low as 45. Among pigs, the neutral lipid of *l.dorsi* muscles in Pietrain has a significantly higher proportion of linoleic acid than that of *l.dorsi* in the Large White breed (Wood and Lister, 1973). Again, when pigs of various breeds are crossed with Hampshires there is a marked tendency for highly unsaturated lipids to be deposited (Lea, Swoboda and Gatherum, 1970).

In respect of sex, males generally have less intramuscular lipid than females, the castrated members of each sex having more than the corresponding entire animals (Hammond, 1932). Thus, whereas the percentage of intramuscular lipid in *l.dorsi* from a year old steer was found to be 3%, it was only 1% in the corresponding muscle of a bull of the same age and breed (Lawrie, 1961). Implantation of hexoestrol in steers is also associated with a lowering of the total intramuscular lipid of the same order. Again, in pigs, the mean content of intramuscular lipid in *l.dorsi* muscles of a group of hogs at 100 kg liveweight was found to be 3.3%; that in corresponding boars of the same group averaged 2.5% (Lawrie, Pomeroy and Gatherum, 1964).

Increasing animal age is associated with increasing content of intramuscular lipid and diminishing iodine number (*Table 12.7*); and it should be pointed out that levels markedly greater than those represented are attained in older animals, according to breed and plane of nutrition.

As with moisture and protein, intramuscular lipid varies characteristically between muscles. Comparative values are given in *Table 12.8,* for corresponding muscles of pigs and cattle. The marked differences in iodine number between intramuscular lipids again reflect differences in their component fatty acids. Thus, whereas in porcine *l.dorsi* (lumbar) linoleic acid represents 11% of the total fatty acids, it represents 16% in porcine *rectus femoris* (Catchpole, Horsfield and Lawrie, 1970).

The effect of plane of nutrition on the lipid content of muscle has already been alluded to incidentally. It

Table 12.7  Effect of animal age on intramuscular lipid

|  | 12 day old calf | 3 year old steer | 5 month old pig | 6 month old pig | 7 month old pig |
|---|---|---|---|---|---|
| Intramuscular lipid (% wet weight) | 0.6 | 3.7 | 2.9 | 3.3 | 4.0 |
| Iodine no. | 82.4 | 56.5 | 57.4 | 55.8 | 55.5 |

Table 12.8  Characteristic intramuscular lipid content of corresponding muscles from pigs and cattle (after Lawrie, et al., 1963; 1964)

| Muscle | Pigs | | Cattle | |
|---|---|---|---|---|
|  | Lipid (%) | Iodine no. | Lipid (%) | Iodine no. |
| l.dorsi (lumbar) | 3.4 | 65.3 | 0.6 | 54.2 |
| l.dorsi (thoracic) | 3.3 | 55.5 | 0.9 | 56.6 |
| psoas major | 1.7 | 62.8 | 1.5 | 52.9 |
| rectus femoris | 1.0 | 71.5 | 1.5 | 67.8 |
| triceps (lateral head) | 1.8 | 67.0 | 0.7 | 62.2 |
| superficial digital flexor | 1.9 | 65.3 | 0.4 | 81.5 |
| extensor carpi radialis | 1.4 | 69.7 | 0.6 | 68.2 |

Table 12.9  Effect of plane of nutrition on intramuscular lipid content of l.dorsi muscles (after McMeekan, 1940; Palsson and Verges, 1952)

| Plane of nutrition | Ewes (41 weeks) | | Pigs (26 weeks) | |
|---|---|---|---|---|
|  | Lipid (%) | Iodine no. | Lipid (%) | Iodine no. |
| High | 5.0 | 51.3 | 4.5 | 59.2 |
| Low | 3.2 | 55.9 | 2.0 | 66.8 |

is a major determinant and, for a given muscle in animals
on diets of similar nature, the higher the plane of
nutrition the greater the percentage of intramuscular
lipid and the lower its iodine number (Callow and Searle,
1956). This relation is to be expected in that the higher
the plane of nutrition, the greater the proportion of
intramuscular lipid likely to be synthesised from
carbohydrate and, therefore, the lower its content of
unsaturated fatty acids. Typical data are given in
*Table 12.9.*

Apart from the *quantity* of nutrients, their *quality*
naturally affects the composition of intramuscular lipid.
Ingestion of unsaturated fatty acids will tend to lead
to the deposition of unsaturated intramuscular lipid,
especially in pigs wherein there is no ruminal
hydrogenation, although such factors as breed and the
individual animal modify the effect. If sufficient whale
oil is fed to pigs their fat may become so unsaturated as
to become rancid *in vivo*.

Apart from the influence of the known factors considered
above, the percentage of intramuscular lipid is subject to
substantial variability of which the nature is not yet
clearly defined. For example, even between litter mates
of the same sex, considerable differences in the percentage
of intramuscular lipid in porcine *l.dorsi* are found.
Indeed differences between the pigs within a given litter
may be greater than those for inter-litter means (Lawrie
and Gatherum, 1964).


4.  CARBOHYDRATE

By far the most prevalent carbohydrate in living muscle
is glycogen. Post-mortem glycolysis normally converts
most of it to lactic acid, whereby the pH falls to
about 5.4-5.5 when the enzymes involved are inactivated.
If reserves of glycogen at death are less than 0.8% the
ultimate pH will be above 5.5. This occurs more
frequently in porcine muscle than in those from sheep or
cattle. In the latter (and in horses) there may be
occasionally 2-2.5% of glycogen at death and 1.2% of
residual glycogen may remain at pH 5.5 (*Table 12.10*).
In certain muscles of the horse, however, substantial
amounts of glycogen may remain unconverted to lactic acid
when ultimate pH values are above 6. This may signify
qualitative differences in glycogen between the species
and indeed there is evidence for differences in the
average number of glucose units forming the chains in

*Table 12.10* Characteristic values for initial and
residual glycogen and ultimate pH in various muscles
(after Lawrie, Manners and Wright, 1959)

| Muscle | Initial glycogen (%) | Residual glycogen (%) | Ultimate pH |
|---|---|---|---|
| Horse heart | 0.7 | 0.1 | 5.86 |
| Horse *psoas* | 1.2 | 0.6 | 5.85 |
| Horse diaphragm | 1.9 | 1.2 | 5.86 |
| Horse *l.dorsi* | 2.2 | 1.2 | 5.45 |
| Ox *psoas* | 0.8 | 0.2 | 5.53 |
| Ox *sterno-cephalicus* | 1.3 | 0.5 | 6.01 |

rabbit, horse, pig and cattle. (Kjølberg, Manners and
Lawrie, 1963).

Occasionally the muscles of pigs which develop pale,
soft, exudative character post mortem form atypically
large quantities of lactic acid, whereby the pH may fall
to 4.7 or below.

The species also differ in the extent to which their
muscles convert glycogen, by a subsidiary pathway, to
glucose. Owing to differing contents of α-amylase the
rates of glucose accumulation post mortem average about
0.06, 0.50 and 0.90 mg/h/g in ox and horse, rabbit and
pig, respectively (Sharp, 1958).

Neither breed nor sex appear to affect glycogen levels
appreciably. The effect of animal age is also ill-defined,
although very high levels (about 7%) have been reported
in the muscles of newly born pigs (McCance and Widdowson,
1959) in which respiratory mechanisms are little
developed. As animals grow older the need for sustained
power increases the dependence on respiratory metabolism
in many muscles and their content of glycogen tends to
fall. Indeed the relative emphasis on respiratory or
glycolytic metabolism is a fairly consistent general
determinant of the glycogen concentrations characteristic
of various muscles. In 'red' muscles *in vivo* energy is
primarily obtained by oxidising glycogen (and fat) to
carbon dioxide and water; in 'white' muscles it is
primarily converted to lactic acid in anaerobic
operations, as already indicated. The latter tend to
have higher concentrations of glycogen and thus to attain
lower ultimate pH values (*Table 12.10*). The complement
of enzymes for metabolising carbohydrate differs
considerably between red and white muscles (Beatty and
Bocek, 1970).

Conditions induced in the animal by extraneous
circumstances - inanition, exhausting exercise, fear and
hypothermia - are liable to alter the glycogen reserves
of muscles more significantly than the above intrinsic
influences, although the latter interact with them. Thus,
while the glycogen reserves of pigs can be readily
depleted, to the point of raising the ultimate pH, through
fasting for only 24 h, by a short walk before slaughter,
or by struggling at the moment of stunning (Callow, 1936,
1938), it is difficult to deplete the glycogen reserves
of cattle by forced preslaughter exercise, even when
applied after 14 days fasting (Howard and Lawrie, 1956).
On the other hand, cattle of an excitable temperament
will have a low level of muscle glycogen even if well fed
and rested preslaughter. Fear and inherent excitability
induce continuous glycogen breakdown leading to low
equilibrium levels *in vivo* and to high ultimate pH values
(Howard and Lawrie, 1957). Exposure of pigs, sheep or
cattle to environments of high cooling capacity, by
establishing a need for enhanced heat production in the
animal, also cause muscle glycogen levels to be severely
depleted.

## 5. MISCELLANEOUS MINOR AND DERIVED COMPONENTS

A distinction can be made between the muscles of the
whale and those of the domestic meat animals in respect
of certain nitrogenous constituents. Thus, because the
whale subsists on krill, its muscles contain trimethylamin
oxide and its breakdown product, trimethylamine. Again,
whereas domestic species contain the dipeptide buffers,
β-alanylhistidine and β-alanyl-1-methylhistidine, whale
muscles appear to be unique in containing β-alanyl-3-
methylhistidine (Dennis and Larkin, 1965).

Regarding differences between the domestic meat species
themselves, there have been indications that the ratio of
$N^\epsilon$-methylysines to 3-methyl-histidine in the myofibrillar
proteins may be characteristic, thus permitting their
distinction even in hydrolysates. These methylated amino
acids were shown to form part of the polypeptide chain in
the major myofibrillar proteins by Perry and co-workers
(Johnson, Harris and Perry, 1967; Johnson and Perry, 1970;
Hardy *et al.*, 1970).

Again, characteristic concentrations of taurine in the
muscles of beef, pork and lamb are reported to be 9.1,
12.6 and 26.3 mg/100g respectively (Macy, Naumann and
Bailey, 1964). In terms of the total amino acid titre of

the muscle proteins generally, beef appears to have
somewhat higher contents of leucine, lysine and valine
than pork or lamb (Schweigert and Payne, 1956).

In respect of vitamins, pork differs from beef and lamb
in having about ten times as much thiamin and beef has
three times as much folic acid as pork and lamb (McCance
and Widdowson, 1960).

The differences in the fatty-acid composition of the
intramuscular lipids of the different species largely
determine the particular pattern of volatiles derived
from each on heating.  Thus octanal, undecanal,
hepta-2:4-dienal and nona-2:4-dienal arise on heating
pork fat but not that of beef and few dienals of any kind
are derived from heated lamb fat (Hornstein, Crowe and
Sulzbacher, 1960, 1963).

The breed and sex of meat animals have not so far been
distinguished on the basis of minor constituents of their
muscles.  Although there is evidence that 3-methylhistidine
is absent from foetal muscle (Johnson, Harris and Perry,
1967; Johnson and Perry, 1970) and that the contents of
methionine, isoleucine, valine and phenylalamine increase
relative to those of other amino acids in older animals
(Gruhn, 1965), age also has not been shown to exert any
major influence on components in this category.

Differentiation between muscles on the basis of their
hydroxyproline content has already been mentioned in
relation to their varying contents of connective tissue.
There may be, however, characteristic differences
between muscles in respect of their contents of free
amino acids.  Thus free tryptophan has been found to
range from 0.01 to 0.02 mg/g, and free lysine from 0.06
to 0.09 mg/g in various porcine muscles (Hibbert, 1973).
Again, the mineral content of different muscles varies.
The sodium content of porcine *l.dorsi* is 0.05% while
that of *extensor carpi radialis* is 0.08%.  In *rectus
femoris* the potassium content is 0.38%, whereas it is
0.29% in *extensor carpi radialis* (Lawrie and Pomeroy,
1963).

## CONCLUSION

As more sophisticated methods of analysis are developed,
it seems likely that the current parameters distinguishing
the muscles of meat animals on the basis of species,
breed, sex, age and anatomical location will be augmented
by others which will be both more subtle and more

unequivocally characteristic. Such differences are not
merely analytical curiosities, they largely determine
differences in the eating quality and nutritive value of
the fresh and preserved commodity.

## REFERENCES

ANON. (1961). *Analyst.*, 86, 557
ANON. (1963). *Analyst.*, 88, 422
BEATTY, C.H. and BOCEK, B.M. (1970). In *The Physiology
and Biochemistry of Muscle as a Food* Vol. II, p.115.
Ed. E.J. Briskey, R.G. Cassens and B.B. Marsh.
University of Wisconsin Press, Madison
BENDALL, J.R. (1967). *J. Sci. Fd Agric.*, 18, 553
BERMAN, M. (1961). *J. Fd. Sci.*, 26, 422
BRISKEY, E.J. (1964). *Adv. Fd Res.*, 13, 90
CALLOW, E.H. (1936). *Ann. Rept. Fd Invest. Bd., Lond.*,
pp.75, 81
CALLOW, E.H. (1938). *Ann. Rept. Fd Invest. Bd., Lond.*,
p.28
CALLOW, E.H. (1948). *J. agric. Sci.*, 38, 174
CALLOW, E.H. (1958). *J. agric. Sci.*, 51, 361
CALLOW, E.H. (1962). *J. agric. Sci.*, 58, 295
CALLOW, E.H. and SEARLE, R.L. (1956). *J. agric. Sci.*,
48, 61
CASSENS, R.G. and COOPER, C.C. (1971). *Adv. Fd Res.*,
19, 2
CATCHPOLE, C., HORSFIELD, C. and LAWRIE, R.A. (1970).
*Proc. 16th Meeting European Meat Res. Workers, Varna*,
p.214
CHAMPION, A., PARSONS, A.L. and LAWRIE, R.A. (1970).
*J. Sci. Fd Agric.*, 21, 7
DAHL, O. (1958). *Acta agric. Scand.*, Suppl. 3
DAVEY, C.L. (1960). *Arch. Biochem. Biophys.*, 89, 303
DENNIS, P.O. and LARKIN, P.A. (1965). *J. Chem. Soc.*,
914, 4968
DICKERSON, J.W.I. and WIDDOWSON, E.M. (1960). *Biochem.
J.*, 74, 747
DUGAN, L.R. (1957). *Amer. Meat Res. Inst. Fdn.*, Circ.
No.36
GILES, B.G. (1962). *J. Sci. Fd Agric.*, 13, 264
GROSS, J. (1958). *Exp. Med.*, 107, 265
GRUHN, H. (1965). *Sond. Nahrung*, 9, 325
HAMM, R. (1960). *Adv. Fd Res.*, 10, 356
HAMMOND, J. (1932). *Growth and Development of Mutton
Qualities in the Sheep*, Oliver and Boyd, London

HARDY, M.F., HARRIS, C.I., PERRY, S.V. and STONE, D.
  (1970). *Biochem. J.*, 120, 653
HIBBERT, I.V. (1973). Ph.D. Dissertation, University of
  Nottingham
HORNSTEIN, I., CROWE, P.E. and SULZBACHER, W.L. (1960).
  *J. Agric. Fd Chem.*, 8, 65
HORNSTEIN, I., CROWE, P.E. and SULZBACHER, W.L. (1963).
  *Nature, Lond.*, 199, 1252
HOWARD, A. and LAWRIE, R.A. (1956). *Spec. Rept. Fd
  Invest. Bd., Lond.*, No.63
HOWARD, A. and LAWRIE, R.A. (1957). *Spec. Rept. Fd
  Invest. Bd., Lond.*, No.65
JOHNSON, P., HARRIS, C.I. and PERRY, S.V. (1967).
  *Biochem. J.*, 105, 361
JOHNSON, P. and PERRY, S.V. (1970). *Biochem. J.*, 119,
  293
KJØLBERG,O., MANNERS, D.J. and LAWRIE, R.A. (1963).
  *Biochem. J.*, 87, 351
LAWRIE, R.A. (1953). *Biochem. J.*, 55, 298
LAWRIE, R.A. (1960). *Brit. J. Nutr.*, 14, 255
LAWRIE, R.A. (1961). *J. agric. Sci.*, 56, 249
LAWRIE, R.A. (1974). *Meat Science* 2nd Edn, p.71,
  Pergamon, Oxford
LAWRIE, R.A. and GATHERUM, D.P. (1962). *J. agric. Sci.*,
  58, 97
LAWRIE, R.A. and GATHERUM, D.P. (1964). *J. agric. Sci.*,
  62, 381
LAWRIE, R.A., MANNERS, D.J. and WRIGHT, A. (1959).
  *Biochem. J.*, 73, 485
LAWRIE, R.A. and POMEROY, R.W. (1963). *J. agric. Sci.*,
  61, 409
LAWRIE, R.A., POMEROY, R.W. and CUTHBERTSON, A. (1963).
  *J. agric. Sci.*, 60, 195
LAWRIE, R.A., POMEROY, R.W. and GATHERUM, D.P. (1964).
  *J. agric. Sci.*, 63, 385
LAWRIE, R.A., POMEROY, R.W. and WILLIAMS, D.R. (1964).
  *J. agric. Sci.*, 62, 89
LEA, C.H., SWOBODA, P.A.T. and GATHERUM, D.P. (1970).
  *J. agric. Sci.*, 74, 279
LOWEY, S. and RISBY, D. (1971). *Nature, Lond.*, 234,
  81
MCCANCE, R.A. and WIDDOWSON, E.M. (1959). *J. Physiol.*,
  147, 124
MCCANCE, R.A. and WIDDOWSON, E.M. (1960). *M.R.C. Spec.
  Rept. No.297*, HMSO, London
MCCLAIN,P.E., PEARSON, A.M., BRUNNER, J.R. and CREVASSE,
  G.A. (1969). *J. Fd Sci.*, 34, 115
MCMEEKAN, C.P. (1940). *J. agric. Sci.*, 30, 276, 287

MACY, R.L., NAUMANN, H.D. and BAILEY, M.E. (1964).
*J. Fd Sci.*, 29, 136

MARUYAMA, K. and EBASHI, S. (1970). In *The Physiology
and Biochemistry of Muscle as a Food* Vol. II, p.373,
Ed. E.J. Briskey, R.G. Cassens and B.B. Marsh.
University of Wisconsin Press, Madison

NEEDHAM, D.M. (1926). *Physiol. Rev.*, 6, 1

PALSSON, H. and VERGES, J.B. (1952). *J. agric. Sci.*, 62,
381

PARSONS, J.L., PARSONS, A.L., BLANSHARD, J.M.V. and
LAWRIE, R.A. (1969). *Biochem. J.*, 112, 673

PERRIE, W.T. and PERRY, S.V. (1970). *Biochem. J.*, 119,
31

ROBERTS, P.C.B. (1972). Ph.D. Dissertation, University
of Nottingham

SCHWEIGERT, B.S. and PAYNE, B.J. (1956). *Amer. Meat Inst.
Fdn. Bull.*, No.30

SCOPES, R.K. (1964). Ph.D. Dissertation, University of
Cambridge

SCOPES, R.K. (1968). *Biochem. J.*, 107, 139

SCOPES, R.K. (1970). In *The Physiology and Biochemistry
of Muscle as a Food* Vol. II, p.471. Eds. E.J. Briskey,
R.G. Cassens and B.B. Marsh. University of Wisconsin
Press, Madison

SEIDEL, J.C., SRETER, F.A., THOMPSON, M.M. and GERGELY,
J. (1964). *Biochem. biophys Res. Commun.*, 17, 662

SHARP, J.G. (1958). *Ann. Rept. Fd Invest. Bd., Lond.*,
p.7

SHARP, J.G. and MARSH, B.B. (1953). *Spec. Rept. Invest.
Bd., Lond.*, No.58

SRETER, F.A. (1964). *Fed. Proc.*, 23, 930

WIDDOWSON, E.M., DICKERSON, J.W.T. and MCCANCE, R.A.
(1960). *Brit. J. Nutr.*, 14, 457

WISMER-PEDERSEN, J. (1969). *Proc. 15th Meeting Europ.
Meat Res. Workers, Helsinki*, p.454

WOOD, J.D. and LISTER, D. (1973). *J. Sci. Fd Agric.*,
24, 1449

# 13

## THE COMPOSITION OF MEAT: PHYSICO-CHEMICAL PARAMETERS

### J. M. V. BLANSHARD

*Department of Applied Biochemistry and Nutrition, University of Nottingham*

### W. DERBYSHIRE

*Department of Physics, University of Nottingham*

### INTRODUCTION

Recent studies on the physiological behaviour of muscle have substantially explained the processes of contraction and relaxation in terms of the four proteins F-actin, myosin, tropomyosin and troponin and the controlled presence of $Ca^{2+}$ (Huxley, 1972). The insight that has been gained thereby has proved of immense value to all those involved in the study of the biochemistry and biophysics of muscle.

Another viewpoint which has particular value for those interested in the consumer acceptability of meat is to consider muscle as a gel system. Such an approach is by no means new; the fact that ATP could convert a hydrated gel of actomyosin into a shrunken pellet of low water content - so-called superprecipitation - has been known for a long time. Indeed a number of workers considered that this was the fundamental process associated with the contraction of muscle. Although with the advance of knowledge this idea has been superceded, the basic concept has considerable merit, particularly in that it is susceptible to a physico-chemical treatment. Furthermore, over the past three decades there has been a substantial increase in our understanding of gel systems. The intention of this paper, therefore, is to examine meat as a gel system from the standpoint of physical chemistry.

According to the theory of Flory and Rehner (1943),
the swelling or syneresis behaviour of macromolecular gel
systems can be construed in thermodynamic terms.   The
free energy change, $\Delta F$, involved in the mixing of a pure
solvent with an initially pure, amorphous, unstrained,
polymeric network consists of two components:   $\Delta F_m$,
the free energy of mixing, and $\Delta F_{el}$, the elastic free
energy consequent to the expansion.   Under non-equilibrium
conditions:

$$\Delta F = \Delta F_{el} + \Delta F_m \tag{1}$$

while the sign of $\Delta F$ decides whether swelling or
syneresis occurs.   At equilibrium:

$$\Delta F_m = \Delta F_{el} \tag{2}$$

When subjected to a more rigorous analysis, and
assuming the validity of the classical theory of rubber
elasticity, Flory and Rehner showed that equation 2 could
be expressed in the form:

$$[\ln(1 - v_2) + v_2 + \chi v_2^2] = 2V_1 \rho (v_2^{1/3} - v_2/2) \tag{3}$$

where $v_2$ is the ratio of the volumes of swollen and non-
swollen structures, $\chi$ the solvent-polymer interaction
coefficient, $V_1$  the molar volume of the solvent and $\rho$
the number of intrastructural cross-linkages per unit
volume.   The term on the left hand side of equation 3
is equivalent to the mixing free energy and contains the
solvent-polymer interaction coefficient, while the term
on the right hand side corresponds to the elastic free
energy and contains the parameter $\rho$ which represents the
number of cross-linkages in the system.   $\rho$ is also
important in that it is clearly inversely proportional
to $M$ (the average molecular weight of the changes
between cross-links), and by the classical theory of
rubber elasticity, $M$ may be related to the rigidity
modulus, $G$, by the following equation:

$$G = \frac{cRT}{M}$$

where $c$ = the concentration of the polymer in the
network.

An examination of these two equations reveals two
important points.   In a gel-type system, increasing the
number of cross-linkages will, on a purely thermodynamic
basis, favour syneresis, while, conversely, reducing the
number will favour swelling.   Secondly, the rigidity
modulus of a system will be proportional to the number
of cross-linkages.

If we consider muscle as a gel, certain similarities
are evident.  A typical gel is composed of a cross-
linked macromolecular matrix in an aqueous system in
which the formation of cross-linkages may be mediated
by inorganic ions.  Such a description could equally
refer to an alginate, pectin or carrageenan gel as well
as to muscle.  However, there are certain important and
significant differences:  muscle is an essentially
anisotropic structure constrained by connective tissue
elements and cell wall structures and of course, pre
mortem, is in a biochemically dynamic state.  Nevertheless the similarities are sufficient to encourage an
examination of muscle behaviour post mortem in this way.


## THE STATE OF WATER IN MUSCLE


This topic has been the subject of considerable
controversy and has been reviewed in Chapter 16.  Until
recently it was suggested that all the water in such
dilute gel systems as those containing only 1%
biopolymer was structured by the presence of the macromolecules.  Woessner, Snowden and Chiu (1970), however,
conclusively demonstrated that even in a 10% w/v agar
gel, in terms of the diffusion coefficient being
different from free water, only 1% of the total water
could be deemed to be structured.  In other words the
loss of water from a gel system on syneresis has very
little to do with a change in 'structuring' of the water
and therefore must involve some change in the network.
If we apply this type of argument to muscle and
initially consider that water is present in either the
bound or free state (i.e. as 'drip' water, either inside
or outside the muscle) then, in a sample undergoing
rigor in a sealed tube, there should be little or no
change in the relative proportions of free and bound
water.  Admittedly, at the end of the rigor process
some of the free water will physically be outside the
muscle sample, but in terms of molecular mobility this
may be no different from that inside the muscle.  At
first sight such a conclusion seems difficult to
reconcile with the widely acknowledged fact that the water-
holding capacity (WHC) of muscle changes on the development of rigor.  In practice, of course, muscle is not a
simple macromolecular system like an alginate gel but a
complex cellular material.

Although the state of water in muscle has been the subject of several investigations by NMR (e.g. Hazlewood, Nichols and Chamberlain (1969), Cooke and Wien (1971), Derbyshire and Parsons (1972), Belton *et al.* (1972)), relatively little attention has been devoted to changes induced during the onset of rigor mortis. However, Pearson, Derbyshire and Blanshard (1972) showed that for porcine muscle, above the freezing point, a non-exponential and rigor-dependent spin-spin relaxation was present. In more detailed studies Pearson *et al.* (1974) found that the spin-lattice relaxation time ($T_1$) remained single exponential throughout the rigor process while the spin-spin relaxation time ($T_2$), though initially single exponential, later displayed two components. It was suggested that the results could be interpreted in terms of a four component phenomenological model. They proposed that the water present could be assigned as existing in either of two environments denoted by subscripts *i* and *e* and characterised by relaxation times $T_b$ or $T_a$ indicative of their being bound or free. It was assumed that exchange between bound and free molecules was rapid within each region, but exchange between the *i* and *e* locations was slow. In the absence of exchange between two such regions, *i* and *e*, the resultant relaxation is the sum of two exponentials:

$$A(t) = (P_{ia} + P_{ib}) \exp \left(- \frac{t}{T_i}\right) + (P_{ea} + P_{eb}) \exp \left(- \frac{t}{T_e}\right)$$

where $P_{ia}$, $P_{ib}$, $P_{ea}$ and $P_{eb}$ represent the populations of the different regions and where the proton relaxation times of water molecules in the *i* and *e* regions may be expressed as:

$$\frac{P_{ia} + P_{ib}}{T_i} = \frac{P_{ia}}{T_a} + \frac{P_{ib}}{T_b} \quad \text{and} \quad \frac{P_{ea} + P_{eb}}{T_e} = \frac{P_{ea}}{T_a} + \frac{P_{eb}}{T_b}$$

It proved possible to calculate both the spin-spin relaxation times associated with these and also the population of each of the species and to monitor any changes during rigor. The changes in populations during the onset of rigor are illustrated in *Figure 13.1*.

Any process which might be expected to destroy the cellular structure as such, e.g. mastication or homogenisation, would naturally facilitate the exchange of water from the *i* to the *e* environments. This is evident in *Table 13.1*.

One important conclusion from this work, however, is that both throughout the rigor process and even during changes induced by homogenisation and mastication, there

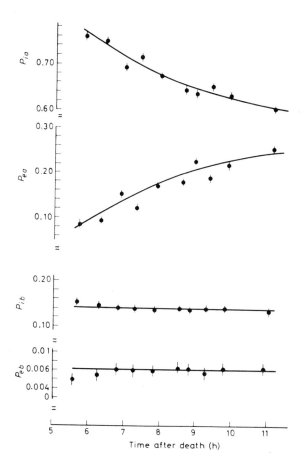

Figure 13.1  Changes in the populations of the four
fractions of water ($P_{ia}$, $P_{ea}$, $P_{ib}$, $P_{eb}$) in porcine
muscle held at room temperature as a function of time
after death (from Biochim. Biophys. Acta, <u>362</u> (1974),
courtesy of Elsevier Excerpta Medica North-Holland)

*Table 13.1* Changes in the observed spin-spin relaxation times $(T_2)$, and the populations of the various fractions of water in muscle, with the progress of rigor and as the result of mastication or homogenisa-tion (from *Biochim. Biophys. Acta*, 362 (1974), p.197, courtesy of Elsevier Excerpta Medica North-Holland)

| Condition | Spin-spin relaxation time (ms) | Populations | | | |
|---|---|---|---|---|---|
| | | $P_{ia}$ | $P_{ib}$ | $P_{ea}$ | $P_{eb}$ |
| 1.5 h after death | Single 55 | 0.88 | 0.12 | – | – |
| 24 h after death | 53±2 and 140±10 | 0.75 | 0.11 | 0.135 | 0.005 |
| Masticated | 41±2 and 76±4 | 0.38 | 0.08 | 0.50 | 0.04 |
| Homogenised | 28±5 and 59±2 | 0.12 | 0.04 | 0.75 | 0.09 |

is no significant overall change in the content of free water nor in the bound water, i.e.

$$P_{ia} + P_{ea} \simeq 87\%; \qquad P_{ib} + P_{eb} \simeq 13\%$$

On the other hand, there has been a significant movement of water from the *i* to the *e* environment effected by presumably progressive interdigitation and cross-linking of the actin and myosin during the onset of rigor. Conversely we may expect that any treatment which will reduce the degree of cross-linking ($\rho$) will tend to increase the WHC. *In vivo* this is accomplished through the involvement of ATP. A similar explanation may account for the beneficial effects of polyphosphates upon WHC when used in certain meat processing operations.

It is important to emphasise that at this stage no positive, morphological identification of the *i* and *e* environments has been attempted by the authors.

## THE ROLE OF $Ca^{2+}$ IONS

Although in gel systems the $Ca^{2+}$ is believed to act largely through the cross-linking of biopolymer chains, a more complex role is attributed to it in muscle. Considerable effort has been devoted to elucidating the mechanism of the contractile and relaxation process in muscle. In recent years particular interest has centred upon the roles of tropomyosin, troponin, actin, myosin, $Ca^{2+}$ and the sarcoplasmic reticulum. A crucial question that has been of major interest to muscle biochemists is the mode of action of $Ca^{2+}$. The majority view at present is that $Ca^{2+}$ is bound to troponin which then causes a conformational change in the tropomyosin resulting in a release of the suppression of the $Mg^{2+}$-activated ATPase of the actin-myosin. The released suppression permits the ATPase site on one of the myosin heads to be activated by $Mg^{2+}$ to hydrolyse ATP. As ATP is hydrolysed the actin-combining site on myosin and the myosin-combining site on actin interact and come closer to generate a movement. The cross-linkages generate a relative sliding movement of one set of filaments past the other. Shortening of muscle fibres is the result of repetitive cyclic changes at the sites of action between actin and myosin. It is presumed, but by no means certain, that the situation which obtains in rigor muscle in terms of cross-linking is the same as that which exists in the reversibly contractile system and Murray and Weber (1974) believe that both so-called rigor and activated complexes are involved in normal contraction. There is general agreement on the importance of the sarcoplasmic reticulum which cyclically effluxes $Ca^{2+}$ to, and then removes these ions from the neighbourhood of the myofilaments and the active sites involved in contraction.

The role of the sarcoplasmic reticulum post mortem is somewhat more obscure. Conversion of glycogen of the muscle to lactic acid through the glycolytic pathway continues in the absence of muscular movement to provide energy primarily for the sarcoplasmic reticular ATPase system of the $Ca^{2+}$ pump. The ultimate pH of the muscle is variable. Nauss and Davies (1966) have suggested that the sarcoplasmic reticulum gradually loses its ability to bind calcium post mortem and that the effect initiates glycolysis. Brostom, Hunkeler and Krebs (1971) further believe that $Ca^{2+}$ release could be a limiting factor in the initiation of glycolysis. The onset of *shortening* in rigor following the release of $Ca^{2+}$ into the sarcoplasm can be accounted for by the cyclic formation and breakage of cross-links between actin and myosin, accompanied by the enzymic hydrolysis of ATP by the actomyosin ATPase. In addition a further breakdown of ATP occurs owing to the action of the ATPase of the sarcoplasmic reticulum since leakage of $Ca^{2+}$ into the sarcoplasm will result in the continuous activation of the calcium pump during the development of rigor until the $ATP_2$ has disappeared.

If the loss of $Ca^{2+}$ accumulating ability of the sarcoreticular membranes causes the onset of *rigor mortis* and variably activates the glycolytic cycle and enhances the degree of interdigitation of the thin and thick filaments post mortem, it is important to know the response of the sarcoplasmic reticulum to events post mortem.  Various possible causes have been suggested:

1.  Post mortem pH decline.
2.  Uncoupling of the $Ca^{2+}$ pump by proteolysis.
3.  Temperature dependent changes in the permeability of the sarcoplasmic reticulum.

Greaser *et al.* (1969) have shown that pH values of below 6.0, a condition that occurs in most post mortem muscle, reduce the ability of sarcoplasmic reticulum membranes to sequester $Ca^{2+}$.  Implicit in the fall in pH and the onset of rigor is a decline in the ATP concentration, though *rigor mortis* frequently occurs at widely different ATP concentrations *in situ*.  Goll *et al.* (1971) and West *et al.* (1974) have demonstrated that the use of proteolytic enzymes either of an exogeneous, e.g. trypsin, or intra cellular nature, e.g. cathepsins, produce a post mortem loss of $Ca^{2+}$ accumulating ability in sarcoplasmic reticular membranes.  Such an explanation accommodates in a natural way the observations that *rigor mortis* can occur at pH values near 7.0 (alkaline rigor).

Temperature has also been suggested as a significant factor by Hay, Currie and Wolfe (1973) along with other parameters as an explanation for the rapid development of rigor in chicken breast muscle.  In this connection it is interesting that Inesi, Millman and Eletr (1973) have recently shown (*Figure 13.2*) that transitions occur in the temperature dependence of $Ca^{2+}$ accumulation and ATPase activity in the sarcoplasmic reticular membranes from rabbit muscle both between $37^{\circ}$ and $40^{\circ}$  and below 20°C.  In comparison to normal muscle temperatures, increasing the temperature to $40^{\circ}C$ results in uncoupling of the processes governing $Ca^{2+}$ accumulation and ATPase activity.  $Ca^{2+}$ accumulation is therefore inhibited while ATPase activity and passive $Ca^{2+}$ efflux proceed at rapid rates.  On the other hand at 20°C and below there is an abrupt change in the activation energy for the cation transport process of the associated enzyme activities.  The values for inactivation energies obtained from the slopes of the Arrhenius plots are

27 to 30 kcal/mol at 20°C and 15-18 kcal/mol above
that temperature. A more detailed thermodynamic analysis
of these results (*Table 13.2*) supported the conclusions
of ESR spin probe studies that below 20°C the reactivity
of activated molecules is limited by a highly ordered
environment. From a practical point of view it is
reasonable to conclude tentatively that above 37°C a
large net efflux of $Ca^{2+}$ will occur while as temperatures
are lowered below 20°C a progressively greater net
efflux of $Ca^{2+}$ will occur, particularly below 10°C.

At this point in time it is difficult to assess the
relative importance of each of the above factors. It
may very well be that the loss of $Ca^{2+}$ accumulating
ability is the result of the compounding of all three
factors, the importance of one or the other reflecting
the state of the animal and musculature at slaughter
and resulting in meat of different quality.


## TEXTURE


If we return to the classical theory of rubber
elasticity then we can define the rigidity modulus *G*
as follows:

$$G = \frac{cRT}{M}$$

where *M* is inversely proportional to the number of
crosslinks ρ. On this basis we should expect that an
increase in the number of crosslinks, a decreased
sarcomere length and increased interdigitation will
result in an increase in the rigidity modulus or
components of the complex Young's modulus (*Figure 13.3;*
Parker, 1972) and toughness. Such crosslink formation
and contraction are encouraged by a high $Ca^{2+}$ concentration.

Although the increase in free calcium ions in the
sarcoplasm undoubtedly causes the contraction during
development of *rigor mortis* (Schmidt, Cassens and
Briskey, 1970), the actual state of myofibrillar
contraction and hence toughness can be affected by a
variety of influences, e.g. the conditioning temperature,
the temperature at which the muscle goes into rigor
(which is also related to the rate of onset of rigor)
and by the degree of slack allowed the muscle by its
skeletal attachments (Bouton *et al.*, 1973). McCrae

Table 13.2 Calculated values for the enthalpy, free energy and entropy of activation, derived from the temperature dependence of the $Ca^{2+}$ pump and associated ATPase activities of rabbit muscle sarcoplasmic reticulum (from Inesi, Millman and Eletr (1973), courtesy of Academic Press Incorporated (London) Ltd.)

| Activity | Temperature range (°C) | $H_a$ (kcal/mol) | $F_a$ (kcal/mol) | $S_a$ (e.u.) |
|---|---|---|---|---|
| $Ca^{2+}$ uptake | 5–20 | 27–29 | 17–19 | +34 to +36 |
| | 20–35 | 16–18 | 15–17 | +3 to +4 |
| ATPase | 5–20 | 28–30 | 17–19 | +38 to +40 |
| | 20–35 | 15–17 | 14–17 | 0 to +3 |

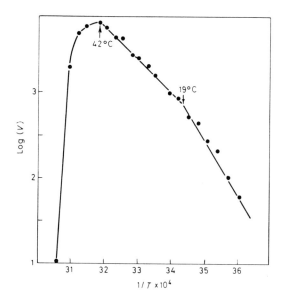

*Figure 13.2  Semi-logarithmic plot of initial rates of Ca²⁺ accumulation by the sarcoplasmic reticulum of rabbit muscle as a function of reciprocal temperature (from Inesi, Millman and Eletr (1973), courtesy of Academic Press Incorporated (London) Ltd.)*

*et al.* (1971) have stated that the entire toughness span can be explained in terms of a uniform attempt to shorten opposed by a far-from-uniform system of skeletal restraints among the muscles of the carcass.

It is possible to apply this type of consideration to certain phenomena in meat handling; the biochemical events surrounding these changes have recently been reviewed by Bendall (1973). We can exclude for the time being the contribution to toughness/tenderness of connective tissue, though the principles enunciated in this section equally apply to this component.

COLD SHORTENING

The excessive shortening that occurs in muscle during glycolysis at temperatures in the range of $0-10^{\circ}C$ is known as cold shortening. Cold shortening is most pronounced in unrestrained muscle (Marsh and Leet, 1966) and is accompanied by a contraction of up to 40% and a pronounced increase in toughness. Cassens and Newbold (1967) attribute this to a release of $Ca^{2+}$ by the sarcoplasmic reticulum and subsequent calcium activation of ATPase and this would certainly accord with the greater net efflux of $Ca^{2+}$ that takes place at this rather than normal, ambient temperatures as reported by Inesi, Millman and Eletr (1973).

THAW RIGOR

A not altogether different situation is that of thaw rigor. This phenomenon which has been known for a number of years occurs when a muscle is frozen prerigor

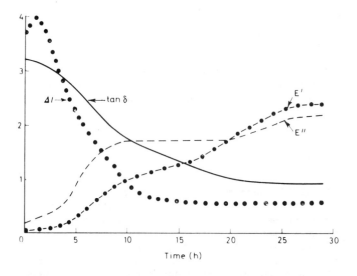

*Figure 13.3 Changes in the components of the complex Young's modulus, the loss tangent and the length of the* sternomandibularis *muscle of ox as it enters rigor. Scales for ordinates are:* $E'$: $10^6$ $N.m^{-2}$ *per unit;* $E''$: $10^5$ $N.m^{-2}$ *per unit;* $\tan\delta$: 0.1 *per unit; and* $\Delta l$: 3% *of maximum observed length per unit*

and subsequently rapidly thawed and is accompanied by a massive contraction leading to a substantial loss of WHC and the development of resultant toughness.  The generally accepted view is that intracellular ice-crystal formation dislocates the sarcoplasmic reticulum resulting in an abundant release of $Ca^{2+}$ on thawing.

## RESOLUTION OF RIGOR

It is also a matter of some interest that the resolution of rigor by conditioning has been observed to be accompanied by a recovery of $Ca^{2+}$ accumulating ability of the sarcoplasmic reticulum (Szabolcs, Kover and Kovacs, 1967) and an increased availability of non-bound $Ca^{2+}$ (Webb *et al.*, 1967).

## PSE (PALE-SOFT-EXUDATIVE) MUSCLE

The formation of PSE muscle is characterised by a fast pH fall within ½-3 h to a final pH of from 5.0 to 5.6 and in general is favoured by a high ambient temperature. In fact it is this condition according to Inesi, Millman and Eletr (1973)  which favours a very rapid $Ca^{2+}$ efflux from the sarcoplasmic reticulum which could well lead to the rapid glycolysis and pH fall.

So far we have merely considered the elastic behaviour in muscle in terms of the number of cross-linkages and completely neglected the nature of the chains between the crosslinks.  Bailey, Mitchell and Blanshard (to be published) have recently shown by computer studies in gel systems that by considering different macromolecules it is possible to compute the relative rigidities of molecules and thereby gain a much more rational under-standing of the interplay of syneresis and elasticity present in such materials (Mitchell and Blanshard, to be published).  There is little doubt too that in a phenomenon such as PSE muscle as the myofibrillar proteins approach a pH at which they are denatured and under temperature conditions too which favour this behaviour, the myosin will change from a relatively rigid to a more flaccid molecule.  Such systems tend to favour syneresis more than those with relatively rigid structures.

## ACKNOWLEDGEMENTS

The authors would like to express their warm thanks for the collaboration of Dr. R.T. Pearson in the development of part of this paper and to many helpful discussions with Professor R.A. Lawrie and Dr. J.R. Mitchell.

REFERENCES

BAILEY, E., MITCHELL, J.R. and BLANSHARD, J.M.V.  To be published

BELTON, P.S., JACKSON, P.R. and PACKER, K.J. (1972). *Biochim, biophys. Acta,* 286, 16

BENDALL, J.R. (1973). *The Structure and Function of Muscle,* 2nd edn. Vol. 2.  p.244, Academic Press, New York and London

BOUTON, P.E., HARRIS, P.V., SHORTHOSE, W.R. and BAXTER, R.I. (1973).  *J. Fd Sci.,* 38, 932

BROSTOM, C.O., HUNKELER, F.L. and KREBS, E.G. (1971). *J. biol. Chem.,* 246, 1961

CASSENS, R.G. and NEWBOLD, R.P. (1967). *J. Fd Sci.,* 32, 269

COOKE, R. and WIEN, R. (1971). *Biophys. J.,* 11, 1002

DERBYSHIRE, W. and PARSONS, J.L. (1972). *J. Magn. Resonance,* 6, 344

FLORY, P.J. and REHNER, J. (1943). *J. Chem. Phys.,* 11, 521

GOLL, D.E., STROMER, M.H., ROBSON, R.M., TEMPLE, J., EASON, B.A. and BUSCH, W.A. (1971). *J. Anim. Sci.,* 33, 963

GREASER, M.L., CASSENS, R.G., HOEKSTRA, W.G. and BRISKEY, E.J. (1969), *J. Fd Sci.,* 34, 1969

HAY, J.D., CURRIE, R.W. and WOLFE, F.H. (1973). *J. Fd Sci.,* 38, 700

HAZLEWOOD, C.F., NICHOLS, B.L. and CHAMBERLAIN, N.F. (1969). *Nature,* 222, 747

HUXLEY, H.E. (1972).  *The Structure and Function of Muscle,* 2nd edn.  Vol. 1, p.301, Academic Press, New York and London

INESI, G., MILLMAN, M. and ELETR, S. (1973).  *J. Mol. Biol.,* 81, 483

MARSH, B.B. and LEET, N.G. (1966).  *J. Fd Sci.,* 31, 450

MCCRAE, S.E., SECCOMBE, G.G., MARSH, B.B. and CARSE, W.A. (1971). *J. Fd Sci.,* 36, 566

MITCHELL, J.R. and BLANSHARD, J.M.V.  To be published

MURRAY, J.M. and WEBER, A. (1974). *Scientific American,* 59

NAUSS, K.M. and DAVIES, R.E. (1966). *J. biol. Chem.*, <u>241</u>, 2918

PARKER, N.S. (1972). *Rheol. Acta*, <u>11</u>, 56

PEARSON, R.T., DERBYSHIRE, W. and BLANSHARD, J.M.V. (1972). *Biochem. biophys. Res. Commun.*, <u>48</u>, 873

PEARSON, R.T., DUFF, I.D., DERBYSHIRE, W. and BLANSHARD, J.M.V. (1974). *Biochim. biophys. Acta.*, <u>362</u>

SCHMIDT, G.R., CASSENS, R.G. and BRISKEY, E.J. (1970). *J. Fd Sci.*, <u>35</u>, 574

SZABOLCS, M., KOVER, A. and KOVACS, L. (1967). *Acta Biochim. and Biophys. Acad. Sci. Hung.*, <u>2</u>, 409. *See:* (1968) *Fleischwirtschaft.*, <u>48</u>, 647

WEBB, N.B., KAHLENBERG, O.J., NAUMANN, H.D. and HEDRICK, H.B. (1967). *J. Fd Sci.*, <u>32</u>, 1

WEST, R.L., MOELLER, P.W., LINK, B.A. and LANDMANN, W.A. (1974). *J. Fd Sci.*, <u>39</u>, 29

WOESSNER, D.E., SNOWDEN, B.S. and CHIU, Y.C. (1970). *J. Colloid and Interface Sci.*, <u>34</u>, 283

# 14

# THE COMPOSITION OF MEAT: ANALYTICAL ASPECTS

R. SAWYER

*Laboratory of the Government Chemist, London*

## INTRODUCTION

The range of analytical techniques employed in the assessment of composition and quality factors of meats has expanded rapidly in the past decade or so. Choice of method or methods used will depend on the purpose for which an analysis is to be carried out. There are distinct contrasts between the traditional methods for assessment of proximate composition and gross nutritive value, the variety used for tissue identification and the sophisticated methods required for the detection of flavours, taints and possible contaminants. Methods will vary according to the nature of the material being assessed, i.e. fresh meat as recognisable pieces or as a processed product. The need for changes in approach and in methodology has stemmed from advances in the technology of producing meat for consumption, typical examples are the use of phosphates, surveyed by Iles (1973), and in the technology of alternative protein materials used as process adjuncts or as substitutes for meat. Thus, the analyst requires to know the question posed and the point of view of the questioner before appropriate methods can be chosen or developed. It is noticeable that in the changing environment the methods developed for histological and biochemical researches are now being exploited by the analyst employed in the control laboratory.

## COMPOSITION

The basic composition of meat varies considerably between different types and cuts. Meat is structurally muscle fibre held by connective tissue through which blood vessels, nerves and fat cells are more or less abundantly distributed. The fibre sheath also encloses soluble protein and other nitrogenous compounds together with mineral salts.

Proteins are second only to water as the most abundant substances in animal tissues and are, without doubt, the most important constituents of the edible portions of meat animals. The edible complex usually consists of the proteins actin and myosin, together with smaller quantities of collagen, reticulin and elastin. There are also quantities of respiratory pigments - myoglobin, nucleoproteins, enzymes and other miscellaneous components The inedible portion of the animal contains a considerable amount of protein such as collagens (skin, tendon, connective tissues); the keratins (hair, horn, hoof); elastin (ligaments); and the blood proteins. Lushbough and Schweigert (1960) give as an overall summary of average values for proximate composition and energy value of the edible portion of fresh meat cuts, the following: protein ($N \times 6.25$), 17%; fat, 20%; moisture, 62%; ash 1%; and calorific value, 250 cal/100 g. Specific values for individual cuts and species of meat vary widely, but the above figures represent 'average' lean meat cuts with a surrounding layer of fat approximately ½" thick.

Individual values are stated to depend on the species, the degree of fattening of the animal prior to slaughter, the degree of cutting and trimming and, for cooked or processed meats, the nature of cooking, curing and other processing methods.

The figures of primary interest for analytical control are the 'lean meat content' and 'meat equivalent' of a product. Thus, for the purpose of comparison the following composition may be quoted from the above source as an average for lean muscle meat: protein ($N \times 6.25$), 20%; fat, 9%; moisture, 70%; ash, 1%; calorific value, 160 cal/100 g. Thus, in the absence of separable fat the gross energy value is reduced but the net protein content and moisture content are increased. Fresh muscle meat contains essentially no carbohydrate (i.e. less than 1%) and no fibre. Organ meats, however, contain glycogen and glucose up to approximately 4%. *Table 14.1* shows proximate compositional data for a number

*Table 14.1* Proximate composition of fresh meats

|      | Cut of meat | Protein (%) | Moisture (%) | Fat (%) | Ash (%) | Cal/ 100 g |
|------|-------------|-------------|--------------|---------|---------|------------|
| BEEF | Chuck       | 18.6        | 65           | 16      | 0.9     | 220        |
|      | Flank       | 19.9        | 61           | 18      | 0.9     | 250        |
|      | Loin        | 16.7        | 57           | 25      | 0.8     | 290        |
|      | Rib         | 17.4        | 59           | 23      | 0.8     | 280        |
|      | Topside     | 19.5        | 69           | 11      | 1.0     | 180        |
|      | Rump        | 16.2        | 55           | 28      | 0.8     | 320        |
| PORK | Ham         | 15.2        | 53           | 31      | 0.8     | 340        |
|      | Loin        | 16.4        | 58           | 25      | 0.9     | 300        |
|      | Shoulder    | 13.5        | 49           | 37      | 0.7     | 390        |
|      | Spare rib   | 14.6        | 53           | 32      | 0.8     | 350        |
| LAMB | Breast      | 12.8        | 48           | 37      | 0.7     | 380        |
|      | Leg         | 18.0        | 64           | 18      | 0.9     | 240        |
|      | Loin        | 18.6        | 65           | 16      | 0.7     | 220        |
|      | Rib         | 14.9        | 52           | 32      | 0.8     | 360        |
|      | Shoulder    | 15.6        | 58           | 25      | 0.8     | 300        |

of cuts and species of animals. Energy values are
calculated on the basis of 4 cal/g for protein and
carbohydrate and 9 cal/g for fat. The effect of various
degrees of finish (fattening) of the animal species are
illustrated in the values shown in *Table 14.2*. As
indicated earlier, the various edible offals differ
somewhat in composition from lean meat and some typical
values are shown in *Table 14.3*.

## PROCESSED MEATS

The cured and processed meats vary widely in their
content of the major nutritive components and this
depends on the specific nature of the meat and processing
practice employed. Thus in products such as sausage,
luncheon meat, frankfurters etc., the fat levels may be
somewhat higher than for fresh muscle and organs, the
level of protein will be thus lowered. This results
from the use of meat trimmings and other ingredients
higher in fat than most muscle cuts. Owing to the
addition of salts - for curing and water retention
purposes - the ash levels are also correspondingly higher
than in fresh meat. The moisture content of products is

*Table 14.2* Proximate composition of fresh meat in relation degree of fatness of the animal

|  | Cut | Protein (%) | Moisture (%) | Fat (%) | Ash (%) | Cal/ 100 g |
|---|---|---|---|---|---|---|
| BEEF | Chuck | | | | | |
| | Thin | 19.2 | 71 | 9 | 0.9 | 160 |
| | Medium | 18.6 | 65 | 16 | 0.9 | 220 |
| | Fat | 17.6 | 60 | 22 | 0.8 | 270 |
| | Loin | | | | | |
| | Thin | 18.6 | 64 | 16 | 1.0 | 220 |
| | Medium | 16.9 | 57 | 25 | 0.8 | 290 |
| | Fat | 15.6 | 53 | 31 | 0.8 | 340 |
| | Topside | | | | | |
| | Thin | 19.7 | 71 | 8 | 1.0 | 150 |
| | Medium | 19.3 | 67 | 13 | 1.0 | 190 |
| | Fat | 18.7 | 63 | 17 | 0.9 | 230 |
| PORK | Ham | | | | | |
| | Thin | 17.2 | 60 | 22 | 0.9 | 270 |
| | Medium | 15.2 | 53 | 31 | 0.8 | 340 |
| | Fat | 13.2 | 46 | 40 | 0.7 | 410 |
| | Loin | | | | | |
| | Thin | 17.9 | 63 | 18 | 1.0 | 230 |
| | Medium | 16.4 | 58 | 25 | 0.9 | 290 |
| | Fat | 14.8 | 52 | 32 | 0.8 | 350 |
| | Hand | | | | | |
| | Thin | 24.4 | 52 | 33 | 0.8 | 360 |
| | Medium | 12.5 | 45 | 42 | 0.7 | 430 |
| | Fat | 10.3 | 37 | 52 | 0.6 | 512 |
| LAMB | Leg | | | | | |
| | Thin | 18.4 | 71 | 9 | 1.0 | 160 |
| | Medium | 18.0 | 64 | 18 | 0.9 | 230 |
| | Fat | 16.7 | 60 | 22 | 0.8 | 270 |
| | Rib | | | | | |
| | Thin | 17.7 | 65 | 16 | 0.9 | 210 |
| | Medium | 14.9 | 52 | 32 | 0.8 | 350 |
| | Fat | 11.2 | 39 | 49 | 0.6 | 490 |
| | Shoulder | | | | | |
| | Thin | 16.7 | 67 | 15 | 0.9 | 200 |
| | Medium | 15.6 | 58 | 25 | 0.8 | 290 |
| | Fat | 13.6 | 51 | 34 | 0.7 | 360 |

*Table 14.3*  Proximate composition of offals

| | | Protein (%) | Moisture (%) | Fat (%) | Ash (%) | Cal/ 100 g |
|---|---|---|---|---|---|---|
| BEEF | Brain | 10.5 | 78 | 9 | 1.4 | 130 |
| | Heart | 16.9 | 78 | 4 | 1.1 | 110 |
| | Kidney | 15.0 | 75 | 8 | 1.1 | 140 |
| | Liver | 19.7 | 70 | 3 | 1.4 | 140 |
| | Lung | 18.3 | 79 | 2 | 1.0 | 90 |
| | Pancreas | 13.5 | 60 | 25 | 1.2 | 280 |
| | Spleen | 18.1 | 77 | 3 | 1.4 | 100 |
| | Thymus | 11.8 | 54 | 33 | 1.1 | 340 |
| | Tongue | 16.4 | 68 | 15 | 0.9 | 210 |
| PORK | Brain | 10.6 | 78 | 9 | 0.7 | 130 |
| | Heart | 16.9 | 77 | 5 | 0.4 | 120 |
| | Kidney | 16.3 | 77 | 5 | 0.8 | 110 |
| | Liver | 19.7 | 72 | 5 | 1.7 | 130 |
| | Lung | 12.9 | 84 | 2 | – | 70 |
| | Pancreas | 14.5 | 60 | 24 | – | 270 |
| | Spleen | 17.1 | 77 | 4 | – | 100 |
| | Tongue | 16.8 | 66 | 16 | 0.5 | 210 |
| LAMB | Brain | 11.8 | 79 | 8 | 1.4 | 120 |
| | Heart | 16.8 | 72 | 10 | 1.0 | 160 |
| | Kidney | 16.6 | 78 | 3 | 1.3 | 110 |
| | Liver | 21.0 | 71 | 4 | 1.4 | 140 |
| | Lung | 17.9 | 79 | 2 | 1.1 | 80 |
| | Spleen | 18.8 | 74 | 4 | 1.6 | 110 |
| | Thymus | 14.1 | 80 | 4 | 1.3 | 90 |
| | Tongue | 13.9 | 70 | 15 | 0.8 | 190 |

highly variable and is dependent on the nature of the product; examples are shown in *Table 14.4*.  It is in this field of processed products that the meat technologist is able to display his skills, additives which find frequent use include simple monosaccharides and disaccharides, honey, liquid glucose, milk solids, cereals, curing salts, emulsifiers and flavouring agents.

*Table 14.4*  Chemical composition of meat products

|  | Protein (%) | Moisture (%) | Fat (%) | Ash (%) |
|---|---|---|---|---|
| Bacon | 10.4 | 25 | 61 | 2.1 |
| (range of values) | 15.3 | 55 | 28 | 3.4 |
| Ham (cured) | 27.6 | 64 | 5 | 3.8 |
| (range of values) | 21.6 | 52 | 23 | 3.4 |
| Corned beef (Argentine) | 23.8 | 54 | 14 | 2.6 |
| (range of values) | 30.7 | 57 | 9 | 2.7 |
| Luncheon meat | 11.4 | 52 | 29 | 2.1 |
| (range of values) | 15.2 | 56 | 23 | – |
| Salt pork | 3.9 | 8 | 85 | 3.5 |
| Frankfurter | 15.2 | 64 | 14 | 3.1 |
| Liver sausage | 16.7 | 59 | 11 | 2.2 |
| Salami | 23.9 | 31 | 37 | 7.0 |
| Beef sausage | 7.5 | 41 | 33 | 2.2 |
| (range of values) | 12.7 | 53 | 17 | 2.9 |
| Pork sausage | 8.6 | 33 | 37 | 1.1 |
| (range of values) | 13.8 | 54 | 23 | 3.2 |
| Roast beef (lean) | 29.6 | 54 | 15 | 1.0 |
| Roast pork (lean) | 28.5 | 57 | 13 | 1.9 |
| Roast lamb (lean) | 26.6 | 57 | 16 | 1.1 |

## CANNED MEATS

When meat and meat products are heat processed and hermetically sealed in a container so as to destroy spoilage organisms and to denature enzyme systems, the changes in proximate composition are limited by ancillary processing stages. A typical example is in the production of corned beef in which a limited amount of fresh meat is added to the cured meat before processing takes place. The nature of such products as canned ham, chicken etc., will depend greatly on the proportions of fat and lean meat included and on the proportion of juices added or removed after cooking and before the closure of the pack. Some typical figures are shown in

*Table 14.4* which also includes analyses of cooked lean meat.

UK legislation incorporates standards for the meat content of the range of products mentioned above and since this subject is to be considered in another paper in this conference, no further detail is given here. However, the control analyst must face the task of assessing the compliance of a product with the require- ments of legislation. The quality control analyst in the producing establishment has the advantage of knowing the recipe for an individual product and if analysis is carried out he has knowledge of the quality of his ingredients. The regulatory analyst, however, has no such knowledge and he is faced with the task of performing the analysis and providing an interpretation of the facts as he may find them.


## ASSESSMENT OF LEAN MEAT CONTENT

The normal analytical method for estimation of the lean meat content of a meat product is based on a measurement of the total nitrogen content as determined by the Kjeldahl method. A correction is applied for the nitrogen content contributed by other ingredients such as vegetable and cereal fillers; calculation of meat content is then based on a 'lean meat factor' determined experimentally. The Meat Products Sub-Committee of the Society for Analytical Chemistry (1961, 1963, 1965) has published various factors recommended for use with different types of fresh food, typical values of the factor are listed in *Table 14.5.*

The lean meat equivalent is obtained by applying the formula $100 \times N/N_F$, where $N$ is the nitrogen content corrected for the contribution from other ingredients and $N_F$ is the factor appropriate to the meat concerned.

Early calculations of this type were applied to determination of meat content of sausage by Stubbs and More (1919); they applied a correction to the total nitrogen on the assumption that the nitrogen equivalent of the cereal content was equivalent to 2% of the total carbohydrate. Pearson (1968, 1969, 1970a) has pointed out that the lean meat content obtained by the Stubbs and More calculation is not necessarily synonymous with the actual lean meat content. He has derived equations of general application in which allowance is made for the intramuscular fat in the 'lean' tissue and for the

*Table 14.5*  Mean conversion factors for various foods recommended by The Society for Analytical Chemistry (1961, 1963, 1965)

| Meat | Recommended factor (fat free basis) |
| --- | --- |
| Beef | 3.55 |
| Veal | 3.35 |
| Pork | 3.45 |
| Tongue | 3.0 |
| Ox liver | 3.45 |
| Pig liver | 3.65 |
| Kidney | 2.7 |
| Blood | 3.2 |
| Chicken (breast) | 3.9 |
| Chicken (dark) | 3.6 |
| Chicken (whole) | 3.7 |
| Turkey (breast) | 3.9 |
| Turkey (dark) | 3.5 |
| Turkey (whole) | 3.65 |

nitrogen content of extramuscular fat.  Pearson has suggested that 'typical' lean contains 10% of 'invisible' fat and that extramuscular fat contains 90% fat.  The general equation derived by Pearson is as follows:

$$\text{Lean meat} = \frac{100 N_T - F_{EXT}\, N_F \left( \dfrac{100-1}{F_F} \right) - 2C}{N_F \left( 1 - \dfrac{F_L}{F_F} \right)}$$

Where $N_T$ = Total nitrogen %
  $F_{EXT}$ = Total fat extracted %
  $N_F$ = Nitrogen factor appropriate to the species
  $F_F$ = Fat in extramuscular fat %
  $F_L$ = Fat in lean meat %
  $C$ = Carbohydrate content %

Allowing for the average proportions quoted above, the equation reduces to:

$$\text{Lean meat} = \frac{112.5\, N_T - A \times F_{EXT} - 2.25\, C}{N_F}$$

Where for pork $A = 0.4132$ and $N_F = 3.45$
  and for beef $A = 0.4437$ and $N_F = 3.55$.

An alternative approach (Pearson, 1970b) is that
favoured on the Continent; the Feder number is calculated
as follows:

% organic not fat = 100 − (% fat + % ash + % moisture)

The Feder Number = $\dfrac{\text{% water}}{\text{% organic not fat}}$

The organic not fat is almost identical with the crude
protein content ($N \times 6.25$). For genuine meat products
the Feder number should not exceed 4. Kroll and Meester
(1962) prefer to use the ratio % water/% protein. It
will be apparent that the above system of calculation is
subject to error owing to the presence of non-meat
protein other than that derived from cereal and also as
a result of the natural variation in the conversion
factor. For example, the Society for Analytical Chemistry
(SAC) committee found a range of 2.96 to 4.53 for beef
(average 3.55), and a range of 2.8 to 4.2 for pork
(average 3.45). The data found for pork suggested that:

$$N_F = 0.0021\ W + 2.97$$

where $W$ = liveweight of pig carcass in lb. Coomaraswamy
(1972) has further examined the problem of the
determination of meat content of comminuted meat products
which contain excessive amounts of skin, rind and
connective tissues. The definition of lean meat above
excludes visible fat but makes no mention of visible
rind or visible connective tissue, hence by implication
lean meat could be held to include the proportions of
skin and connective tissue normally associated with that
meat since these are specifically mentioned in the
definition of meat. Thus by definition additional rind
and connective tissue which may be used in sausage and
similar products as part of the filler should not count
towards the meat content. Coomaraswamy makes the point
that the Stubbs and More method allows for additional
fat to be used in the manufacture of sausage products
since total meat content = lean meat + fat (as determined).
However, he makes the following points in justification
of the practice:

(a) Fat confers some organoleptic quality.
(b) The use of fat is a long standing practice.
(c) The consumer is aware of the practice.
(d) Fat has nutritive value.

   (e)   The amount of fat added is limited by a
          requirement for lean meat content.

He contends that none of the arguments applies to the use
of rind or connective tissue; the main contention is that
the nutritive value of connective tissue is less than
that of muscle protein and permission to use excessive
amounts of these tissues would result in a lowering of
the nutritive value of the foodstuff.  The basis for
assessment of connective tissue is the collagen and the
fact that collagen contains a significant proportion of
hydroxyproline (13-14.5%); typical analyses for muscle
protein are shown in *Table 14.6*.  From the data it can
be seen that hydroxyproline may be used as an index of
the collagen content of the meat product in question.
In his paper, Coomaraswamy presents an analytical method
for hydroxyproline content and a modification of the
Stubbs and More method for calculation of total meat
content.  Preliminary experimental data in support of the
method are given and calculations are performed on the
basis of 10, 15 and 20% connective tissue in lean meat.
The author proposes the method for regulatory purposes
and concludes that there is a need for more definitive
data to establish a reasonable value for connective
tissue in lean meat.


## QUALITY ASSESSMENT


The assessment of quality of meat and meat products may
be subdivided into two broad headings, the first relates
to the quality of the meat itself in terms of consumer
appeal, whilst the second relates primarily to the nature
of the material which is presented as 'meat' in a
processed product.  Much of the evaluation of quality in
fresh meats is based on subjective tests and in this case
it is obvious that much reliance is placed on the
experience of specialist meat inspectors.  The main
emphasis on evaluation of fresh meat quality is placed
upon relating factors such as weight, finish, quality and
age of the live animal and the relative quantities of
external fat, marbling, muscle distribution, appearance
and colour of the carcass to tenderness, flavour and aroma
of the cooked product.  As a first order assessment of
beef for example, the purchaser has become accustomed to
associate dark colour with toughness and green, brown or
grey tints with poor or 'off' flavour.  Visual inspection

*Table 14.6* Average amino-acid content ot
pork and lamb (values expressed as % of cr

| Amino acid | Beef (%) | Pork (%) | |
|---|---|---|---|
| Arginine | 6.6 | 6.4 | |
| Histidine | 2.9 | 3.2 | |
| Isoleucine | 5.1 | 4.9 | |
| Leucine | 8.4 | 7.5 | 7.4 |
| Lysine | 8.4 | 7.8 | 7.6 |
| Methionine | 2.3 | 2.5 | 2.3 |
| Phenylalanine | 4.0 | 4.1 | 3.9 |
| Threonine | 4.0 | 5.1 | 4.9 |
| Tryptophan | 1.1 | 1.4 | 1.3 |
| Valine | 5.7 | 5.0 | 5.0 |
| Alanine | 6.4 | 6.3 | 6.3 |
| Aspartic acid | 8.8 | 8.9 | 8.5 |
| Cystine | 1.4 | 1.3 | 1.3 |
| Glutamic acid | 14.4 | 14.5 | 14.4 |
| Glycine | 7.1 | 6.1 | 6.7 |
| Proline | 5.4 | 4.6 | 4.8 |
| Serine | 3.8 | 4.0 | 3.9 |
| Tyrosine | 3.2 | 3.0 | 3.2 |
| Hydroxyproline | 0.1 | 0.1 | 0.1 |

will also help in the assessment of the proportions of
edible to inedible material. Meanwhile the assessment
of bacterial quality, freedom from disease and parasites
together with proper handling are the qualities for which
the consumer looks elsewhere for support. The subjective
assessments may be assisted by the use of such techniques
as measurement of pH, light reflectance, conductivity,
juiciness by (expression of liquor under pressure) and
tenderness by tenderometer testing. Reviews of the use
of the latter two techniques have been published by
Tannor, Clark and Hankins (1943) and Schultz (1957). If
nutritional value is assessed in terms of protein quality,
vitamin and mineral content, there may be little to choose
between the flesh of prime steak and the so-called
'stewing steak'. As indicated earlier, there are
differences in the amino acid composition of fresh and
other tissue.

In the case of processed meats the problem is of a
different nature, by processing the manufacturer is able
to blend and adjust the product by incorporation of fat,
less attractive tissues, flavour adjuncts, mineral salts

and other protein material; there is also the possibility of mixing in meat from other species of animal. Thus it may be seen that the analyst is faced with a range of problems which require the use of a variety of techniques and analytical approaches.

Microscopy provides a means of detecting the kind of tissues or adulterants which may be present in many foods and which may not be detected by chemical means. As a supporting technique the use of the microscope is an often neglected analytical tool. Identification of skeletal and non-skeletal tissues by microscopy presupposes an adequate knowledge of histology and perhaps this is an excuse for its apparent neglect by analytical chemists. Biological methods suffer by virtue of the effects of comminution and heat treatment during processing. However, in uncooked ground meat products a semiquantitative evaluation of the amount of non-skeletal muscle tissue present may be obtained in certain cases. Use of these methods has been discussed for sausage products by Holey, Roberts and Jones (1964). Useful results may be obtained by combination of microscope techniques with phase contrast and polarising illumination together with enzyme digestion processes, Birkner and Auerbach (1960) describe useful combined digestion and staining methods. This approach has been described as a means of identifying the presence of textured vegetable protein in uncooked comminuted meats (Linke, 1969). The method was based on the fact that many commercial vegetable proteins contain reasonable levels of carbohydrate; in some cases 30% may be present, Smith and Wolfe (1961) and Wood (1971). The presence of the carbohydrate fraction may be demonstrated by the McManus (1946) modification of the periodic acid-Schiff method; Linke (1969) modified the McManus procedure by using chromic acid instead of periodic acid. Jewel (1973) has reported a drawback to the use of the Linke method and reported that iodine staining allows for a differentiation between rusk filler and textured vegetable protein in luncheon meat formulations. Coomaraswamy and Flint (1973) also describe differential staining techniques.

Serological techniques for species identification based on the principles developed by Uhlenhuth (1901) were described for detection of raw horseflesh in comminuted meats in the classic work on Food Inspection and Analysis by Leach and Winton (1920). The method has been applied by Castledine and Davies (1968) for a range of fresh meats in raw sausages; variations for identification of

vegetable proteins and protein concentrates have been
reported by Munsey (1947), Hale (1945) and Glynn (1939).
A precipitin test for detection of egg albumin in
hamburger was reported by Victor *et al.* (1973). Further
modifications using antisera produced from the heat
stable fraction of soya-bean globulin have been described
by Catsimpoolas and Meyer (1968) and by Kagenkolb and
Hingerle (1967). The method is applicable to bean meals
and protein concentrates but the texturised protein shows
no reaction - presumably this effect is due to further
denaturing of the protein during processing. A significant
advantage is that the method is applicable to mildly
processed products since the active globulin may be
extracted at 80°C in physiological saline at pH 7.5.
A further advantage is that under these conditions most
meat proteins are coagulated and so in the case of non-
heat-treated products significant inactivation of the
meat proteins is also effected. Dagenkolb and Hingerle
(1967) claimed to be able to detect 0.2% of foreign
protein in a meat product.

Combinations of immunological techniques and
electrophoresis have been reported by Kamm (1970) and
Gunther (1969); Peter (1970) combined immunology with
diffusion whilst de Hoog, van den Reek and Brouwer (1970)
used a density gradient combination for examination of
denatured soya.

Electrophoresis techniques were applied to water-soluble
proteins of fresh fish for the purpose of species
identification by Connell and co-workers at Torry Research
Station; application of the method has been extended,
modified and adapted by numerous workers, e.g. Mackie
(1969). Protein identification following various degrees
of denaturation was found to be possible by the use of
progressive solubilisation of the heat-treated protein
by use of concentrated solutions of urea. Initially the
electrophoretic methods employed were the classical
moving boundary methods, but the introduction of gel
techniques in the last decade has reduced it to such a
level that it may be carried out in a moderately equipped
laboratory. Electrophoresis of polypeptides in the
presence of sodium dodecyl sulphate was reported by
Shapiro, Vinuela and Maizel (1967); Scopes and Penny
(1971) used the method to determine sub-unit sizes of
muscle protein. Hofmann and Penny (1971) used the method
for detection of soya protein in meat heated to 100°C.
They found that the patterns of soya and meat proteins
were different, and that one soya band could be used to
give quantitative results for protein concentrate in a

meat/soya mixture. Mattey, Parsons and Lawrie (1970)
had reported a similar method using a laser densitometer
for quantitative detection of meat species heated to
120°C for 3 minutes. These workers used 10M urea for
solubilisation and later claimed that a 10% admixture of
soya protein with 90% meat could be detected even after
heating at 100°C for 1 h 20 min had taken place (Parsons
and Lawrie, 1972).

Various chemical procedures have been proposed for the
assessment of foreign protein admixtures in meats.
McVey and McMullin (1940) described methods for calcium
and lactose in sausages in order to assess the extent of
the use of milk powder. They make the point that 6% of
milk powder is equivalent to 10% lean meat if assessed
on the basis of nitrogen content. Kutscher, Nagel and
Pfaff (1962) and Thalacker (1963) described methods for
milk protein by determination of the phosphoprotein
fraction. Zwetkova (1968) used a TLC method for
caseinates in sausage products.

A number of parameters have been proposed for the
assessment of soya protein: Illings and Whittle (1944),
sucrose, salt-free ash and manganese; Hendry (1939),
Lythgol, Ferguson and Racicot (1941), Fredholm (1967),
insoluble non-fermentable sugars; Kent-Jones and Amos
(1957), lipid phosphorus; Hayward (1939), nitrogen free
extract; and Kerr (1936), alcoholic potash insolubles.
Recent work by Formo, Honold and MacLean (1973) has
suggested that the combination of magnesium, manganese
and fibre content may be used as an index of soya-meal
content in soya-meat blends. On the basis of their
preliminary work they conclude that magnesium provides
the most precise index.

It is of interest to note that Leach and Winton (1920)
describe the use of glycogen as an index of horseflesh
content of sausage products, this in combination with
the precipitin test and the characterisation of the fat
by iodine value and butyro-refractometer reading was the
method of choice for many years. Dalley (1950) reported
favourably on the use of tests on the fat for such species
identification. Further work on the characterisation of
species by fatty-acid analysis using gas liquid
chromatography (GLC) has been reported. Cook and
Sturgeon (1966) have shown that horse, pork and beef fat
can be identified by the GLC pattern of unsaponifiable
fractions. Castledine and Davies (1968) employed trans
esterification of the fat in methanol to prepare methyl
esters of the fatty acids. Horse fat is identified by
pronounced linoleate and linolenate content, whilst pork

fat may be identified by the characteristic levels of linoleic and arachidonic acids when compared with beef and lamb fat. However, it has been demonstrated that the feeding regime may significantly affect the fatty-acid composition of a particular species (Hubbard and Pocklington, 1968; Hubbard, Pocklington and Thomson, 1969).

Whilst by no means exhaustive, the above summary indicates the versatility required of the analytical chemist in the task of assessing meat for quality, quantity and type in meat and meat products.

REFERENCES

BIRKNER, M.L. and AUERBACH, E. (1960). In *The Science of Meat and Meat Products,* American Meat Institute Foundation, W.H. Freeman and Co., San Francisco and London, p.10

CASTLEDINE, S.A. and DAVIES, D.R.A. (1968). *J. Assoc. Public Analysts,* 6, 39

CATSIMPOOLAS, N. and MEYER, E.W., (1968). *Arch. Biochem Biophys.,* 125, 742

CONNELL, J.J. (1953). *Biochem. J.,* 54, 119, and 55, 378

COOK, M.R., and STURGEON, J.D. (1966). *J. Assoc. Offic. Analyt. Chem.,* 49, 877

COOMARASWAMY, M. (1972). *J. Assoc. Public Analysts,* 10, 33

COOMARASWAMY, M. and FLINT, F.O. (1973). *Analyst, Lond.,* 98, 542

DALLEY, R.A. (1950). *Analyst, Lond.,* 75, 336

DEGENKOLB, E. and HINGERLE, M. (1967). *Arch. Lebensmittelhyg.,* 18, 241

FORMO, M.W., HONOLD, G.R. and MACLEAN, D.B. (1973). Proc. 87th Annual Meeting AOAC

FREDHOLM, H. (1967). *Fd Technol.,* 21, 93

GLYNN, J.H. (1939). *Science,* 89, 444

GUNTHER, H. (1969). *Arch. Lebensmittelhygiene,* 20, 97 and 128

HALE, M.W. (1945). *Fd Res.,* 10, 60

HAYWARD, J.W. (1939). *J. Assoc. Offic. Agric. Chem.,* 22, 552

HENDRY, W.B. (1939). *Anal. Chem.,* 1, 611

HOFMANN, K. and PENNY, I.F. (1971). *Fleischwirtschaft.,* 51, 577

HOLEY, N.H., ROBERTS, H.P. and JONES, K.B. (1964). *Analyst, Lond.,* 89, 332

DE HOOG, P., VAN DEN REEK, S. and BROUWER, F. (1970).
  *Fleischwirtschaft.*, 50, 1663
HUBBARD, A.W. and POCKLINGTON, W.D. (1968). *J. Sci. Fd
  Agric.*, 19, 571
HUBBARD, A.W., POCKLINGTON, W.D. and THOMSON, J. (1969).
  *Chem. and Ind.*, 977
ILES, N.A. (1973). *Scientific and Technical Surveys*,
  Number 81, BFMIRA
ILLINGS,E.T. and WHITTLE, E.G. (1944). *Food*, 13, 32
JEWEL, G.G. (1973). Technical Circular No.528, BFMIRA
KAMM, L. (1970). *J. Assoc. Offic. Anal. Chem.*, 53, 1248
KENT-JONES, D.W. and AMOS, A.J. (1957). *Modern Cereal
  Chemistry*, 5th edn., Northern Publishing Co., Liverpool
KERR, R.H. (1936). *J. Assoc. Offic. Agric. Chem.*, 19, 409
KROLL and MEESTER, (1962). *Conserva*, 11, 163
KUTSCHER, W., NAGEL, W. and PFAFF, W. (1961).
  *Z. Lebensmitt Untersuch*, 115, 117. In (1962) *Analyt.
  Abstr.*, 9, 868
LEACH, A.E. and WINTON, A.L. (1920). *Food Inspection and
  Analysis*, John Wiley & Sons, N.Y.; Chapman & Hall, Londc
LINKE, H. (1969). *Fleischwirtschaft.*, 49, 469
LUSHBOUGH, C.H. and SCHWEIGERT, B.S. (1960). In *The
  Science of Meat and Meat Products*, American Meat
  Institute Foundation, W.H. Freeman and Company,
  San Francisco and London, p.185
LYTHGOL, H.D., FERGUSON, C.S. and RACICOT, P.A. (1941).
  *J. Assoc. Off. Agric. Chem.*, 24, 799
MACKIE, I.M. (1969). *J. Assoc. Public Analysts*, 7, 85
MATTEY, M., PARSONS, A.L. and LAWRIE, R.A. (1970).
  *J. Fd  Technol.*, 5, 41
MCMANUS, J.F.A. (1946). *Nature, Lond.*, 158, 202
MCVEY, W.C. and MCMULLIN, H.R. (1940). *J. Assoc. Off.
  Agric. Chem.*, 23, 811
Meat Products Sub-Committee, Society for Analytical
  Chemistry (1961). *Analyst, Lond.*, 86, 557
Meat Products Sub-Committee, Society for Analytical
  Chemistry (1963). *Analyst, Lond.*, 88, 422
Meat Products Sub-Committee, Society for Analytical
  Chemistry (1965). *Analyst, Lond.*, 90, 579
MUNSEY, V.E. (1947). *J. Assoc. Off. Agric. Chem.*, 30,
  187
PARSONS, A.L. and LAWRIE, R.A. (1972). *J. Fd Technol.*,
  7, 455
PEARSON, D. (1968). *J. Assoc. Public Analysts*, 6, 129
PEARSON, D. (1969). *Fd Manufacture*, 44, 42
PEARSON, D. (1970a). *J. Assoc. Public Analysts*, 8, 22
PEARSON, D. (1970b). *The Chemical Analysis of Foods*,
  p.384, J. and A. Churchill, London

PETER, M. (1970). *Arch. Lebensmittelhyg.*, 21, 220

SCHULTZ, H.W. (1957). National Livestock and Meat Board USA, Proceedings 10th Annual Reciprocal Meat Conference, p.17

SCOPES, R.K. and PENNY, I.F. (1971). *Biochem. Biophys. Acts.*, 236, 409

SHAPIRO, A.L., VINUELA, E. and MAIZEL, J.R. (1967). *Biochem. and Biophys. Research Comm.*, 28, 815

SMITH, A.K. and WOLFE, W.J. (1961). *Fd Technol., Champaign*, 15, 4

STUBBS, G. and MORE, A. (1919). *Analyst*, 44, 125

TANNOR, B., CLARK, N.G. and HANKINS, O.G. (1943). *J. Agr. Research*, 66, 403

THALACKER, R. (1963). *Deutsch. Lebensmitt Rdsch.*, 59, III

UHLENHUTH, P. (1901). *Deutsch. Med. Wochs.*, 27, 780

VICTOR, F.M., DUFFETT, N.F. and RIGHT, E. (1970). *Q. Bull. Assoc. Food and Drug Office*, 34, 241 John Wiley & Sons, N.Y.; Chapman & Hall, London

WOOD, J.C. (1971). *Fd Manufacture*, 46, 37

ZWETKOVA, (1968). *Fleischwirtschaft.*, 48, 1486

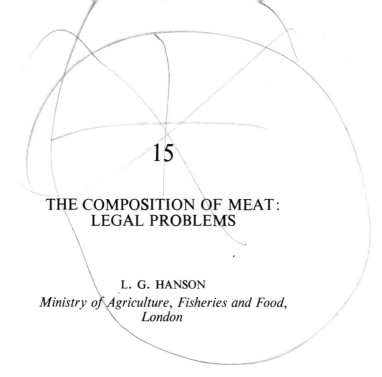

# 15

# THE COMPOSITION OF MEAT: LEGAL PROBLEMS

L. G. HANSON

*Ministry of Agriculture, Fisheries and Food, London*

## INTRODUCTION

This chapter deals with the food standards aspects of meat and meat products so meat hygiene, meat inspection and control of disease are not covered, nor are price and weight. Problems involved in devising, drafting and enforcing legislation on meat and meat products are described. The period 1956 to 1966 was one of analysis and discussion; 1967 saw the regulations on meat products and the period since has seen significant developments in home and international food standards, in consumer protection generally and in food technology which will undoubtedly influence any new or revised regulations.

## FOOD STANDARDS LEGISLATION

Food is an essential of life and purchases represent about a third of the weekly budget. Food not only sustains life but it can be hazardous; food poisoning can have immediate effects and there can be other longer term effects if precautions are not taken to prevent them. It is necessary therefore to protect the purchaser of food from being harmed or deceived. It is also in the interests of the honest trader if laws are made accordingly.

British food laws have been developed since 1860.  The basic aims are to require that food is:

1.  Not injurious to health.
2.  Not unfit.
3.  Not falsely described or described so as to be misleading.
4.  Of the nature, substance or quality demanded or of the prescribed composition.

Two things should be noted.  First, the law (now in the Food and Drugs Act 1955) lays an absolute duty on those engaged in the food trades so to conduct their affairs as to avoid the committing of an offence - it is not necessary for the prosecution to show a criminal intention and there are defences which enable the offence to be passed back to the real offender together with a system of warranties to safeguard traders.  Second, the law is a complex of general and particular rules; the particular rules (usually in regulations) may be horizontal - applying to all foods, or vertical - applying to a category.  In addition to the main body of food law in the Food and Drugs Act and the 40 or so regulations made under it, the Trade Descriptions Acts 1968 and 1972 provide rules about false trade descriptions and origin marking.

The period from 1960 onwards saw the development of modern food manufacture and distribution with more complex and sophisticated foods sold prepacked in supermarkets.  The purchaser is now faced with a difficult choice from a host of competing products with only the information on the label and what he can remember from advertisements to help him.  Food legislators have tried to keep pace by making detailed rules with the aim of keeping the right balance between trader and trader, and trader and purchaser.  Most of the important manufactured foods are subject to compositional regulations which prescribe a minimum composition and statutory names. Since the Labelling of Food Regulations 1970 came into force on 1 January 1973 the purchaser can see on all labels of prepacked food a specific description of the food; a declaration of ingredients including additives in descending order of proportion (except for a few foods); the name and address of someone responsible and basic information, for example about nutrition, if certain claims are made.  For most foods when sold loose a specific description is required together with a declaration of certain classes of additives, if present.

There is strict control over pictures and over the manner in which the information is given on labels and in advertisements.

In respect of one of the central features of the food laws for the past 100 years – that it is an offence to sell food not of the nature, substance or quality demanded – the position now is that the statutory information is designed to enable the purchaser to evaluate the nature, substance and quality and thus to make an informed purchase. Before most foods were prepacked, the purchaser was more likely to invoke the protection of the law by asking for or about the product. It was – and still is – an offence to say anything false or misleading about the product.

## THE UNDERLYING QUESTIONS

How far is it necessary or desirable for the legislator to intervene in the market place? Is it practicable? Is it economic? Can the law be made specific enough to help without restricting unduly the development of 'new and improved' products? Can the rules be expressed in legal form; can they be enforced? These are the underlying questions for food standards legislators.

## THE BRITISH SYSTEM

The British system for tackling these questions has been seen at its most active since 1960. It rests on the proposition that a committee of independent experts should examine the subject in depth with the help of evidence (oral and written) from all concerned and that its report should be published as a basis for Ministers to decide, in the light of comments from all concerned, what should be done. When their proposals for regulations are issued all may comment again before the law is finally made to come into force after a suitable period.

The Food Standards Committee (FSC) with Professor Ward as Chairman and three members with scientific and technical experience, three with experience of the food trades and three of general experience (two of whom are women) deals with matters of composition, labelling and advertising. The Food Additives and Contaminants Committee under Professor Weedon has appropriate scientific, technical and trade experts.

The reports of these two committees are recognised as
authoritative and they have been found invaluable not only
in the United Kingdom but also in many other countries and
in the Codex Alimentarius Commission.

The food trades and the enforcement authorities have
played a very active part in discussions in the Committees
and in Whitehall.  Until recently it was possible to
maintain that the consumer voice was too muted but now it
is directly expressed (or is it?) by consumer associations
The media and Parliament are now very sensitive and there
is wide interest in consumer protection.  This is all to
the good.  All views can now be made known and the balance
of the whole system is about right, resting as it does on
the sound and expert foundation of a published committee
report.

## FOOD STANDARDS IN MEAT AND MEAT PRODUCTS

"Of course everyone knows what meat is even if the
sausage is something of a mystery; so if there are any
'legal problems' they must relate to meat products rather
than meat".  Neither proposition is true; there are
problems in meat and in meat products and they arise
partly because not everyone knows what the food is and
partly because many think they know and are mistaken
consciously or unconsciously.

### MEAT:  APPEARANCE AND ADDITIVES

The appearance of a piece of meat is an important means
of assessing its worth.  During 1963 some dusting
powders were applied to raw meat to maintain its fresh
red colour.  Not only was the effect to deceive but it
could mask deterioration and also, in some consumers the
result was an unpleasant, tingling irritation in the face
and neck.  Sections 1 (injurious to health) and 2 (nature
substance or quality demanded) could have been invoked
but after a special review by the Food Standards Committe
the Meat (Treatment) Regulations 1964 were made to ban
the addition of these substances and the sale of meat so
treated.

The addition of a colour or preservative to raw
unprocessed meat has never been allowed except for a
concession in Scotland to allow sulphur dioxide on mince
in the days when refrigeration facilities in the home and
elsewhere were not what they are today.

More subtle problems have arisen from the use of
proteolytic enzymes and polyphosphates neither of which
are at present subject to detailed control.  The Food
Standards Committee noted (paragraph 104 Food Labelling)
that enzymes were being used to make meat and meat
products more tender.  Regulation 18 of the Labelling of
Food Regulations 1970 now requires the word 'tenderised'
to appear on the label or the ticket as part of the
statutory name and this must be done whether the enzyme
treatment has been applied to the meat or to live
animals.  Two points should be noted:  firstly, treatment
of the meat would make the meat subject to the rules about
foods with more than one ingredient; secondly, the other
provision is unique in requiring a declaration or
description relating to treatment or feeding practices
applied to the animal.

One of the functions of polyphosphates is to hold
water and their use in preparation of meat and poultry
has the effect of making the product moister (which the
consumer might like) but also of increasing the retention
of natural and added water which would, in excess, appear
to be an adulteration and an offence under Section 2.  As
yet there is no detailed legislation not least because of
the difficulties in deciding what would be the 'right'
amount of water in variable circumstances (meat like
other natural products is also variable itself).  The
addition of polyphosphates would require appropriate
designations on tickets and labels (and in declarations
of ingredients when prepacked).


MEAT:  WHAT TO TELL THE CONSUMER

The Food Standards Committee recommended (paragraph 26,
Food Labelling) that for meat and edible offal when sold
loose, a ticket should state 'the type of meat or edible
offal and where appropriate, the joint or cut of meat'.
At that time (1964) an origin-marking order required
imported meat and offal to be specially marked, e.g. as
'imported' or 'Empire'.  The other characteristics which
might be considered relevant are whether the meat is
fresh, chilled or frozen and for what purposes the meat
can best be used.

In some ways there has been no more controversial area
of food labelling than meat in butchers' shops.  The
butcher has objected to a multiplicity of labels and
tickets while others have maintained that more information
must be given to the consumer who is alleged to be

ignorant and innocent when buying meat.  The increased
sale of prepacked meat in supermarkets has been a factor
for change even without legislation as the label has
often given much more information.  The Meat and Livestock
Commission also developed a very good Code of Meat
Labelling although not everyone followed it.  The position
now is that a declaration of the country of origin is
required under the Trade Descriptions Act 1972 if any UK
mark is applied and the label of prepacked meat must
carry an appropriate designation '... a name or
description or a name and description sufficiently
specific to indicate to an intending purchaser the true
nature of the food ...'  However it is still very much
*caveat emptor*, and the housewife is faced with a
bewildering number of cuts and very little to go on in
judging variations in qualities such as flavour and
tenderness.

Earlier this year advertisements making offers of
'roasting beef or frying steak' were criticised as
being an inadequate guide and potentially misleading.
If they were false or misleading they would offend
against Section 6 of the Act which applies to advertisements as well as to labels.  But inadequate guidance or
incomplete information is not easy to deal with.  Leaving
aside publicity, education and voluntary measures there
are two ways in which the legislator can tackle it.  The
subject is covered in the FSC Report on Claims and
Misleading Descriptions.  The terms which must be used
on labels or advertisements can be defined and everyone
made to use them whether they want to or not - a remedy
to be reserved for the essential cases - or the use of
certain terms voluntarily can be stopped unless they are
accompanied by the necessary information or restricted
to carefully defined criteria.  However, the unspecific
claim like the advertising 'puff' is difficult to counter
except by an informed public.


MEAT:  DATE MARKING

Finally the growth of prepacking of perishable foods has
led to calls for date marking and the FSC Report on Date
Marking recommended that vacuum packed bacon (or similar
products) should be marked conspicuously with an 'open
by' date in order to avoid hazard (note that the trade
already did this where they considered it necessary or
prudent in view of Sections 1 and 8 of the Act.)  As to
the 'sell by' date which the FSC recommended for general

use, the FSC said of meat:   "For a limited period after
killing most fresh meat improves in quality with age.
Once, however, it has been enclosed in a package, as in
film wrapping, it seems reasonable to require a 'sell by'
date ...".   The Committee also considered it important
that the perishable meat products, e.g. sausages and
pies, should be clearly marked with a 'sell by' date.


### FOOD STANDARDS FOR MEAT PRODUCTS: FOUR MAIN QUESTIONS

There are four main questions about a meat product which
are relevant to food standards legislation.   They are not
separate but must be considered together.   They are:

1.  What is it?
2.  What sort of meat and what type?
3.  How much meat?
4.  What control should be placed over the information
    given to the consumer?

However, it will be convenient first to consider food
additives.


### FOOD ADDITIVES

The use of polyphosphates in meat products presents the
same sort of problems as for raw meat.   The British
system requires that the use of an additive should be
necessary and have no short or long term harmful effects.
Preservatives are *expressly permitted* in ham, bacon,
cured meats and sausages; colours are *not forbidden* in
meat products (although as already noted they are
*expressly prohibited* in raw or unprocessed meat).   The
distinction which reflects the different system of
control of preservatives and colours means that whether a
particular meat product may be coloured is left to the
general laws.
  Colours are used in a wide range of meat products to
standardise appearance or to retain colour after the
product has been taken out of its container as its
unattractive appearance would otherwise put the consumer
off.   In the Codex Committee those countries which do not
allow colour have argued that they are not necessary if
the right types of quality of meat are used.   Legal

problems of the use of colour and the type of meat in a
product are thus interrelated and there is advantage in
considering them product by product (as in the Codex).


## WHAT IS IT?

'Meat product means any product intended for sale for
human consumption containing meat'.  Examples are
sausages, meat pies, sausage rolls, corned meat, cured
meat, meat with jelly, meat with cereal, pressed meat,
meat with gravy, meat with vegetables, curried meat,
meat roll, meat loaf, meat paste, meat spread, pâté,
meat soup and others including many consisting mainly of
other ingredients with meat.

   Regulations cover all meat products in four main
groups;  Meat Pie and Sausage Roll; Canned Meat Product;
Sausages and other Meat Products; and Meat Spreadables.
They cover all types, fresh and frozen, canned and loose,
prepacked and those sold in catering establishments.  As
yet, there are no regulations on meat soups although the
FSC recommended control.  The use of offals has been
closely controlled since 1953 and the FSC has recommended
certain important changes particularly to give more
information to the purchaser.  The regulations contain
detailed definitions.  The general meat products group
are categorised according to the next largest ingredient
by weight after meat and each has a prescribed minimum
meat content; there are special labelling rules and
control of descriptions and pictures for products with
less than 35% of meat by weight.  The statutory names or
descriptions laid down for each product or group of
products must now always be used so the consumer has
detailed protection although, as common or usual names
may sometimes be used as an alternative, the list of
ingredients must often be consulted to correct or
supplement first impressions based only on the name
itself.


## WHAT SORT OF MEAT AND WHAT TYPE?


### *Definitions of meat*

A scientific description might be:  'the anatomically
definable skeletal muscles of the carcass, including
their connective tissue and such fat as is clearly

deposited in the connective tissue associated with the muscles' (*Offals in Meat Products* (1972) based on Professor Lawrie's advice). The Codex Committee in its Second Session found two possibilities in elaborating definitions for its work '... to draw up a strict and fairly detailed definition of meat, and to list in each individual commodity standard ... such additional ingredients as might be considered necessary' or '... a broad and fairly brief definition ... listing ... the exclusions ... in each standard.' The FSC in its early reports (Sausages, 1956; Canned Meat, 1962; Meat Pies, 1963) concentrated more on the type of meat, e.g. pork or beef, and on the quantity of lean meat although they dealt with the question of the inclusion of offals.

During the discussions leading up to the 1967 Regulations it became clearer that manufacturing practices required the inclusion of a detailed definition. The difficulty was to reflect good manufacturing practice while protecting the consumer. Definitions of 'meat' were varied as necessary but the basis was 'the flesh, including fat, and the skin, rind, gristle and sinew in amounts naturally associated with the flesh used, of any animal or bird which is normally used for human consumption and includes cured meat and permitted offal ...'. More recently questions of reformed meat (slices or pieces reformed by advanced technology and the use of certain additives) have arisen which need to be looked at in relation to these definitions.

Any skin etc. used in excess of that naturally associated may not be counted towards the meat content although this may be difficult to detect. The consumer 'is in a much better position to judge them [quality factors such as the proportion of fat and gristle, the texture of the fibres, the flavour, the tenderness and the freshness .....] than, for example, to judge the quantity of meat' (Sausages, 1956). Since 1956, however, technology has improved and comminuted products in particular are now very difficult to assess. Should there be greater control of the use of anything but the flesh? Can the consumer be said to be content? Does he accept the product for what it is - or what he thinks it is? Of the inclusion of offals the Food Standards Committee observed '... it is doubtful whether it could be said that all consumers find present usage acceptable since, in the absence of a labelling provision, they are ignorant ...' (*Offals in Meat Products*, 1972). The Committee recommended that excess skin be declared as

'detached skin' or 'added skin' and that the consumer
always be informed of the presence of offal.

Meat pie, meat with cereal, luncheon meat or meat loaf,
sausage, all imply the use of meat as generally understood
by the consumer, e.g. the flesh but perhaps not the skin
etc. associated with it, and the declaration of ingredients
might therefore need to be a detailed source of information
for inclusion of anything but the flesh. Although there
is no permitted generic term for 'meat' (except in very
restricted circumstances for a few comminuted products
such as sausages) and therefore nearly all lists of
ingredients should show the specific name of each
ingredient, the labels of meat products rarely contain
references to added skin or to offal (unlike the labels
of some canned pet foods) even though there is some use
of them. Are there analytical or enforcement explanations

'Luncheon meat' is an example of a name which is used
in several countries for a product which may look the
same but which is significantly different. Like the use
of colour it gave delegates to the Codex Committee many
exciting moments in Copenhagen (and so did the differences
between 'chopped meat' and 'comminuted meat' which like
many food standard differences were quality and money
differences). In the USA the product is all meat but in
UK it is 80% meat with cereal added. The question in
the UK is whether the consumer realises the difference
between chopped meat, luncheon meat and meat loaf
without the inclusion in the name of the category name,
e.g. meat with cereal, and an indication of the amount
of meat.


WHAT TYPE OF MEAT?

The regulations require the term meat in a description
but leave open whether the type is given. The usual
rules are that a named meat should either represent all
the meat or that if more than one is named the greater
should be placed first. The regulations lay down,
exceptionally, certain rules which allow, for example
pork sausage or pork loaf (i.e. 65% meat with cereal)
to be used even if other meats are present. In both
cases the rules reflect previous manufacturing practice
and they illustrate some of the most difficult - and
controversial - aspects of food standards. How far to
interfere, change and inform? Does the consumer accept
does he know - does he care? Now that a declaration of

ingredients is required any consumer who reads it will
be able to identify the presence of other meat, but should
he have to look further than the name and can he in a
restaurant or when the product is sold loose?

The Food Standards Committee's only failure in a long
history to reach a consensus was over the use of other
meat in a pork or beef sausage and the issues are clearly
set down in paragraphs 24-27 of the Report on Sausages.
The regulations allow 'pork sausage' if at least four-
fifths of the meat used is pork. The Report records that
in the absence of express permission there might be an
offence under Sections 2 and 6 of the Food and Drugs Act
1955 (and now also under Section 2 of the Trade
Descriptions Act 1968) as the purchaser would expect to
find only pork in a product described as pork and not,
for example, pork with beef. Meat spreadables are
separated into pastes and spreads and 'spread' may only
be used if the prescribed minimum of 70% meat consists
of the named meat although other meat in excess may be
included but rarely is. Here again, the declaration of
ingredients must be consulted in order to help interpret
the name.


HOW MUCH MEAT?

How much meat should be included in a meat product is a
critical question. Perhaps the essence of the food
standards problem (and not only on meat products) is
contained in paragraphs 3 to 6 of the Food Standards
Committee Report on Meat Pies which, unfortunately,
cannot be reproduced here because of limitations of
space. The Committee did not accept that the consumer
knew that many meat pies had hardly any meat in them or
that consumers could be protected without detailed
standards.

In these days of advanced technology and in view of
the use of flavourings and colours, taste and appearance
cannot be an accurate guide to the amount of meat in a
product. All the Regulations therefore exercise control
by a prescribed minimum meat content and, where
appropriate, a lean meat content to control the fat for
nutritional and quality reasons. The minimum meat
content for each product had to be decided in the light
of all the evidence of manufacturing practices and
consumer expectancy and account must be taken of natural
variations in the meat, of problems of getting the same

amounts into each product, of analytical methods, of
technical factors such as the transfers between meat and
pastry and meat and brine and of sampling questions. The
standard is based on each product complying, not on average
compliance.

When the amount of meat falls below 35% the Food
Standards Committee elaborated a principle that (except
for soups), unless it was clear that a product was not
being described or claimed as a meat product, it should
be clearly distinguishable from meat products with at
least 35% meat. The Regulations did this by a control of
descriptions and pictures but the need to allow for a
diversity of legitimate products below 35% with at least
two vegetables ('Ready Meal') and for common or usual
names like Irish Stew or Lancashire Hot Pot (which imply
meat but how much?) and the difficulties (real and
imagined) of interpreting and enforcing when it is
being implied that 'meat is a major ingredient' have not
had entirely satisfactory results. The march of
technology and the diversity of consumer demands have
added further difficulties. In 1962 the Committee
concluded that a series of categories each with its
description was preferable to a declaration of meat
content. Perhaps now both would be best for trader and
consumer.

The determination analytically of the meat content is
not easy and it has been made more difficult by the
legislation not being used restrictively to stop, for
example, the inclusion of other animal or soya proteins
which can be confused with meat when the product is
analysed. The view taken is that the standard should
reflect what the product should be, not what would be
easy to analyse.

There is a greater possibility of confusion - of the
consumer and the analyst - in the developing use of
vegetable and other sources of protein which are textured
and flavoured to resemble and taste of meat. The Food
Standards Committee's Report on Novel Proteins will no
doubt throw light on this interesting subject. The
general provisions of the Food and Drugs Act, reinforced
by the Labelling Regulations and the Trade Descriptions
Act, should effectively prevent these new products from
being passed off as meat but if they are to be used for
the benefit of the consumer there is a need for detailed
rules to control their names and their inclusion in meat
products. Pictures can be even more misleading. This
is an area where the absence of detailed rules may well
have inhibited a new development and where it will be

particularly important to get the right balance in
framing those detailed rules.

## MEAT DISHES IN RESTAURANTS AND OTHER CATERING ESTABLISHMENTS

There is scope for confusion - deliberate and accidental -
in catering establishments too. The general provisions
of the Act apply and so does the 1973 requirement for an
appropriate designation or common or usual name. The
compositional regulations do not necessarily apply to
sales to catering establishments but this does not
absolve the establishment from complying with
compositional requirements when selling the product.
Here again, the problem of novel proteins has been
causing some anxiety and the Committee's report should
be helpful in clarifying the issue and in pointing the
way to a more detailed control. However, the basic
problem is that the restaurant whether the Ritz or Joe's
cafe is not a shop and the detailed rules have to be
drawn up *mutatis mutandis* if they are to be fair,
sensible and practicable.

## NUTRITIONAL CONSIDERATIONS

Meat and meat products are nutritionally important. Any
legal action is therefore directed towards distinguishing
other products which are being confused with meat or meat
products. There is little abuse in relation to meat or
meat products themselves and the general safeguards in
the Labelling Regulations provide adequate protection and
they facilitate nutritional claims for unprocessed meat
by allowing claims to be based on average values so as to
overcome the practical difficulties associated with
natural and seasonal variations. In fact nutritional
claims for meat and meat products are rarely made
perhaps because the consumer knows how good they are!

## CONCLUSION

The legal problems relating to meat and meat products
are numerous and complex.  Such an important part of the
nation's diet cannot be left to the general protection
of the food laws.  The detailed control of the manufacture
and sale of meat and of meat products requires an expert
analysis and evaluation of the problem, thorough
examination and discussion by all interested parties, an
accurate assessment of the scientific, practical,
commercial and legal effects of detailed rules and
continuous review.  There are philosophical and political
considerations to be kept in mind as well as international
ones (now including the European Economic Community where
the approach to the subject in other Member States is by
no means similar).  It is well worth trying to get the
right balance between trader and trader, and between
trader and consumer; between flexibility of formulation
and confusion of differing products of similar
descriptions; between licence to compete and freedom of
competition based on detailed, sensible rules and
consumers' knowledge and awareness of what they are
buying.

The aims, like in all food standards, are first that
the consumer should get only safe, sound, wholesome food,
accurately and informatively described and, second, to
enable him to make an informed choice in the market
place.  'Give me neither poverty nor riches; feed me
with food convenient for me' (Proverbs 30:8).  And what
better than meat or meat products so long as the consumer
knows what he wants and gets what he asks for?

## ACKNOWLEDGEMENTS

The views in the paper are my own and not those of the
Ministry.  My understanding and enjoyment of the subject
owes much to the knowledge, wit and good sense of the
experts with whom I have worked – especially Professor
Alan Ward, Chairman of the Food Standards Committee (FSC)
Dr V. Enggaard, Chairman of the Codex Committee on
Processed Meat Products; John O'Keefe, editor of the
standard work, *Bell and O'Keefe's Sale of Food and Drugs,*
and member of the FSC; and, perhaps most of all, John
Davies who was Secretary of the FSC at the time of the
reports on 'Canned Meat' and 'Meat Pies' and Head of Food

Standards Division in 1967 when the meat products
regulations were made.

REFERENCES

Codex Committee on Processed Meat Products Reports (not
   available)
Food Standards Committee Reports (HMSO):
   Sausages                          (1956)
   Canned Meat                       (1962)
   Meat Pies                         (1963)
   Food Labelling                    (1964)
   Fish and Meat Pastes              (1965)
   Claims and Misleading Descriptions (1966)
   Soups                             (1968)
   Offals in Meat Products           (1972)
   Date Marking of Food              (1972)
O'KEEFE, J.A. (1968). *Bell and O'Keefe's Sale of Food
   and Drugs*, 14th edn. Butterworth, London
The Labelling of Food Regulations (1970) SI 1970 No.400
The Labelling of Food (Amendment) Regulations (1972)
   SI 1972 No.1510
The Meat Product Regulations (HMSO):
   The Meat Pie and Sausage Roll Regulations 1967 SI 1967
      No.860 (as amended)
   The Canned Meat Product Regulations 1967 SI 1967 No.861
      (as amended)
   The Sausage and other Meat Product Regulations 1967
      SI 1967 No.862 (as amended)
   The Fish and Meat Spreadable Products Regulations 1968
      SI 1968 No.430
The Offals in Meat Products Order (1953) SI 1953 No.246

# IV

## EATING QUALITY

# 16

# WATER-HOLDING CAPACITY OF MEAT

R. HAMM

*Bundesanstalt für Fleischforschung Kulmbach,*
*West Germany*

## INTRODUCTION

The cross-striated muscles of meat animals contain about
75% water, and under certain conditions, e.g. by addition
of water to the minced tissue, even more water can be
taken up. The power with which this water is bound by
the muscle proteins is of great importance for the
quality of meat and meat products. Almost all procedures
for the storage and processing of meat are influenced by
the water-holding capacity (WHC) of the tissue; and,
vice versa, these procedures can change the WHC of muscle.
Such procedures include transportation, storage, ageing,
mincing, salting, curing, canning, cooking, freezing and
thawing or drying. It is well known that WHC is of
particular importance for the quality of sausages of the
frankfurter type and canned ham. The great economic
problem of weight losses during storage, cooking or
freezing and thawing of meat is related to the binding of
water within the muscle. Thus, investigation of the WHC
of meat is of considerable economic interest. Moreover
it gives information not only on WHC of meat itself but
also on changes in meat proteins. Changes in WHC are a
very sensitive indicator of changes in the charges and
structure of muscle proteins (Hamm, 1960; 1963; 1972).

## THE STATE OF WATER IN THE MUSCLE CELL

There is no doubt that the myofibrillar proteins are
primarily responsible for the binding of water in muscle.
It is also obvious that different types of water-binding
exist in the tissue.  Water is of tremendous importance
for the structure and function of the living cell and,
therefore, recently great effort has been made in order
to elucidate the state of water in biological systems,
mainly by means of NMR studies.  In agreement with earlier
results it is generally accepted that a relatively small
part of the tissue water (4-5% of the total water) is
tightly bound on the surface of the protein molecules as
hydration water which has an ice-like structure and
properties different from those of free water, e.g. lower
freezing point (Cooke and Wien, 1971; Dydynska, 1970;
Elford, 1970; Finkh, Harmon and Muller, 1971; Ling and
Negendank, 1970).  Non-polar amino-acid residues of the
muscle proteins seem to be of importance for the binding
of this fraction of cellular water (Karmas and Dimarco,
1970).  The data on the binding of this hydration water by
muscle proteins vary between 15 and 36 g $H_2O/100$ g
protein.  Belton, Jackson and Paker (1972), however, in
an NMR study found a much higher proportion of bound
(non-freezing) water, namely about 20% of the total
water; 15% of the total water could be located in the
extracellular space, and the remainder of water was
present in the myofibrils and the sarcoplasmic reticulum.

This tightly bound hydration water is hardly influenced
by changes in the structure and charges of the muscle
proteins.  This is true for example, in regard to
muscular contraction (Cooke and Wien, 1971) and even for
the cooking of meat (Palnitkar and Heldman, 1971).
Therefore, the remarkable changes in WHC of meat caused
by changes of protein charges (pH), by rigor mortis, by
heating etc. cannot be due to any changes in the
hydration water.  The changes in WHC of meat of practical
importance must be related to the remaining 95% of muscle
water.  Apparently this water can be more or less
immobilised within the system of myofibrillar proteins.
In my first review on WHC of meat I mentioned:  "Most of
the 'free' water present in muscle must be considered as
'immobilized' water.  We do not know exactly which
forces cause this immobilization."  After 15 years we
still are uncertain about the state of water in the
muscle cell.  Hazlewood, Nichols and Chamberlain (1969)
concluded from NMR studies that most of the skeletal

muscle water has restricted motional freedom; they
suggested that most of the cell water is highly ordered.
These water molecules must be much more ordered than
those of ordinary water.  Other authors confirmed this
hypothesis (Chang *et al*., 1972; Cope, 1969; Ling and
Negendank, 1970; Tait and Franks, 1971).  The
interpretation of NMR data, however, is still controversial
because other investigators suggested that most of the
muscle water (except the tightly bound hydration water)
is not bound but freely movable like ordinary water
(Cooke and Wien, 1971; Finkh, Harmon and Muller, 1971;
Hansen, 1971; Hansen and Lawson, 1970).

According to Karmas (1973) dividing water in biological
systems into two major phases as bound and free water
seems to be inadequate.  It would be more meaningful to
use the term 'biological water activity' which ranges
from biologically active water to structural water.
Biological active water is needed by biosystems at the
peak of biological activity.  The structural or
protective water ('hydration water', *see* above) is vital
for survival of the living organism and it cannot be
frozen or evaporated without destroying the organism.
As the controversial interpretation of NMR studies
reveals, the term 'biological water activity' is not well
defined and at present we know less than ever about the
relationship between biological water activity and the
differences in WHC of meat.  It should be mentioned,
that great differences in WHC of beef and pork could be
obtained at the same water activity of the tissue,
measured by relative humidity (Vrchlabsky and Leistner,
1970).


## IMMOBILISED WATER:   WATER-HOLDING AND
## SWELLING CAPACITY


The remarkable changes in WHC occurring during storage
and processing of meat are determined by the extent to
which the non-tightly bound water is immobilised within
the microstructure of the tissue.  There seems to be a
continuous transition from the water strongly immobilised
within the tissue, which can be expressed with difficulty,
to the 'loose' water which can be squeezed out by very
low pressure.  It is not possible, therefore, to get any
absolute figures for the immobilised part of water
because the amount of immobilised water determined depends
on the method used.  Consequently, we must define WHC in

terms of the method of measurement. Using the same
method one can measure relative differences in WHC quite
exactly.   In most of the methods for determination of
WHC, the loose water is released by pressing the sample
between two plates (e.g. the 'filter-paper press method')
or by centrifuging.  A new type of method has been
proposed by Hofmann (1971).  A gypsum diaphragm is put on
the surface of the intact or minced tissue with a
relatively low pressure.  The loose water is sucked up
into the diaphragm by the effect of capillary forces.  The
air displaced from the capillaries by the meat juice goes
into a U-shaped calibrated glass tube, closed by a
coloured fluid.  The volume of the displaced air, read
from the shift of this fluid, is equal to the volume of
loose water and is inversely proportional to the WHC of
the sample.  According to these methods, WHC is defined
as the ability of meat to hold fast its own and added
water during application of any force.

Swelling capacity of meat is defined as the spontaneous
uptake of water by meat from any surrounding fluid
resulting in an increase of weight and volume of muscle.
Swelling capacity usually shows a close correlation with
WHC.


## FACTORS INFLUENCING THE IMMOBILISATION
## OF WATER


As I have already mentioned, we do not know exactly
which forces restrict the mobility of water in the tissue,
but we know much more about the factors which influence
changes in the immobilisation of water, i.e. changes in
the WHC of meat.  The amount of water immobilised within
the tissue is influenced by the spatial molecular
arrangement of the myofibrillar proteins or filaments,
myosin and actin (*Figure 16.1*).  By decreasing the
cohesion between adjacent molecules or filaments, as is
caused by increasing the electrostatic repulsion between
similarly charged groups or by weakening of hydrogen
bonds, the network is enlarged, the swelling increases
and more water can be immobilised within the larger
meshes, i.e. there is an increase in WHC (*Figure 16.1,*
A→B).  As the intermolecular cohesion becomes more and
more loose, the network finally collapses and the gel
becomes a colloid solution of the myofibrillar proteins
(B→C).  On the other hand, by increasing the attraction
between adjacent molecules, as happens when the

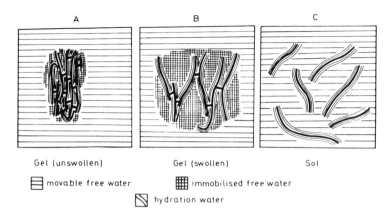

A                       B                       C

Gel (unswollen)         Gel (swollen)           Sol

▤ movable free water         ▦ immobilised free water

◩ hydration water

*Figure 16.1   Influence of cross-linking of proteins or
filaments on the WHC or swelling of meat.
A: Strong cross-linking causes low WHC.   The bulk of
'free' water is not immobilised. B: Few cross-links
allow a high WHC.   A great part of the 'free' water is
immobilised within the protein network.   C: No cross-
linkages and, therefore, no WHC or swelling.   The protein
molecules are freely movable forming a colloidal solution*

electrostatic attraction between oppositely charged
groups increases or by the effect of interlinking bonds,
less space is available for the retention of immobilised
water.   Thus, during tightening of the network
unshrinking occurs, a part of the immobilised water
becomes free and flows out at low pressure (*Figure 16.1,*
B→A).

   The protein network is represented by the three-
dimensional arrangement of myosin and actin filaments.
In addition to these two proteins, tropomyosin also seems
to contribute to the WHC of muscle tissue.   This could be
shown with a model system, namely, with a gel of the
myofibrillar protein from poultry muscle (Nakama and
Sato, 1971).   It is also possible that the sarcoplasmic
reticulum has some effect on WHC because it promotes the
gelatinisation of the extract of sarcoplasmic proteins
(Ivanov *et al.*, 1969).   If WHC or swelling decreases, the
water is probably shifted from the fibrils into the
interfibrillar spaces.   It could be that it is not
necessary to explain the phenomenon of water-binding in
muscle by the existence of special forces lowering the
mobility of the bulk of cellular water; perhaps water-
binding is caused just by the mechanical fixation of

water in the network of the protein gel which is greater
as more water finds its place in the network, i.e. as the
meshes of the network become larger.

## INFLUENCE OF pH ON WHC

A good example of the importance of protein-protein
interactions on the WHC and swelling of muscle tissue,
according to the scheme in *Figure 16.1,* is the influence
of pH on WHC and swelling of meat *(Figure 16.2).*
A loosening of the microstructure and, consequently an
increase of immobilised water is caused by raising the
protein net charge by the addition of acid or base
*(Figure 16.3).*
   The pH at which the WHC is at a minimum (pH 5.0)
corresponds to the iso-electric point (IP) of the
actomyosin which makes up the bulk of the structural
muscle proteins.  At the IP the net charge of a protein
is at a minimum; at this pH we should expect a maximum

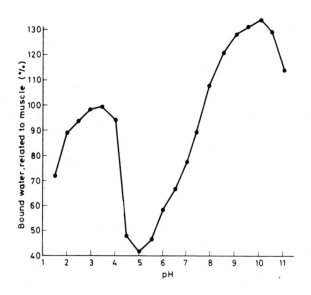

*Figure 16.2   Influence of pH on WHC of ground beef
(filter-paper press-method)*

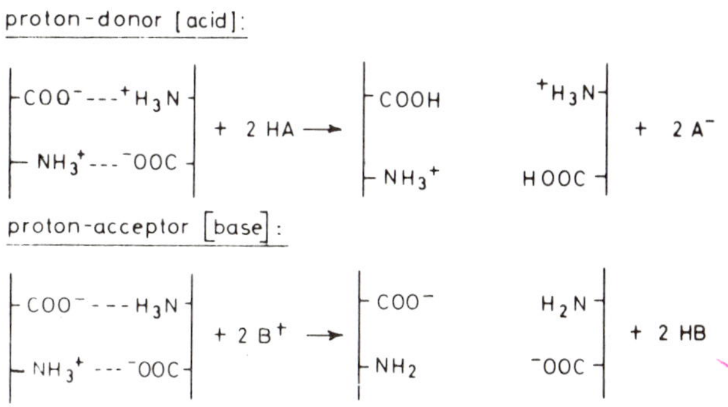

proton-donor [acid]:

$$\left.\begin{array}{l} -COO^-\text{---}^+H_3N- \\ \\ -NH_3^+\text{---}^-OOC- \end{array}\right] + 2\,HA \longrightarrow \left.\begin{array}{l} -COOH \\ \\ -NH_3^+ \end{array}\right. \quad \left.\begin{array}{l} ^+H_3N- \\ \\ HOOC- \end{array}\right] + 2\,A^-$$

proton-acceptor [base]:

$$\left.\begin{array}{l} -COO^-\text{---}H_3N- \\ \\ -NH_3^+\text{---}^-OOC- \end{array}\right] + 2\,B^+ \longrightarrow \left.\begin{array}{l} -COO^- \\ \\ -NH_2 \end{array}\right. \quad \left.\begin{array}{l} H_2N- \\ \\ ^-OOC- \end{array}\right] + 2\,HB$$

*Figure 16.3  Influence of acid and base on the inter-action between protein charges. Left: Isoelectric protein (low WHC)*

of intermolecular salt linkages between positively and negatively charged groups which explains the minimum WHC at the IP.

Differences in WHC of meat between animals of the same species can be related to pH differences; with an increasing pH the WHC increases provided that the variation in pH is not too small. This relationship was confirmed for red meats, but not for poultry meat (Nakamura, 1970), by more recent work (Herring, Haggard and Hansen, 1971; Pagano Toscano and Autino, 1970; Partmann, Frank and Gutschmidt, 1970). During the past several years efforts have been made to increase the ultimate pH of meat (e.g. from 5.4 to 7.0) by injection of adrenaline ante mortem, which lowers the glycogen content of the muscle just before slaughter. An increase in the pH of meat caused by this treatment results in a significant increase of WHC of beef (Bouton *et al.,* 1973a and b), lamb and/or mutton (Bouton, Harris and Shorthose, 1971; 1972) and pork (Hatton *et al.,* 1972).

## POST-MORTEM CHANGES IN WHC

The development of rigor mortis in muscle, which is induced by the break-down of adenosine triphosphate (ATP) post mortem, is caused by a strong cross-linking between actin and myosin filaments. This tightening of

the protein network leads to a considerable decrease in
the WHC of meat during the first hours post mortem
(*Figure 16.1,* B→A).  During further ageing, the WHC
increases somewhat, probably by a loosening of the
myofibrillar system caused by the attack of proteolytic
enzymes on the region of the Z line.  Earlier
observations which showed a loss of WHC during rigor
mortis particularly in the basic range of IP and an
increase of WHC in a larger range of pH during ageing
were confirmed by El-Badawi, Anglemier and Cain (1971).
With broilers, the WHC increased during ageing between 3
and 24 h post mortem (Wardlaw, McKaskill and Acton,
1973).  Brendl and Klein (1970) demonstrated a coincidence
in the post-mortem course of changes in WHC, pH and the
solubility of myofibrillar proteins (*Figure 16.4*).

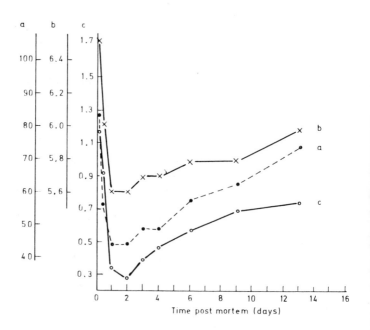

*Figure 16.4  WHC:   (a) % bound water, (b) pH value and
(c) soluble myofibrillar protein-N of beef at different
times post mortem.   (From Brendl and Klein, 1970, by
courtesy of Inst. Chem. Technol, Prague)*

An extremely fast rigor mortis and, therefore, a
strong drop of WHC occurs if pre-rigor frozen meat is
thawed ('thaw rigor'). For this reason the 'one phase'
freezing can be disadvantageous with regard to WHC and
tenderness of meat. Such adverse high WHC losses and
toughness of thawed beef or lamb can be prevented if the
temperature of the meat, frozen at low temperature, is
raised to just a few degrees below its freezing point for
several hours before thawing. Under such conditions
normal glycolytic changes and rigor mortis will slowly
go to completion while enough ice exists to physically
prevent the muscle from shortening (Behnke and Fennema,
1973b; Behnke, Fennema and Cassens, 1973; Marsh,
Woodhams and Leet, 1968). Drip losses during thawing
increase rapidly as shortening exceeds a critical value
of about 35% (*Figure 16.5*).

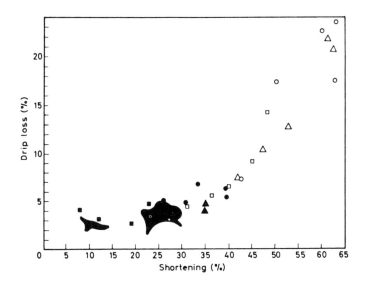

*Figure 16.5 Effect of per cent shortening on per cent
drip loss of excised* sternomandibularis *muscles
(irregular areas represent range of 30 control values)
(From Behnke and Fennema, 1973b, by courtesy of the
Institute of Food Technologists)*

## WHC AND TENDERNESS

The WHC of meat is closely related to tenderness,
juiciness and colour.  Recently particular attention has
been drawn to the relationship between WHC and
tenderness.  An increase in WHC is associated with a
loosening of the network of the protein gel (*Figure 16.1,*
A→B) which results in an increase in tenderness
(Tyszkiewicz, 1969).  The tenderness of ovine muscle has
been shown to increase with rising WHC, owing to
increasing pH (Bouton, Harris and Shorthose, 1971).
Increased shortening of muscle fibres is accompanied by
a decrease in WHC (e.g. increase in drip loss) as was
observed during the freezing and thawing of poultry
(Behnke and Fennema, 1973a).  Contracted muscles have a
significantly reduced WHC when compared with their
stretched counterparts (Bouton, Harris and Shorthose,
1972).  On the other hand, a large increase in WHC can
counteract the increased shear force which is associated
with the cold-shortening of bovine and ovine muscles
(Bouton *et al.,* 1973a).

## 'BINDING' AND WHC

The phenomenon of 'binding' refers to the knitting of
chunks of meat together to produce a unit system.  The
binding of chunks of meat together to form rolls or
loaves has received considerable attention especially by
the poultry industry.  According to Vadehra and Baker
(1970), binding appears to be related to WHC of meat as
well as to the cell disruption and release of intracellular material.  The same might be true for the effect
of tumbling on the quality of canned ham.

## CHANGE OF WHC DURING HEATING

The considerable decrease of WHC during the heating of
meat, which results in the release of juice, is due to a
tightening of the myofibrillar network by heat-
denaturation of the proteins (*Figure 16.1,* B→A).  The
biggest change in total juice loss (equal to the sum of
cooking loss and expressible liquid) during the cooking
of beef occurs between the temperature of the raw sample

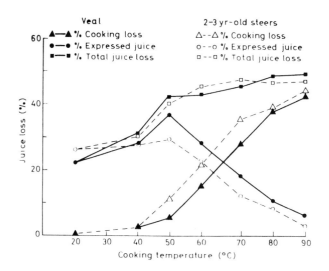

*Figure 16.6  Changes in cooking loss, centrifugally
expressible juice and total juice loss for deep
pectoralis muscles from veal and 2-3 yr-old steers (pH 5.4
5.6) cooked at different temperatures for 1 h (From
Bouton and Harris, 1972, by courtesy of the Institute of
Food Technologists)*

and about 50°C   (Bouton and Harris, 1972; *Figure 16.6*).
According to Karmas and Dimarco (1969; 1970), determination of the heat denaturation thermoprofiles revealed two
peaks at 65 and 82°C.  The protective, semi-crystalline
water structure, surrounding the non-polar amino-acid
radicals, may collapse as a result of the heat, this is
followed by the formation of hydrophobic bonding which
results in an aggregated, denatured state.


INFLUENCE OF SALTS ON WHC


Besides the effect of pH, the influence of salts is a
typical example of the importance of protein charges on
WHC of meat.  The effect of NaCl on WHC or swelling
depends on the pH of the tissue.  NaCl increases the WHC
at pH>IP and decreases it at pH<IP (*Figure 16.7*)

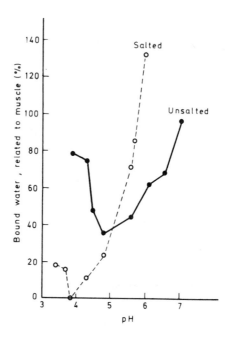

*Figure 16.7   Influence of pH on the WHC of salted (2% NaCl) and unsalted ground beef (filter-paper press-method*

This effect is predominantly due to the $Cl^-$ ion of NaCl, which causes a weakening of the interaction between oppositely charged groups at pH>IP (*Figure 16.1,* A→B) and a strengthening at pH<IP (*Figure 16.1,* B→A), as is shown in the scheme of *Figure 16.8*.

It is in agreement with this concept that at pH values greater than IP, increases in ionic strength or in pH increase the electrostatic repelling forces between the muscle filaments and, consequently, the interfilamental spacing of the fibre (April, Brandt and Elliott, 1972). Ohashi, Sera and Ando (1972) suggested that the addition of NaCl may lead to liberation of bivalent cations ($Ca^{++}$, $Mg^{++}$) from muscle proteins and thus also a loosening of the microstructure of the tissue may take place.

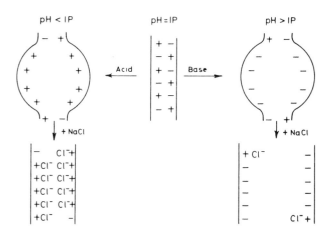

*Figure 16.8  Schema of the influence of NaCl on swelling or WHC of meat at pH values above and below the isoelectric point*

## WHC WITH REGARD TO SAUSAGES OF THE FRANKFURTER TYPE

Mincing of meat for the production of sausages of the frankfurter type (frankfurters, wieners, bologna etc.) destroys the sarcolemma.  For this reason, the myofibrillar system is transformed from one of limited swelling to one of unlimited swelling, whereby the WHC is increased.  The effect of the removal of the sarcolemma on the swelling of fibre (*Figure 16.9*) shows that the skinning of the muscle fibres increases the effect of pH on the interfilamental spacing (April, Brandt and Elliott, 1972).

Most of the factors which increase the WHC of meat also improve the distribution of fat in the sausage. Under normal conditions during sausage production, solid or semi-solid particles of added fat (fat tissue) are mixed with the system of fibre fragments, sarcoplasma and added water.  During heating of the sausage mixture, the coagulating network of proteins or filaments surrounds the melting fat particles, which cannot coalesce because of a mechanical fixation within the meshes of the coagulated network.  The larger those meshes are the less coalescence of the fat can occur and

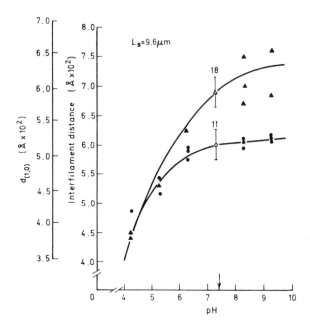

*Figure 16.9  Interfilament spacing as a function of pH.
The lattice spacing and interfilament spacing of living
single muscle fibres (●) and skinned fibres (▲) is
plotted against the pH of the medium.  The open symbols
(o,Δ) represent mean values, with the numbers of
experiments and the standard deviation from the mean
indicated (From April et al., 1972)*

the better is the WHC of the mixture (Hamm, 1973).  I
feel that such an effect might be more important for the
desired distribution of the fat in a sausage than the
so-called emulsion of fat, which is caused through coating
fat droplets by layers of dissolved muscle proteins (or
added proteins).

Processing of pre-rigor meat results in sausages of
excellent quality because of the high WHC of this
material owing to the presence of ATP.  It is possible
to prevent the strong loss of WHC after slaughter by
salting the ground beef within the first hours after
death, i.e. before the breakdown of ATP starts.  In
this way, the high WHC of the pre-rigor tissue can be

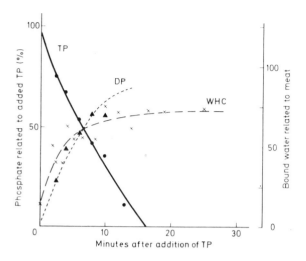

*Figure 16.10  Effect of sodium tripolyphosphate (TP) on WHC of ground beef at different times after addition of TP.  The amount of TP decreases and the diphosphate (DP) concentration increases because of the enzymatic breakdown of TP to DP*

kept for several days in spite of the breakdown of ATP and glycogen which cannot be prevented by the salt (van Hoof and Hamm, 1972).  It could be shown, moreover, that the WHC in pre-rigor salted meat does not decrease, because the onset of rigor mortis in the fibre fragments is prevented (Rede and Hamm, 1972).  This is probably due to a strong electrostatic repulsion between adjacent protein molecules caused by the initial combined effect of ATP, NaCl and high pH value.  By the freezing of pre-rigor salted ground beef the high WHC can be preserved for several months and by mincing the frozen material in a cutter excellent products can be obtained which show the same quality as sausages made from fresh pre-rigor beef ('warm' meat) (Hamm, 1973; Dimitrijevic, Panin and Miljevic, 1970).

The effect of phosphates on WHC of meat will not be discussed here because this topic has been recently reviewed (Hamm, 1971).  It should be only mentioned here that according to recent results triphosphate (TP) added to minced, salted beef is quickly hydrolysed to

diphosphate (DP) by a tissue triphosphatase. TP itself
seems to have no effect on the WHC of the meat but seems
to increase the WHC to the same extent as it is broken
down to DP. The enzymatic breakdown of DP to monophosphat
by tissue diphosphatases occurs at a much slower rate and
does not result in a decrease of WHC (Neraal and Hamm,
1973; *Figure 16.10).*

*General reviews, encompassing many practical and
theoretical aspects of the water-holding capacity of
meat, were published by the author earlier (Hamm, 1960
and 1972). In this chapter only such references are
given which were not quoted in both of these earlier
reviews.*

## REFERENCES

APRIL, E.W., BRANDT, P.W. and ELLIOTT, G.F. (1972).
    *J. Cell Biol.,* 53, 53
BEHNKE, J.R. and FENNEMA, O. (1973a). *J. Fd Sci.,* 38, 275
BEHNKE, J.R. and FENNEMA, O. (1973b). *J. Fd Sci.,* 38, 539
BEHNKE, J.R., FENNEMA, O. and CASSENS, R.G. (1973).
    *J. Agric. Fd Chem.,* 38, 539
BELTON, P.S., JACKSON, R.R. and PAKER, K.J. (1972).
    *Biochim. Biophys. Acta,* 286, 16
BOUTON, P.E. and HARRIS, P.V. (1972). *J. Fd Sci.,* 37,
    140 and 218
BOUTON, P.E., HARRIS, P.V. and SHORTHOSE, W.R. (1971).
    *J. Fd Sci.,* 36, 435
BOUTON, P.E., HARRIS, P.V. and SHORTHOSE, W.R. (1972).
    *J. Fd Sci.,* 37, 351
BOUTON, P.E., CARROLL, F.D., HARRIS, P.V. and SHORTHOSE,
    W.R. (1973a). *J. Fd Sci.,* 38, 404
BOUTON, P.E., CARROLL, F.D., HARRIS, P.V. and SHORTHOSE,
    W.R. (1973b). *J. Fd Sci.,* 38, 816
BRENDL, J. and KLEIN, S. (1970). *Sb. Vys. Sk. Chem.-
    Technol. Praze, Potravny,* E 28, 117
CHANG, D.C., HAZLEWOOD, C.F., NICHOLS, B.L. and BORSCHACH
    H.E. (1972). *Nature,* 235, 170
COOKE, R. and WIEN, R. (1971). *Biophys. J.,* 11, 1002
COPE, F.W. (1969). *Biophys. J.,* 9, 303
DIMITRIJEVIC, M., PANIN, J. and MILJEVIC, M. (1970).
    *Tehnologija mesa,* 11, 341

DYDYNSKA, M.D. (1970). *Acta Biochim. Polon.*, <u>17</u>, 209

EL-BADAWI, A.A., ANGLEMIER, A.F. and CAIN, R.F. (1971). *Alex. J. Agric. Res.*, <u>19</u>, 89

ELFORD, B.C. (1970). *Nature*, <u>227</u>, 282

FINKH, E.D., HARMON, J.F. and MULLER, B.H. (1971). *Arch. Biochem. Biophys.*, <u>147</u>, 299

HAMM, R. (1960). *Adv. Fd Res.*, <u>10</u>, 355

HAMM, R. (1963). *Recent Adv. Fd Sci.*, <u>3</u>, 218

HAMM, R. (1971). *Symposium: Phosphates in Food Processing* pp.65-82. Ed. J.M. DeMan and P. Melnychin. AVI Publishing Co., Westport, Conn.

HAMM, R. (1972). *Kolloidchemie des Fleisches.*, Parey, Hamburg, Berlin,

HAMM, R. (1973). *Fleischwirtschaft*, <u>53</u>, 73

HANSEN, J.R. (1971). *Biochim. Biophys. Acta,* <u>230</u>, 482

HANSEN, J.R. and LAWSON, K.D. (1970). *Nature,* <u>225</u>, 542

HATTON, M.W.C., LAWRIE, R.A., RATCLIFF, P.W. and WAYNE, N. (1972). *J. Fd Technol.*, <u>7</u>, 443

HAZLEWOOD, C.F., NICHOLS, B.C. and CHAMBERLAIN, N.F. (1969). *Nature,* <u>222</u>, 747

HERRING, H.K., HAGGARD, J.H. and HANSEN, L.J. (1971). *J. Anim. Sci.*, <u>33</u>, 578

IVANOV, I.I., BELYAVTSEVA, L.M., IVANTEEVA, E.P. and MATVEEVA, I.M. (1969). *Biokhimiya*, <u>34</u>, 1184

HOFMANN, K. (1971). *Jahresbericht der Bundesanstalt für Fleischforschung, Kulmbach, 1971,* 106

KARMAS, E. (1973). *J. Fd Sci.*, <u>38</u>, 736

KARMAS, E. and DIMARCO, G.R. (1969). *Bull. Inst. Internat. Froid*, Suppl. 1969, 117

KARMAS, E. and DIMARCO, G.R. (1970). *J. Fd Sci.*, <u>35</u>, 615 and 725

LING, G. and NEGENDANK, W. (1970). *Physiol. Chem. Phys.*, <u>2</u>, 15

MARSH, B.B., WOODHAMS, P.R. and LEET, N.G. (1968). *J. Fd Sci.*, <u>33</u>, 12

NAKAMA, T. and SATO, Y. (1971). *J. Texture Stud.*, <u>2</u>, 475

NAKAMURA, R. (1970). *Jap. J. Zootech. Sci.*, <u>41</u>, 471

NERAAL, R. and HAMM, R. (1973). *Proceed. 19th Meat Europ. Res. Workers Meeting, Paris,* Vol. IV, 1419

OHASHI, T., SERA, H. and ANDO, N. (1972). *Bull. Fac. Agric., Miazaki Univ.,* <u>19</u>, 261; ref. *Fd Sci. Technol. Abstr.* (1973), 5, 9S1061

PAGANO TOSCANO, G. and AUTINO, C. (1970). *Atti della Soc. Ital. della Sci. Veter.*, <u>24</u>, 405

PALNITKAR, M.P. and HELDMAN, D.R. (1971). *J. Fd Sci.*, <u>36</u>, 1015

PARTMANN, W., FRANK, H.K. and GUTSCHMIDT, J. (1970).
  *Fleischwirtschaft,* 50, 1205
REDE, R. and HAMM, R. (1972). *Fleischwirtschaft,* 52,
  331
TAIT, M.J. and FRANKS, F. (1971). *Nature,* 230, 91
TYSZKIEWICZ, I. (1969). *Rocznicki Inst. Przemyslu
  Miesnego,* 6, 75
VADEHRA, D.V. and BAKER, R.C. (1970). *Fd Technol.,* 24,
  No. 7, 42
VAN HOOF, J. and HAMM, R. (1973). *Z. Lebensmittelunter-
  such.u.-Forsch.,* 150, 282
VRCHLABSKY, J. and LEISTNER, L. (1970). *Fleischwirtschaf*
  50, 967
WARDLAW, F.B., MCKASKILL, L.H. and ACTON, J.C. (1973).
  *J. Fd Sci.,* 38, 421

# 17

# TENDERNESS

## B. B. MARSH

*Muscle Biology Laboratory,
Department of Meat and Animal Science,
University of Wisconsin, Madison, USA*

## INTRODUCTION

The Carib Indians, whose unorthodox food preferences
gave us the word *cannibal*, were sensitive to meat
toughness and astute enough to do something about it.
According to Rouse (1963), their victims were roasted
very soon after death, the extremities then being
allocated to the women. We may perhaps surmise that the
Carib had discovered both the significantly decreased
toughness of meat cooked in a pre-rigor condition and a
simple distribution system assuring the males of the more
tender cuts.

Neither of these methods is open to us in our more
complex society. Nor can we hope any longer that
tenderness will be markedly improved merely by adding
to the existing catalogue of factors sometimes appearing
to be associated with it, an approach frequently
demonstrating only that the practical man was right all
along, though perhaps for the wrong reason (E.H. Callow,
personal communication). There remains but one way to
go if we are to make real progress in this refractory
area: to acquire a fuller understanding of the nature
of toughness itself by studying directly the components
responsible for it.

## MUSCLE STRUCTURE

Meat is muscle, and the primary function of muscle is to move. Movement is initiated and maintained by the contractile proteins, but would be confined to the myofibrillar level were it not for the supportive, adhesive and force-transmissive abilities of the connective tissue. Since the striated muscles of the mammalian body are each designed for a specific role in the living animal, it might be expected that their composition would vary, just as the great differences in their size, shape and anatomical complexity reflect the wide range of tasks they are required to perform. Systematic intermuscle differences are indeed found, and in particular the content of connective tissue varies appreciably with specialised function; they all conform to the same general morphological pattern, however, regardless of variations in the proportions of their components.

CONNECTIVE TISSUE

A muscle can be physically divided or teased into successively smaller longitudinal units, each of which is surrounded by a sheath of connective tissue; the muscle itself being surrounded by the epimysium, the fibre bundle by the perimysium, and the fibre by the endomysium and sarcolemma (*Figure 17.1*). The connective tissue around and within a muscle is much more than a supporting framework; it is also a protective wrap, a unit-divider, a bed within which the vascular and neural pathways reach every cell, and a restraint preventing over-extension and damage of the delicate contractile structure.

*Collagen*

The connective-tissue component of most interest to us is collagen, for this protein is a very significant contributor to the toughness of meat. Its basic structur unit is tropocollagen, a thin rod-like molecule of weight about 300,000, length 280 nm, and an axial ratio giving it, in Piez's (1966) appropriate analogy, the shape of a seventeen-foot length of one-inch hose. Tropocollagen is made up of three peptide chains wound round each other to form a very compact and strong unit (Bailey, 1968).

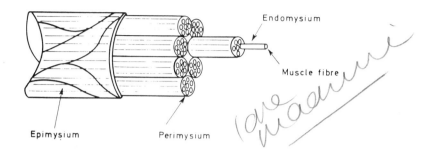

*Figure 17.1  Idealised diagram of muscle, fibre bundles and fibres, each level encased in a sheath of connective tissue (after Bailey (1972), by kind permission of the author)*

Collagen fibres appear to be composed of precisely aligned molecules of tropocollagen in a quarter-stagger form of construction, the units bonded together at intervals to prevent sliding when tension is applied. The crosslinks (*Figure 17.2*) responsible for this added strength are formed following the enzymic oxidation of specific lysine residues on adjacent but off-set molecules (Bornstein and Piez, 1966), and in the young animal most of them are sufficiently labile that they break easily when the tissue is heated (Bailey, Peach and Fowler, 1970). They are stabilised, however, as the

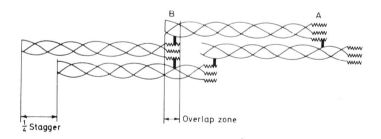

*Figure 17.2  Suggested architecture of a collagen fibre, showing the helical structure of the three peptide chains, the quarter-stagger formation of the tropocollagen molecules and the crosslinks preventing slip (after Bailey (1972), by kind permission of the author)*

animal ages, being gradually replaced by bonds which are
much more resistant to thermal rupture (Bailey and
Shimokomaki, 1971; Shimokomaki, Elsden and Bailey, 1972)
The nature of these crosslinks is of particular
significance to meat quality, not only in explaining some
of the curious empirical observations of past studies but
also in offering a possible means of eventual toughness
limitation.

*Elastin*

Connective tissue contains many components besides
collagen, but on the basis of quantity and structure the
only one of potential toughening significance is the
rubber-like protein, elastin (Partridge, 1966).  There i
still some uncertainty about the absolute amounts of
elastin present in various muscles (Bendall, 1967; Cross
Carpenter and Smith, 1973), but it is generally agreed
that elastin is present in smaller (perhaps much smaller
quantity than collagen, and that only the latter protein
contributes significantly to panel awareness of
connective tissue.  It may be safely assumed, therefore,
that it is the collagen rather than the elastin which
merits the greater attention in current tenderness
investigations.

CONTRACTILE SYSTEM

*Structure*

The three-dimensional network of connective tissue
pervades the muscle only to the level of the fibre
boundary.  To confront the other architectural feature
of significance in meat tenderness studies, a further
penetration must be made, since the contractile apparatu
is entirely intracellular.
   Within the fibre are found the long, thread-like
myofibrils, about 2000 of them, each about one micron in
diameter.  They are separately enwrapped in the
sarcoplasmic reticulum, a highly specialised mesh of
tubules concerned with calcium-ion control and hence
with the initiation and arrest of contraction.  Inside
the myofibril and still strictly in line with the fibre
direction, the filaments are arrayed in a configuration
(*Figure 17.3*) which has become increasingly familiar to

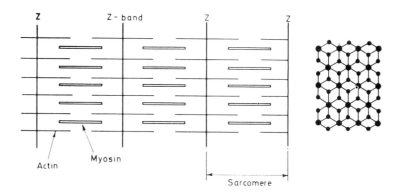

*Figure 17.3  The interdigitating array of thick (myosin)
and thin (predominantly actin) filaments and their
positions relative to the Z-line.  Left:  longitudinal
section.  Right:  transverse section at a point where
thick and thin filaments overlap  (after Bailey (1972),
by kind permission of the author)*

muscle and meat investigators since its first description
nearly 20 years ago (Hanson and Huxley, 1955).  Transverse
Z-lines divide the myofibril into (usually) regular units
or sarcomeres.  Extending longitudinally in both directions
from the Z-lines, and attached to them, are thin (or I)
filaments composed principally of the protein actin, their
free tips reaching somewhat less than half way along the
sarcomeres when the muscle is at its rest length.  The
thick, A, or myosin filaments span the gap between the
tips of opposing actin units, interdigitating with their
free ends; being appreciably shorter than the rest-length
sarcomere, they fail by some distance to reach the
Z-lines.

The contraction of muscle is essentially a shortening
of the sarcomere produced by the relative movement of the
two filament types.  The actin tips advance toward each
other, sometimes to the point of overlap and beyond, as
a result of a co-ordinated making and breaking of cross-
bridges between the thick and thin filaments; energy for
the contractile process being supplied by the calcium-
activated enzymic dephosphorylation of adenosine
triphosphate (ATP).  Despite the great and still increasing
complexity of this area of investigation (A.F. Huxley,

1971; H.E. Huxley, 1972; Needham, 1973), it is
unnecessary for our present purposes to go beyond a
recognition that contraction, relaxation and stretch
involve the longitudinal sliding of two discontinuous
filament types, the extent of their overlap determining
the length of the sarcomere and hence of the muscle
itself.

*Rigor mortis*

The slow rate of ATP dephosphorylation in living,
resting muscle continues into the post-mortem phase, but
restoration cannot keep pace for long with breakdown
under the anaerobic conditions prevailing in the dying
tissue. After a delay (varying among species from a few
minutes to many hours) during which the nucleotide level
remains almost constant, a fairly rapid ATP decline takes
place. Accompanying the fast phase, and in fact because
of it, cross-bridges form between the thick and thin
filaments. The bridges effectively prevent the sliding
of actin and myosin filaments past each other, just as
they must have done in life during contraction, for a
living contracted muscle would be of little value if it
were readily extensible. This time, however, in the
continued absence of ATP, the bonds are permanent, and
the muscle is in rigor mortis.

The time course of rigor onset in relation to
temperature is of great practical significance,
particularly in beef and lamb. Using the *l.dorsi* muscle
of a well-fed and relatively passive ox as an example,
we may note total times to rigor of about 4 h at 37°C,
16 h at 17°C, and 20 h at 7°C; for lamb the corresponding
times are about two-thirds of these values.

Our knowledge of rigor onset is built almost entirely
upon the studies of Bate-Smith and Bendall (1947, 1949)
and Bendall (1951), whose investigations of the process
in rabbit muscle were a model for later extensions to
the whale (Marsh, 1952), horse (Lawrie, 1953), ox (Marsh,
1954), lamb (Marsh and Thompson, 1958) and pig (Lawrie,
1960). The whole subject has been reviewed recently by
Bendall (1973).

*Post-mortem shortening*

Of all the post-mortem changes taking place before or
during rigor onset, it is the extent of shortening which
is of over-riding importance to meat tenderness.  Muscle
has a natural tendency to shorten during the rapid phase
of ATP breakdown, but because this *rigor shortening*
declines with falling temperature, the length change is
insignificant in the intermediate temperature range
reached by the cooling carcass at the time of its
occurrence.

A much more spectacular event is observed in *thaw
shortening*, produced when pre-rigor muscle is rapidly
frozen and later thawed, and first recorded over a
century ago (Walker, quoted by Perry, 1950).  If optimal
conditions are selected for its demonstration, a muscle
strip may shorten within minutes by 80% or more while
exuding almost 40% of its weight as 'drip' fluid (Marsh
and Thompson, 1957).  Thaw shortening has been observed
in species as far apart as frog (Moran, 1929), rabbit
(Szent-Györgyi, 1950), whale (Sharp and Marsh, 1953) and
lamb (Marsh and Thompson, 1958).  It appears to be due to
a rapid flux of calcium ions during thawing, these
stimulating the tissue to shorten in much the same way
as they provoke the living muscle to contract (Bendall,
1960; Kushmerick and Davies, 1968).

The third type of length change which muscle can
undergo is *cold shortening*, discovered by Locker and
Hagyard (1963).  Excised bovine muscles if cooled to
$0-5^{\circ}C$ well before rigor onset, shorten very appreciably –
sometimes by more than 50% – during a period of a few
hours (*Figure 17.4*).  The great bulk of the change
occurs quite quickly, before the rapid phase of ATP
breakdown (Cassens and Newbold, 1967).  The effect is
more species-selective than thaw shortening; beef
(Locker and Hagyard, 1963) lamb (McCrae *et al.*, 1971)
and the red muscles of the rabbit (Bendall, 1966) display
it well, and pork (Henderson, Goll and Stromer, 1970)
cold shortens to some extent, but rabbit white muscles
fail to respond at all (Locker and Hagyard, 1963).  As
with thaw shortening, this phenomenon declines with post-
mortem delay before cold application and is totally
absent if rigor is established before the muscle
temperature reaches about $5^{\circ}C$.

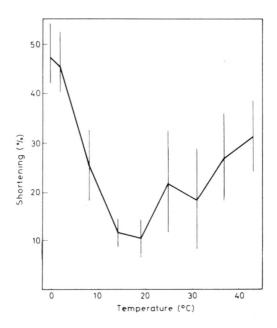

*Figure 17.4  The effect of temperature during rigor onset
on the extent of shortening in bovine neck muscles
excised shortly after slaughter.  Shortening as per cent
initial excised length.  Vertical lines:  standard
deviations (after Locker and Hagyard (1963), by kind
permission of Dr. R.H. Locker)*

## STRUCTURE AND TOUGHNESS

The structural elements, connective tissue and contractile
apparatus, affect meat tenderness in entirely different
ways:  the former by a slow and age-dependent increase in
the stability of intercollagen bonds during the life of
the animal, the latter by a rapid shortening-dependent
increase in the number of interfilament bonds following
its death, and during the dying of its musculature.

THE COLLAGEN CONTRIBUTION

It is almost 80 years since K.B. Lehmann established
that the toughness of meat is related to its content of
connective tissue (Mitchell, Zimmerman and Hamilton, 1927).
Although many investigators have since supported
Lehmann's observations, several studies have led to the
conclusion that little or no correlation exists. The
conflicting views up to the mid-sixties are well
summarised by Szczesniak and Torgeson (1965). Particularly
significant are the reports that veal contains more
connective tissue than beef, despite its greater tenderness (Bate-Smith, 1948; Wilson, Bray and Phillips, 1954).
More recently, it has been shown (Bendall and Voyle, 1969)
that the *l.dorsi* muscles of calves younger than 6 weeks
contain about three times the collagen percentage found
in those of older (6-24 months) animals.

These observations do not necessarily indicate that
collagen is unrelated to meat quality. Rather, they
suggest that the collagens of young and older animals
may make quite different contributions to toughness.
Studies in other fields provide a precedent for this
concept; collagen fibres prepared from the tails of young
rats, for instance, are prevented from heat shortening
(at 65°C) by quite light loading, but those prepared from
older rats are restrained only when the load is more than
trebled (Verzar, 1963). Several meat-tenderness
investigations have indeed indicated just this sort of
age influence. Thus the percentage of intramuscular
collagen solubilised by various treatments - collagenase,
or water at 70-100°C - is several times greater in veal
than in mature steers (Goll *et al.*, 1964a,b,c), and a
similar age effect has been demonstrated using the
muscles of sheep and pigs (Hill, 1966). It has also been
shown that the collagen forms which are soluble in neutral
salt solutions or dilute acids fall to low levels between
birth and 1-2 years of age (Carmichael and Lawrie, 1967).

The recent work of Bailey and his group has established
a firm chemical and structural foundation for these
observations relating animal age to ease of collagen
extractability. It is now clear that many of the
covalent bonds linking tropocollagen molecules are
relatively labile in the young animal, being easily
ruptured by pH changes, heat or denaturing agents; they
change with age, however, to a much more thermostable
form (Bailey, 1972). The three major reducible cross-
links of the collagen of both muscle and tendon decline

in amount until they are virtually absent at maturity
(Bailey and Shimokomaki, 1971; Shimokomaki, Elsden and
Bailey, 1972), a result supporting the suggestion
(Bailey, 1969) that these more labile bonds are
intermediates in the cross-linking process.

It is this decrease in the proportion of labile to
stable bonds - the 'quality' of the collagen - which is
of very real consumer concern, for it directly determines
the increasing resistance of the collagen fibre to
physical breakdown during cooking. This knowledge does
much more than merely explain declining tenderness with
increasing maturity; the structural information on which
it is based opens the way to possible modification in
life or manipulation during post-mortem storage. The
life of the 'youthful' form of collagen bonding might be
extended into maturity, for instance, or it may prove
possible to inhibit or retard cross-link information by
use of additives or a controlled nutritional deficiency.
With increasingly detailed knowledge of the offending
structures becoming available, an enzyme might be found
(or, eventually, constructed) which could modify the
properties of collagen either just before or shortly
after slaughter, so keeping the tenderising process under
the control of the processor. The recent demonstration
that hydroxyproline is essential to the integrity of the
collagen molecule (Jiminez, Harsch and Rosenbloom, 1973)
could suggest some means of fibre limitation or
modification .

THE SHORTENING CONTRIBUTION

Toughness has been associated with connective tissue
since the last century, but its dependence on length
changes in the contractile machinery was first
suggested little more than a decade ago. Locker (1959)
observed that the muscles of a hanging ox carcass enter
rigor mortis in widely differing states of contraction,
and in the following year he reported that relaxed
muscles are more tender than those which have shortened
during rigor onset (Locker, 1960). The later discovery
of the cold-shortening phenomenon (Locker and Hagyard,
1963) has been discussed earlier. Present knowledge of
the length/toughness relationship is built entirely on
these three remarkable contributions, which have introduced a new and exciting concept into tenderness research

Experiments on excised muscles have revealed the
magnitude of the toughening caused by cold-stimulated

shortening in beef (Marsh and Leet, 1966) and lamb (McCrae *et al.*, 1971). The rate of toughening with length decrease is slow at first, but rises with increasing shortening until, at about 40% change, the meat may be four times as tough as its unshortened control. With still further shortening, toughness declines because of a major rupturing of the structure produced by alternating zones of supercontraction and fracture (Marsh, Leet and Dickson, 1974).

A recent study (Marsh and Carse, 1974) has unexpectedly revealed a small but significant rise in toughness when the muscle is held at a slightly extended length during rigor onset. No advantage is to be gained, therefore, by stretching the muscle during the few hours following death - indeed, it could be slightly diadvantageous - unless a quite impractical extension almost doubling the muscle length is achieved. The same investigation has also provided support for the supposition that it is solely the configuration of the contractile proteins which determines the extra toughness over that caused by the 'background' connective tissue, for the entire complex shape of the length/tenderness relationship can be explained in terms of the varying overlap or stretch-fracture of thick and thin filaments (*Figure 17.5*).

If this pre-rigor shortening and its associated toughening were confined to excised muscles, it would be of little more than academic interest, for hot boning of beef or lamb is rarely practised. Unfortunately several major muscles are free to shorten even though still firmly attached to the skeleton, since suspension of the carcass by the hind leg allows them to develop a slack which can be taken up easily when a cold stimulus provokes a length change. That the phenomenon of muscle shortening is of considerable practical concern on a fairly large scale was demonstrated in 1966, when a promising trade in New Zealand frozen lamb was halted by the US importer because of excessive toughness. The defect was due to the well intentioned elimination of the customary cooling-floor delay, with consequent pre-rigor chilling and freezing of the carcass (Marsh, Woodhams and Leet, 1968).

Because of its smaller size, a lamb carcass cools much more rapidly than a beef side under the same ambient conditions, and so displays cold-induced toughening to a greater extent. Nevertheless a practical problem exists in that species also (Davey, 1970; Smith, Arango and Carpenter, 1971), and a worsening must be expected as faster and earlier post-mortem chilling methods are

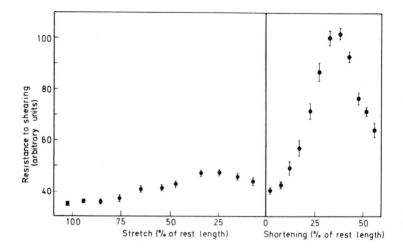

*Figure 17.5  The length/tenderness relationship for bovine neck muscles.  The muscles were either held in an extended position during rigor onset ('stretch'), or permitted to shorten in response to cold exposure applied at various times post mortem ('shortening').  Vertical lines:  standard errors.  Filament-length calculations indicate that:  (1) the minor peak occurs at about the point (24% stretch) at which the tips of the thin filaments first contact each other, and (2) the major peak coincides approximately with the point (35% shortening) at which interaction could occur between thin filaments and those thick filaments successfully penetrating the Z-line from the adjacent sarcomere. (after Marsh and Carse (1974), by kind permission of the Editor, Journal of Food Technology)*

introduced in the interests of improved hygiene (Ingram, 1972), reduced weight loss (Taylor, 1971), or accelerated processing.  Since the extent of ageing declines with shortening and becomes negligible if the earlier length change was high (Davey, Kuttel and Gilbert, 1967), the damage done by muscle shortening cannot be fully reversed by later treatment.

The full impact of shortening-provoked toughness is probably still to be felt, but some of its implications are already clear.  First, it is a major defect currently being 'built in' quite unwittingly to some of our meat, and likely to become more widespread if counter-measures are not introduced. Second, recognition of the effect clears the way for a much more precise approach to the whole subject of tenderness, since studies can now be designed to eliminate the confounding effect of shortening. In this connection we may note that the sometimes-observed beneficial effects of marbling and fat cover on tenderness may be little more than a reflection of the slower cooling rate (and hence reduced muscle shortening) of a larger and fatter carcass. Finally, there is a very real hope of appreciable tenderness improvement in the foreseeable future, for muscle is much more amenable to modification while it is dying than it was in life or will be in its final rigor condition.

Within the limitations of present knowledge, four general approaches to the problem of cold-induced shortening can be envisaged. It should be possible:

1. To delay chilling until rigor onset has 'locked' the primary filaments into a relaxed configuration.
2. To restrain the more seriously affected muscles during normal early post-mortem chilling.
3. To freeze so rapidly that cold shortening has no opportunity to take place, and then to maintain a slightly sub-freezing temperature allowing glycolysis to proceed in the presence of a restraining ice matrix.
4. To accelerate glycolysis and rigor onset so that only a relatively brief pre-chilling delay would be necessary.

It is an indication of the intense concern aroused by this problem that, in the few years since its recognition, all four of these possibilities have been examined and at least two of them are in full commercial operation.

The first, a pre-rigor delay, has been used in New Zealand for some years, its introduction being responsible for re-opening the American market for frozen lamb very soon after its enforced closing. The procedure, which has been described by Haughey and Frazerhurst (1972), results in an entirely satisfactory product, but it is a space-wasting, time-consuming and costly process, and hopefully is to be regarded as an interim measure (albeit a prolonged one) pending the application of further knowledge. Its extension to beef *in toto* is impractical because the much slower cooling rate of the larger carcass, although fast enough to cause toughening, would permit a too ready development of deep-spoilage organisms.  A modified conditioning treatment has been devised for beef, however, and market response has been favourable (Davey, 1970 and 1971).

The second possibility, physical restraint to prevent shortening, has led to two alternative methods.  In the first (Stouffer, Buege and Gillis, 1971), the *l.dorsi* muscle is maintained in a slightly stretched condition by an 'extenderiser', an extendible rod which is applied to the beef side or lamb carcass soon after slaughter to hold the relaxed configuration until rigor is established. In the second, the side of beef is suspended by the hip instead of the leg (Hostetler *et al.*, 1970), the hind limb then taking up a more natural posture which maintains the muscles at lengths approximating those in the living relaxed state.  The bunching-up and consequent ready shortening which occur in some muscles of the normally hung side are effectively avoided by this quite simple modification.  The general concept of altered posture was originally advanced by Locker (1960), and its validity was demonstrated by Herring, Cassens and Briskey (1965) with beef sides laid horizontally on their cut surfaces during rigor onset.  The hip-suspended process developed by the Texas group is now in large-scale commercial operation in Australia (Anon., 1973).  A similar method based on postural change has been tried experimentally for lamb in New Zealand, with encouraging initial reports (Davey, Gilbert and Curson, 1971).

The third approach, extremely rapid freezing followed by a holding period just below freezing point, stems from the early observations of Moran (1929) and Smith (1929), and has been used to prevent thaw shortening in several laboratory studies (Marsh and Thompson, 1958; Marsh, Woodhams and Leet, 1968; Behnke *et al.*, 1973a,b).  Although effective use of this method is limited at present to small samples, future developments in ultra-rapid freezing technology might permit eventual extension to commercial meat operations.

The final method by which the problem might be over-come, acceleration of rigor onset, offers several possibilities. The most obvious and simple way is to elevate the temperature, but bacterial proliferation and the possibility of *heat* shortening would make this a hazardous course.  The recent studies of MacFarlane (1973) in Australia indicate a potentially more practical method; a brief application of a pressure of about 1000 atmospheres to pre-rigor ox and sheep muscles greatly hastens glycolysis and rigor onset, and the meat is very significantly tenderised despite immediate post-pressure chilling. Explanations of both the accelerating and tenderising mechanisms are awaited with interest. In New Zealand, rigor acceleration in lamb is being attempted by post-mortem electrical stimulation. Carse (1972) found that a pH decline normally taking 16 h would occupy only 3 h when 250 volts were applied, and Chrystall and Hagyard (1973) have since reported a total conditioning time of only 2-3 h for a lamb carcass stimulated immediately after slaughter with a 3000-volt capacitive discharge, compared with 16-24 h for a control carcass. If these results can be obtained consistently, and provided high-voltage hazards can be eliminated, the Carse process would be a major advance, the more so since the rate appears to be readily controllable through voltage manipulation; problems arising from extremely rapid acid production such as may occur in porcine muscle are thus less likely to be encountered. It is of interest that a process for tenderising meat by early post-mortem electrical stimulation was developed some years before the shortening/toughening relationship was suspected (Harsham and Deatherage, 1951); the highly beneficial results were accredited at that time to the rapid development of 'an acid medium which aids or favors the action of certain enzymes upon both the connective tissue and the muscles'.


## CONCLUSION


Very considerable progress has been made during the past few years in our understanding of meat tenderness. The earlier oblique approach to the subject, with its emphasis on general influencing treatments, has given way to a direct confrontation with the specific influenced components collagen and actomyosin. This frontal attack has been responsible already for major advances in both areas of concern, and we may confidently expect that further progress will follow. For the connective-tissue component, the point of greatest vulnerability has been detected and largely characterised, and several ways in which the final assault might be mounted have been indicated. For the shortening component, a very large and previously unrecognised form of toughness has been revealed, and a number of methods by which the problem can be overcome have been devised.

This change in approach to tenderness research has come
only just in time. The introduction of new, faster, or
more efficient processes into meat plants is proceeding
at an increasing pace, frequently with little or no
consideration of their effects on eating quality. If
empirical studies have been so unproductive in the
relatively static conditions of the past, we can hardly
expect their record to improve significantly in response
to the much more dynamic situations of the present and
future.

REFERENCES

ANON. (1973). *Meat Producer and Exporter (Aust.)*, 28(5),
  5
BAILEY, A.J. (1968). In *Comprehensive Biochemistry*,
  1st edn. 26B, 297. Elsevier, Amsterdam
BAILEY, A.J. (1969). *Gerontologia*, 15, 65
BAILEY, A.J. (1972). *J. Sci. Fd Agric.*, 23, 995
BAILEY, A.J., PEACH, C.M. and FOWLER, L.J. (1970).
  *Biochem. J.*, 117, 819
BAILEY, A.J. and SHIMOKOMAKI, M. (1971). *FEBS Letters*,
  16, 86
BATE-SMITH, E.C. (1948). *J. Soc. Chem. Ind.*, 67, 83
BATE-SMITH, E.C. and BENDALL, J.R. (1947). *J. Physiol.*,
  106, 177
BATE-SMITH, E.C. and BENDALL, J.R. (1949). *J. Physiol.*,
  110, 47
BEHNKE, J.R., FENNEMA, O. and CASSENS, R.G. (1973a).
  *J. Fd Sci.*, 38, 539
BEHNKE, J.R., FENNEMA, O. and HALLER, R.W. (1973b).
  *J. Fd Sci.*, 38, 275
BENDALL, J.R. (1951). *J. Physiol.*, 114, 71
BENDALL, J.R. (1960). In *The Structure and Function of
  Muscle*, 1st edn., 3, 227, Academic Press, New York
BENDALL, J.R. (1966). In *The Physiology and Biochemistry
  of Muscle as a Food*, 1st edn. p.257. University of
  Wisconsin Press, Madison
BENDALL, J.R. (1967). *J. Sci. Fd Agric.*, 18, 553
BENDALL, J.R. (1973). In *The Structure and Function of
  Muscle*, 2nd edn. 2, 244. Academic Press, New York
BENDALL, J.R. and VOYLE, C.A. (1969). *J. Fd Technol.*,
  4, 275
BORNSTEIN, P. and PIEZ, K.A. (1966). *Biochemistry*, 5,
  3460
CARMICHAEL, D.J. and LAWRIE, R.A. (1967). *J. Fd Technol.*,
  2, 299

CARSE, W.A. (1972). *Ann. Res. Rep., Meat Ind. Res. Inst. N.Z.*, p.32

CASSENS, R.G. and NEWBOLD, R.P. (1967). *J. Fd Sci.*, 32, 269

CHRYSTALL, B.B. and HAGYARD, C.J. (1973). *Ann. Res. Rep., Meat Ind. Res. Inst. N.Z.*, p.38

CROSS, H.R., CARPENTER, Z.L. and SMITH, G.C. (1973). *J. Fd Sci.*, 38, 998

DAVEY, C.L. (1970). *Proc. Res. Conf., Meat Ind. Res. Inst. N.Z.*, p.73

DAVEY, C.L. (1971). *Proc. Res. Conf., Meat Ind. Res. Inst. N.Z.*, p.54

DAVEY, C.L., GILBERT, K.V. and CURSON, P. (1971). *Ann. Res. Rep., Meat Ind. Res. Inst. N.Z.* p.39

DAVEY, C.L., KUTTEL, H. and GILBERT, K.V. (1967). *J. Fd Technol.*, 2, 53

GOLL, D.E., BRAY, R.W. and HOEKSTRA, W.G. (1964a). *J. Fd Sci.*, 29, 622

GOLL, D.E., HOEKSTRA, W.G. and BRAY, R.W. (1964b). *J. Fd Sci.*, 29, 608

GOLL, D.E., HOEKSTRA, W.G. and BRAY, R.W. (1964c). *J. Fd Sci.*, 29, 615

HANSON, J. and HUXLEY, H.E. (1955). *Symp. Soc. Expt. Biol.*, ix, 228

HARSHAM, A. and DEATHERAGE, F.E. (1951). *U.S. Patent* 2, 544, 681

HAUGHEY, D.P. and FRAZERHURST, L.F. (1972). In *Meat Chilling - Why and How?* Symp. No. 2, Meat Res. Inst., Langford 32·1

HENDERSON, D.W., GOLL, D.E. and STROMER, M.H. (1970). *Am. J. Anat.*, 128, 117

HERRING, H.K., CASSENS, R.G. and BRISKEY, E.J. (1965). *J. Fd Sci.*, 30, 1049

HILL, F. (1966). *J. Fd Sci.*, 31, 161

HOSTETLER, R.L., LANDMANN, W.A., LINK, B.A. and FITZHUGH, H.A. (1970). *J. Anim. Sci.*, 31, 47

HUXLEY, A.F. (1971). *Proc. Roy. Soc.*, B, 178, 1

HUXLEY, H.E. (1972). In *The Structure and Function of Muscle*, 2nd edn. 1, 301. Academic Press, New York

INGRAM, M. (1972). In *Meat Chilling - Why and How?* Symp. No. 2, Meat Res. Inst., Langford 1·1

JIMINEZ, S., HARSCH, M. and ROSENBLOOM, J. (1973). *Biochem. Biophys. Res. Comm.*, 52, 106

KUSHMERICK, M.J. and DAVIES, R.E. (1968). *Biochim. Biophys. Acta*, 153, 279

LAWRIE, R.A. (1953). *J. Physiol.*, 121, 275

LAWRIE, R.A. (1960). *J. Comp. Path.*, 70, 273

aroma and flavour during cooking and eating, but meat
from the entire boar, particularly animals several years
old, sometimes has a characteristic odour called boar
taint or sex odour. When sex odour is referred to in
relation to the eating quality of the meat, it is the
odour which can sometimes be detected during cooking or
less frequently during eating, that is intended, and not
the smell of the live animal in the pen, nor that of the
preputial sac of a discarded service boar (Patterson,
1972).

The substance principally responsible for sex odour in
the cooked flesh is 5α-androst-16-en-3-one (Patterson,
1968), and it occurs mainly in the adipose tissue at low
concentrations. It is closely akin to the steroidal
hormones and is inescapably associated with the male
metabolism. Its physiological function is pheromonic,
imparting a characteristic odour to the boar's breath,
which enables him to elicit maximum response from an
oestrous female during courtship (Perry, Patterson and
Stinson, 1972).

In the young, six month old male pig, the concentration
of androstenone in the fat is approximately 1 µg/g
(1 ppm), and at this level the odour is not detectable
by many people either during cooking or eating of the
meat. The belief that the odour would be detectable and
objectionable to the consumer if male pigs were left
entire to pork or bacon weight has been shown to be
fallacious in extensive trials with consumer panels
(Rhodes, 1971c and 1972).

AGE

The flavour of cooked meat from a young animal is related
to the age of the animal at slaughter but not rigidly so.
This arises partly because of the changes in metabolism
which occur as the animal develops and because of the
natural variation between individuals, resulting in some
beasts reaching maturity earlier than others.

Calves up to about 6 months of age have neither the
typical beef flavour nor the intensity of flavour of
animals over 12 months. Various studies carried out
over the last 30 years have shown that full beef flavour
is developed up to about 18 months of age; thereafter,
between 18 and 30 months, age has no further significant
effect on flavour. For example, steaks from 30 month
steers were rated lower in all palatability aspects
except for flavour and juiciness when compared to steaks

LOCKER, R.H. (1959). *J. Biophys. Biochem. Cytol.*, <u>6</u>,
   419
LOCKER, R.H. (1960). *Fd Res.*, <u>25</u>, 304
LOCKER, R.H. and HAGYARD, C.J. (1963). *J. Sci. Fd Agric.*,
   <u>14</u>, 787
MCCRAE, S.E., SECCOMBE, C.G., MARSH, B.B. and CARSE, W.A.
   (1971). *J. Fd Sci.*, <u>36</u>, 566
MACFARLANE, J.J. (1973). *J. Fd Sci.*, <u>38</u>, 294
MARSH, B.B. (1952). *Biochim. Biophys. Acta*, <u>9</u>, 127
MARSH, B.B. (1954). *J. Sci. Fd Agric.*, 5, 70
MARSH, B.B. and CARSE, W.A. (1974). *J. Fd Technol.*,
   <u>9</u>, 129
MARSH, B.B. and LEET, N.G. (1966). *J. Fd Sci.*, <u>31</u>, 450
MARSH, B.B., LEET, N.G. and DICKSON, M.R. (1974).
   *J. Fd Technol.*, <u>9</u>, 141
MARSH, B.B. and THOMPSON, J.F. (1957). *Biochim. Biophys.
   Acta*, <u>24</u>, 427
MARSH, B.B. and THOMPSON, J.F. (1958). *J. Sci. Fd Agric.*,
   <u>9</u>, 417
MARSH, B.B., WOODHAMS, P.R. and LEET, N.G. (1968).
   *J. Fd Sci.*, <u>33</u>, 12
MITCHELL, H.H., ZIMMERMAN, R.L. and HAMILTON, T.S. (1927).
   *J. Biol. Chem.*, <u>71</u>, 379
MORAN, T. (1929). *Proc. Roy. Soc.*, B, <u>105</u>, 177
NEEDHAM, D.M. (1973). In *The Structure and Function of
   Muscle*, 2nd edn. <u>3</u>, 364. Academic Press, New York
PARTRIDGE, S.M. (1966). In *The Physiology and
   Biochemistry of Muscle as a Food*, 1st edn. p.327.
   University of Wisconsin Press, Madison
PERRY, S.V. (1950). *J. Gen. Physiol.*, <u>33</u>, 563
PIEZ, K.A. (1966). In *The Physiology and Biochemistry
   of Muscle as a Food*, 1st edn. p.315. University of
   Wisconsin Press, Madison
ROUSE, I. (1963). In *Handbook of South American Indians*,
   1st edn. <u>4</u>, 547. Cooper Square Publishers, New York
SHARP, J.G. and MARSH, B.B. (1953). *Spec. Rep. Fd Invest.
   Bd No. 58*, HMSO, London
SHIMOKOMAKI, M., ELSDEN, D.F. and BAILEY, A.J. (1972).
   *J. Fd Sci.*, <u>37</u>, 892
SMITH, E.C. (1929). *Proc. Roy. Soc.*, B, <u>105</u>, 198
SMITH, G.C., ARANGO, T.C. and CARPENTER, Z.L. (1971).
   *J. Fd Sci.*, <u>36</u>, 445
STOUFFER, J.R., BUEGE, D.R. and GILLIS, W.A. (1971).
   *US Patent* 3, 579, 716
SZCZESNIAK, A.S. and TORGESON, K.W. (1965). *Adv. Fd Res.*,
   <u>14</u>, 33
SZENT-GYORGYI, A. (1950). *Enzymologia*, <u>14</u>, 252

TAYLOR, A.A. (1971). *Proc. 17th Europ. Mtg. Meat Res. Workers,* p.357

VERZAR, F. (1963). *Sci. Amer.,* 208, (4), 104

WILSON, G.D., BRAY, R.W. and PHILLIPS, P.H. (1954). *J. Anim. Sci.,* 13, 826

# 18

# THE FLAVOUR OF MEAT

R. L. S. PATTERSON

*Meat Research Institute, Langford, Bristol*

## INTRODUCTION

Flavour, whether of meat or any other food, embraces
sensations arising from two distinct responses, those of
taste and smell, as well as less clearly defined
contributions from the pressure and heat sensitive areas
of the mouth. It is generally accepted that man can
recognise only four basic tastes, those of sweet, sour,
salt and bitter. In contrast, innumerable odours can be
distinguished by the human nose, deriving from volatile
chemical substances which, for reasons not yet fully
understood, have the special ability to stimulate the
olfactory receptors in the nasal cavity.

Uncooked meat has little odour and only a blood-like
taste, and cooking is necessary to develop the flavour.
Mastication breaks down the fibre matrix and releases
flavorous juices and volatile aroma compounds into the
mouth. Methodology for the separation and identification
of the inorganic ions and polar organic compounds present
in the juices, and the effects that they cause, is less
well developed than for their volatile counterparts, with
the result that the majority of studies on meat flavour
have been concerned with the aroma compounds. This is
perhaps unfortunate because the fullsome and satisfying
feel of the juice in the mouth plays an important part
in the appreciation of flavour, and omission of its
contribution may explain in part the lack of marked

success in attempts to fully define meat flavour on the
basis of volatile compounds alone.  There can be no doubt,
however, that odoriferous volatiles play a major role in
the formation of flavour in cooked meat; this can be
demonstrated easily by pinching the nose whilst chewing
meat, causing much of the 'flavour' to disappear.

Most research on meat flavour follows one or other of
two clearly defined paths.  One is concerned with the
effect on flavour of the history of the animal prior to
slaughter, its physiological condition at slaughter and
the post-mortem treatment of the meat.  In this type of
work, flavour quality is judged usually by taste panels.
The other approach is that of the analytical chemist using
an array of expensive and sophisticated equipment to
analyse flavour extracts  Lists of chemicals, with
occasional comments on the relevance of their odours,
result from this work.  It is only recently that serious
attempts have been made to combine these two approaches;
Swedish workers have analysed the aroma of canned beef
by physico-chemical techniques and have related it
successfully to sensory assessments (Persson and von
Sydow, 1972; Persson, von Sydow and Åkesson, 1973a and
1973b).

## THE INFLUENCE OF THE ANIMAL ON MEAT FLAVOUR

The pre-slaughter factors affecting the flavour include
species, breed, sex, age, fatness and feed.  Other
factors, such as induced stress at slaughter, post-
mortem ageing, storage and cooking procedure may also
affect the final flavour, but are not further discussed
in this paper.

### SPECIES

Since animal muscle has universally the same active
proteins and biochemical mechanism, any inter-species
differences in muscle composition are small, and hence
the basic flavour precursor substances present are also
very similar.  It is perhaps not surprising, therefore,
that it is very difficult to differentiate between the
flavour of cooked beef, lamb and pork if pieces of very
lean meat are cooked identically and tasted (Pearson
*et al.*, 1973).  Similarly, if water-extracts of the three

meats are heated, the odours are again very similar
(Hornstein and Crowe, 1960 and 1963). When pieces of
fatty tissue are heated in the same way, different
odours are evolved which are typical of the three species
and can be readily identified; if the lipids are
extracted and washed free of water soluble material the
identification is less exact (Wasserman and Spinelli,
1972). The gas chromatographic separation of volatiles
from the fats of beef, pork and lamb, however, shows
qualitatively similar patterns for all three, and no
individual peak in any case has an odour typical of the
species.


BREED

Relatively few comparative studies on the effect of breed
on flavour have been reported, largely because of the
difficulties of obtaining suitable groups of animals and
of ensuring uniform treatment during raising.  However,
in the United States in 1939, data on eating quality
were collected from over 700 heifers and steers of the
Hereford, Angus, Shorthorn and Brahman breeds, and the
flavour quality of the lean was found to be significantly
related to breed (Barbella, Tannor and Johnson, 1939).
Later Branaman *et al.* (1962) compared beef-type Hereford
with the dairy-type Holstein cattle raised under similar
conditions, and whilst he found no significant difference
in the general desirability of the flavour between the
beef and dairy breeds, the intensity of the flavour was
rated significantly greater for the beef breed.
    More recently, Rhodes (1969a) has compared the eating
quality of Hereford x Friesian with pure Devon steers
and found no differences in flavour or odour.  Similarly,
Aberdeen Angus steers and heifers, raised in Scotland,
were paired with Hereford x Friesian steers raised in
Somerset. Although the Scotch beef was found to be
significantly more tender, no significant differences
were recorded for flavour (Rhodes, 1971a).


SEX

Traditionally in the United Kingdom, male animals not
required for breeding are castrated at an early age
prior to fattening, and females and castrates are
considered to produce meat of comparable eating quality.
While these practices still continue, some of the more

progressive breeders have abandoned castration of pigs
and cattle in the last year or two, and are now raising
males as entires.  Experience is showing that this
change in practice can be of considerable advantage to
the producers in terms of faster growth rates, improved
feed conversion, reduced overall consumption of feedstuffs, enhanced lean to fat ratios and, consequently,
cuts of meats which more closely provide, without trimming,
the low fat, high lean requirements of today's housewives.
But such action almost invariably meets with resistance
from both the wholesale and retail elements of the British
meat trade on the grounds that meat from male animals,
whether bull, boar or ram, always possesses an unpleasant
'taint' in the flavour, characteristic of the male.
Whilst pigs are a special case, the results of experimenta
work on cattle and sheep lends little support to this
allegation.

Bryce-Jones *et al.* (1964) compared the eating quality
of ten pairs of bull and steer twins, and found that,
although the steer meat was more flavoursome than that of
the bulls, there was no suggestion of taint or off-odour
in the bull meat. / In the United States, Reagen *et al.*
(1971) confirmed these results, finding that the flavour
of 12 and 16 month bull beef was less intense than that
of equivalent steers; however, they found that the meat
from the older bulls was less palatable than that from
the equivalent steers or the younger bulls.  Joints from
Hereford x Friesian bulls and steers, slaughtered at 15
months of age, have been cooked by two different methods
and compared (Crystall, 1971).  Flavour scores were not
found to differ significantly between bulls and steers
and, while there was an overall acceptability in favour
of the steers, there was no evidence of any undesirable
odour or flavour associated with the bull meat.  One
conclusion of the study was that variation was often
greater within than between the sex groups.

With sheep, even fewer comparisons have been reported
but there is no firm evidence to support allegations that
meat from uncastrated males has less acceptable flavour
than that from females and castrates.  In one trial,
involving a controlled comparison between sibling rams
and wethers, slaughtered to give a carcass weight of
30 lb, panellists were quite unable to identify the meat
from the ram when tasting roasted joints (Rhodes, 1969b).

In the case of the pig, it is known that the sex of the
animal can influence the flavour, or more accurately, the
odour of the hot meat.  Meat from the females and
castrated males displays what is considered to be normal

from 18 month old steers (Simone, Carroll and Chichester,
1959).  Similarly, taste-panel studies revealed no
significant differences in the flavour of meat from 18,
42 and 90 month female beef animals (Tuma, 1963; Tuma
*et al.*, 1962).  Evidence therefore suggests that once
maturity has been achieved flavour remains largely
unchanged in the normal animal.


FATNESS

A good covering of fat on a beef carcass and a high
degree of intramuscular (marbling) fat are considered by
many to be indicative, if not almost a guarantee, of good
flavour and good eating quality in cooked beef.  For
example, the fat content of beef carcasses has long been
an important factor in the official United States
Department of Agriculture grading scheme which is overtly
intended to predict eating quality; a very high fat level
is necessary to qualify for the highest quality grade.
The eating quality of meat of differing levels of
fatness has been compared by taste-panel evaluation of
flavour, juiciness, texture  etc.  The results, however,
are conflicting.  On the one hand, studies by Tuma *et al.*
(1962) involving 24 Hereford heifers demonstrated the
absence of a significant relationship between degree of
marbling and taste-panel scores for flavour and juiciness
Similarly other workers found that lean flavour did not
differ between the well-marbled 12th rib and the poorly-
marbled 10th rib steaks from either top or poor grade
carcasses (Doty and Pierce, 1961).  In contrast,
Henricksen and Moore (1965), in a comprehensive study of
steer beef derived from 6 to 92 month old beasts, found
fatness had a significant influence on flavour and
juiciness in older animals aged 42 to 92 months but was
unimportant up to 18 months; similarly, in 60 heifer or
cow carcasses, marbling level was found to affect flavour
(Breidenstein *et al.*, 1968).

   Jeremiah *et al.* (1970) have summarised and analysed
the results of fifteen studies carried out in the United
States involving 1136 beef carcasses.  Correlation
coefficients for the studies between intramuscular fat
and flavour were:  1 in the range 0.6 to 0.4, 3 between
0.4 and 0.2, 9 between 0.2 and -0.2, and 2 between -0.2
and -0.4.  They calculated an overall correlation
coefficient on a weighted basis for all data, and its
value was 0.11.  Similarly, the values for the overall
weighted correlation coefficient between flavour and USDA

quality grade was 0.07 and between flavour and marbling
score, 0.10.  Although objections can be raised to these
latter methods, the data as a whole show no clear
evidence of the influence of fat content on flavour.

Supporting results have been obtained recently at the
Meat Research Institute.  Carcass meat from 24 bulls,
212 steers and 53 heifers, covering a range of fatness
from 10 to 50 on an arbitrary scale (correlating closely
with fat as a percentage of carcass weight), has been
examined and analysis within these sexes showed no
significant relationship between taste-panel assessment
of flavour or of flavour intensity with fatness (Rhodes,
unpublished results).  In a carefully controlled
experiment (Rhodes, 1973) 14 Hereford x Friesian steers
were raised to produce different fatness levels at age
17 months.  They were graded on the MLC fatness scale,
six as Grade 1, two as Grade 2, four as Grade 3 and
three as Grade 4, with total fat contents (excluding
kidney knob) of 11 to 28%.  When individual joints were
tasted, not more than 25% of the variability in flavour
scores was accounted for by the level of fatness.
Excessive fat, therefore, is not necessary in a carcass
for the lean meat to have good flavour.  However, it is
obvious from tasting fat separately that it does have
flavour, and as has been suggested by several studies
(Hornstein and Crowe, 1964; Wasserman and Talley, 1968;
Wasserman and Spinelli, 1972) it is specific to the
animal species concerned and is essential for the
formation of the full, characteristic flavour of the
meat.  Thus while fat is essential for flavour, it
appears that in most cases there is sufficient lipid
material available in lean meat to allow development of
normal flavour, and that quantities of fat in excess of
this are of little consequence to flavour development.

FEED

The traditional method of raising cattle in the United
Kingdom is on pasture with a winter period in yards
using hay or silage and various supplements.  A variety
of forage is fed to beef animals and only rarely do
complaints about flavour arise; odoriferous compounds
occurring naturally are generally lipophilic, unsaturated
and usually contain one or more of the principal hetero
atoms, e.g. oxygen, nitrogen or sulphur.  Such compounds
are readily modified in the powerful reducing conditions
of the rumen and the products are then subjected to the

normal digestive processes. These compounds, and those
which survive unchanged from the forage and are deposited
in the carcass fat, are presumably part of the normal
chemical spectrum and subsequently may play either an
active or passive role in flavour formation.

In modern raising techniques, animals are kept indoors
and supplied with a high energy diet designed to sustain
maximum growth rate. These intensively fed animals are
slaughtered at younger ages and tend to have lean
carcasses devoid of excessive fat. Both the fat and the
lean are often pale in appearance and this has probably
given rise to the prejudice that modern beef is insipid
compared with the traditional material. This criticism
has been extended to flavour but comparisons both in the
United States and in this country have produced little
support for the belief. For example, Hereford x Friesian
steers raised on a barley concentrate ration for twelve
months were compared with a similar group raised on grass
for twenty-four months and with a group of Devon steers;
no significant differences in flavour were detected
(Rhodes, 1969a). Similarly, groups of Friesian x Ayrshire
heifers were raised intensively or on grass and were
slaughtered at either 18 or 24 months, again no
significant differences in flavour were detected in
direct comparisons (Rhodes, unpublished results).

In other trials, the use of field beans (Rhodes, 1970)
and 25% poultry waste (Rhodes, 1971b) as substitutes for
conventional protein supplements in the fattening ration
has been studied in separate experiments. In both cases,
panellists were unable to distinguish flavour or odour
differences between the cooked meat from the treated and
control animals. The use of 20% brewers' grains, however,
did result in small differences in flavour (Rhodes, 1971d)

A positive effect of feed on flavour can be clearly
seen in lambs. Evidence from studies in Australia and
New Zealand shows that compounds in the forage are
rapidly deposited, either native or modified, in the fat
and can greatly affect the flavour acceptability of the
cooked meat. Shorland *et al.* (1970) showed that lambs
fed white clover had a stronger flavour in the fat and
lean and a greater intensity of odour in the casseroled
12th rib chop than lambs fed on perennial ryegrass; these
differences in flavour became apparent about three weeks
after starting the separate treatments. Significant
differences were found in the flavour characteristics of
lamb raised on either lucerne or phalaris pastures
(Park, Corbett and Furnival, 1972). Rape in forage
produced meat with a nauseating aroma and flavour, oats

gave a pungent note and vetch a more intense meaty
flavour compared with meat from similar sheep grazed on
ryegrass and white clover or native pastures; the oats
and rape both gave meats with significantly lower flavour
acceptability (Park, Spurway and Wheeler, 1972).
A highly significant difference was recorded for the
intensity of flavour between meat from Finnish x Scottish
half-bred wethers raised indoors on barley concentrate
plus hay, and grass-fed controls; however, since the
results bracketed closely the 'ideal' intensity point of
the scale and the difference between them was less than
required to predict a significant effect in terms of
consumers eating in domestic circumstances, the result
was considered not to be of commercial importance
(Rhodes, 1971e).

   The use of feed additives can introduce abnormal
flavours, a recent example being the proprietary growth
promoter containing the hormones diethylstilbestrol and
methyltestosterone.  Studies have shown that this
combination can give rise to an undesirable flavour in
pork meat, described variously as oily, burnt, rubber or
faecal (Elliott, 1971; Patterson, 1972).  However, while
this particular disadvantage of the product has been
recognised, there is little evidence to suggest that
this odour difference has caused significant adverse
consumer reaction.


## THE CHEMISTRY OF MEAT FLAVOUR

### DISTRIBUTION OF PRECURSORS

Some of the earliest flavour work showed that lean meat
flavour originated in the water-soluble fraction of
muscle (Kramlich and Pearson, 1958) and, in the absence
of fat, that lean meat extracts of beef and pork
(Hornstein and Crowe, 1960), lamb (Hornstein and Crowe,
1963) and whale (Hornstein, Crowe and Sulzbacher, 1963)
had virtually the same basic meaty aroma when heated to
100°C.  The presence of the corresponding fat was judged
to be necessary for the development of the full species-
specific aroma and flavour (Hornstein and Crowe, 1963;
Wasserman and Talley, 1968).  Pepper and Pearson (1971)
found some differences in the proteins present in lean
and adipose tissue and suggested that the water-soluble
fraction from adipose tissue might make a distinct and
characteristic contribution to meat flavour in addition

to any effects arising from the lipids *per se*.

Wasserman and Spinelli (1972) extracted adipose tissue
(which produced the characteristic aroma of the species
on heating) with chloroform-methanol and then washed the
extract with water.  The original extract possessed the
species-specific aroma but after water-washing, the
characteristic aroma was either missing or greatly reduced.
The water soluble material from the lipid contained amino
acids and at least one sugar, glucose, in the carbohydrate
fraction, and developed a typical lean meat aroma on
heating.  Preliminary gas chromatographic analyses of the
extracted fats and water-washes showed primarily
quantitative differences among the meat species.  These
findings are largely in agreement with those of Pepper
and Pearson (1971), and with later findings by Pearson
*et al.* (1973) who concluded, from taste-panel studies on
beef and lamb, that the species differences in flavour
were extremely subtle and difficult to distinguish and
appeared to originate in the fatty tissue.


THE NATURE OF THE PRECURSORS

Flavour precursors are distributed between the lean and
the fat, the species-specific elements most probably
residing in the fat.  The adipose tissue has been shown
to contain, in addition to triglycerides, amino acids,
glucose, (Wasserman and Spinelli, 1970 and 1972),
ribonucleic acids, nucleoproteins and glycoproteins
(Pepper and Pearson, 1971).  Extensive studies by Macy,
Naumann and Bailey (1970) have shown that the water-
extractable components of muscle include amino acids,
nitrogen bases, carbohydrates, inorganic phosphate and
various nucleotides.

Separation of raw beef constituents by gel permeation
chromatography produced 12 fractions, seven of which
produced recognisable broiled beef aroma of different
intensities when heated.  The two fractions with the most
intense aroma represented nearly 80% of the diffusate
and contained the sulphur-containing amino acids,
methionine and cysteic acid (cystine + cysteine), and a
reducing sugar, 2-deoxy-D-ribose (Mabrouk, Jarboe and
O'Conner, 1969).  The procedures were sufficiently
refined to demonstrate differences in precursor contents
in various muscles.

## CHEMICAL REACTIONS OF PRECURSORS

When raw meat is heated, a series of physical and
chemical changes take place, involving both the lean
and the fat. The fat melts, thereby becoming more liable
to chemical reaction; in the lean, collagen structures
disintegrate, soluble proteins coagulate as the
temperature increases above 60°C, but myofibrillar
structures remain largely intact. Browning reactions
commence at about 90°C, and increase with time and
temperature, and with the concentration of sugar in the
meat (Pearson *et al.*, 1962). With further increase in
temperature, the surface of the meat dehydrates and
pyrolytic reactions commence. The reactions which occur
between and within the major classes of compounds present
in meat have been reviewed recently by Wasserman (1972)
and, together with the reaction products can be considered
under the following headings: amino acids; carbohydrates;
amino-acid - carbohydrate interactions; lipids; nitrogen,
oxygen and sulphur compounds and nucleotides.


## AMINO ACIDS

In a quantitative study of the amino acids in raw and roasted
beef, Macy, Naumann and Bailey (1970, Pt 5) found that
most free amino acids increased in concentration during
cooking of roasts to an internal temperature of 170°F
(77°C) with the exception of threonine, serine, glutamic
acid, histidine and arginine. They attributed this
general increase in amino-acid content to hydrolysis of
the protein, probably involving cathepsins or other
proteolytic enzymes in the tissue, since they had shown
previously that most free amino acids decreased during
heating when isolated from the tissue by dialysis (Macy,
Naumann and Bailey, 1964). This increase therefore
probably represents the net difference between amino
acids formed and destroyed during cooking at these
temperatures. Wasserman and Spinelli (1970) however,
found that no aroma developed in model systems containing
only amino acids after drying at 125°C and that 80-90%
of the amino acids remained unchanged. Temperatures of
about one hundred degrees in excess of normal roasting
temperatures are required to initiate pyrolytic
decarboxylation and deamination reactions. Merritt and
Robertson (1967) have listed the major products derived
from 17 amino acids in a study using pyrolysis-gas
chromatography and mass spectrometry; for example,

3-methyl- and 2-methyl-butanal derive from leucine and iso-leucine, 2-methylpropanal from valine, and benzene, toluene and ethylbenzene from phenylalanine. Such compounds have been identified in cooked meat volatiles, but may of course arise also by other reaction mechanisms. An interesting finding was that the products formed during pyrolysis of dipeptides and dipeptide pairs depended on the sequence of the amino acids; if this pertains for longer peptides, it may explain in part why aromas derived by heating mixtures of single amino acids do not reproduce faithfully those formed by heating meat extracts of apparently similar constitution.

## CARBOHYDRATES

Carbohydrates are important precursors of flavour in many foods; at high temperatures (*c*. 300°C), sugars undergo caramelisation, resulting in the formation of many highly odoriferous and pungent substances, among which furan derivatives, carbonyl compounds and aromatic hydrocarbons are prominent (Fagerson, 1969). At lower temperatures, (100-130°C) bound water is lost without alteration of the molecular structure; at temperatures between 150 and 180°C, a molecule of water cleaves from the sugar leaving the anhydride, followed by a second molecule of water at about 200°C, yielding oxygenated furans. Macy, Naumann and Bailey (1970, Pt 5), in a comparison of eight different cuts of beef, found that the total carbohydrate content after roasting was significantl reduced ($p < 0.05$, 10 replicates) in all but one cut.

## AMINO ACID-CARBOHYDRATE INTERACTIONS

Of reactions leading to the development of meat flavour probably the most important occur in this category. Wasserman and Spinelli (1970) showed that, in the absence of sugars, no aroma or colour formation occurred when a solution of amino acids was dried *in vitro* at 125°C. Addition of glucose to the initial solution, however, resulted in large losses of amino acids and the development of odour. The same authors found that, after boiling the diffusate from dialysed meat extracts for 30 min, the concentration of ribose decreased by about 20% although only small changes occurred in the concentration of the majority of amino acids (except arginine and cysteic acid which were each reduced by

approximately 60%). When the same diffusate was dried
at 125°C for 15 min, meaty aroma and brown colouration
developed rapidly with nearly complete elimination of
ribose, glucose and fructose, and 40-60% losses of most
amino acids. Therefore, although the amino acids and
carbohydrates can each be shown to undergo reactions
separately which yield odoriferous products, relatively
mild heating only is required for interaction when the
reactants are mixed together at the correct relative
humidity.

Browning is the general term used to describe the
reaction between amino acids or protein material and
sugars, first described by Maillard (1912). The reaction
parameters have been studied in detail and the subject
reviewed extensively (Hodge, 1953 and 1967; Reynolds,
1963 and 1965; Spark, 1969; Stewart, 1969).

Other routes may lead to the formation of flavour and
browning products, for example, aldol condensation of
aldehydes formed as products of Maillard reactions, or
polymerisation of $\alpha,\beta$-unsaturated aldehydes and amines
which occurs readily at temperatures about -10°C to yield
resins and complex condensation products. Although all
these reactions are known to occur in model systems, the
extent of the contribution of each to flavour production
and browning in meat and other foods is still not known.
However, it is clear that innumerable intermediate
products can be formed in the reactions between amino
acids and sugars before final production of melanoidins,
and it is well established that aliphatic aldehydes and
ketones, heterocyclic compounds, pyrazines, furan and
pyrrole derivatives, contribute substantially to the
aroma and flavour of foods.


CONTRIBUTION OF LIPIDS

The role of lipids in meat flavour has already been
mentioned; it appears that the species-specific flavours
may originate from subtle differences in the amino-acid
and carbohydrate content of the adipose tissue (Wasserman
and Spinelli, 1970 and 1972). Apart from constitutional
differences, lipids may contribute to flavour in other
ways, for example, by acting as solvents for extraneous
odorants derived from the feed; the effect is particularly
marked in the case of sheep. Products of the metabolism
may accumulate in the fat and give rise to characteristic
odours in certain animals, for example, androstenone may
occur in the fat of entire male pigs but not in that of

the females or castrated males. The presence of such
compounds becomes noticeable when they are volatilised
by heat; no chemical change in the fat is necessary for
their release.

Forss (1969) has reviewed the role of lipids in flavours
and their effect on flavour perception, and lists
*n*-alkanals, *n*-2-alkenals, *n*-2,4-alkadienals, ketones,
lactones, alcohols and lower fatty acids as the most
frequently occurring classes of compounds. Saturated and
unsaturated aldehydes and ketones were identified in the
volatiles from beef, pork and lamb fats heated to 100°C
in air (Hornstein and Crowe, 1960 and 1963), but the
classes and concentrations of the various carbonyls
differed between the species. Deca-2,4-dienal, which has
a pronounced oily aroma, was prominent in both beef and
pork fats, but was absent in lamb fat. Volatile
monocarbonyl compounds derived from lamb, beef and pork
fats were present in the ratio 1:5:14.

Aldehydes and ketones have also been reported in beef
fat by Sanderson, Pearson and Schweigert (1966), and by
Yamato *et al.* (1970), although it was concluded in the
latter paper that none of the carbonyls possessed
characteristic beef aroma. In a series of investigations
Watanabe and Sato reported the identification of gamma-
and delta-lactones in beef-fat volatiles (1968) and in
pork fat (1969). Later, the same authors (1971b)
studied beef-fat volatiles using gas chromatography and
mass spectrometry and identified 70 non-acid volatiles
using gas chromatography and mass spectrometry and
identified 70 non-acid volatiles comprising aldehydes
and ketones, lactones, alcohols, esters, hydrocarbons
and pyrazines. Many strong odours were recognised
during analysis of the fractionated material, and the
authors suggested tentatively that normal beef-fat aroma
probably resulted from an appropriate blend of aldehydes,
ketones, esters and (unspecified) sulphur compounds;
they did not include the pyrazines.


COMPOUNDS CONTAINING NITROGEN, OXYGEN AND SULPHUR

Of the many substances identified in the volatiles of
cooked meat, the nitrogen-containing pyrazines and
several sulphur- or oxygen-containing compounds have very
low odour thresholds and are believed to play a significant
part in the formation of meaty aroma.

*Pyrazines*

The pyrazines are a class of unsaturated 6-membered ring
compounds containing nitrogen atoms at positions 1 and 4.
Maga and Sizer (1973) have reviewed their flavour
properties and occurrence in foods.  They may arise from
Maillard reactions, but recently Wang and Odell (1973)
found that the presence of reducing sugars was not
essential; aliphatic aminohydroxy compounds yield
pyrazines when heated individually.

The identification of pyrazines in meat volatiles has
been reported only recently by Watanabe and Sato (1971a,b)
who found 9 alkyl-substituted pyrazines in the volatiles
from beef fats and shallow fried beef, three of which
were also found by Liebich *et al.* (1972) in boiled beef.
17 pyrazines were identified in the pyrolysed aqueous
extract of beef (Flament and Ohloff, 1971), and Mussinan,
Wilson and Katz (1973) identified a total of 33 pyrazines
in pressure-cooked beef.  Extraction of the volatiles of
shallow fried beef with hydrochloric acid reduced
considerably the roasted odour of the total extract.
Pyrazines would be removed by this treatment and it seems
probable that they are responsible to a large degree for
the 'roasted' element of dry-cooked meat odour.  Their
occurrence in boiled beef, however, suggests that their
flavour role is not exclusively the provision of
empyreumatic odours.


*Oxygen and sulphur compounds*

The oxygen- and sulphur-containing compounds identified
in meat volatiles have been discussed by Herz (1968);
Tonsbeek, Plancken and van de Weerdhof (1968); and Herz
and Chang (1970).  Prior to 1968, mainly carbonyls,
sulphides and disulphides had been described, none of
which was considered to possess characteristic meaty
aroma.  Since then, many compounds containing one or
more hetero atoms have been identified, including
thiophenes, thiazoles and polysulphur heterocyclics.
Wilson *et al.* (1973) found 46 sulphur-containing substances
in pressure-cooked beef.

Tonsbeek, Plancken and van de Weerdhof (1968)
identified 4-hydroxy-5-methyl-3(2H)-furanone and its
2,5-dimethyl homologue in beef broth.  Neither had a
meaty aroma, the former was described as roasted chicory
root and the latter as caramel.  Later, Tonsbeek,
Copier and Plancken (1971) isolated 2-acetyl-2-thiazoline

which had an intense smell of freshly baked bread.  Chang
*et al.* (1968) identified four substances in the volatiles
of boiled beef which had meat-like odours at low
concentration:  2,4,5-trimethyl-3-oxazoline; 3,5-dimethyl-
1,2,4-trithiolane; 5-thiomethyl-2-furaldehyde; and
thiophene-2-carboxaldehyde.

Various furans and pyrrole derivatives have been
reported by Liebich *et al.* (1972) and Watanabe and Sato
(1972).  Although these compounds were identified in
fractions which possessed meaty aromas, no individual
compound had a pronounced meaty odour.

1-methylthioethanethiol and 2,4,6-trimethyl-perhydro-
1,3,5-dithiazine (thialdine) have been isolated and
identified in the headspace gas of simmering beef broth
(Brinkman *et al.*, 1972).  Odours were described as 'fresh
onion' and 'dry camphor' respectively in concentration,
however, in very dilute aqueous solution 1-methylthio-
ethanethiol has a meaty odour.  Herz and Chang (1970)
described a similar change from sulphury to meaty with
change in concentration of 3,5-dimethyl-1,2,4-trithiolane,
although Brinkman described both the *cis* and *trans* isomers
of the compound as onion-like.


5'-RIBONUCLEOTIDES

The nucleotides are a group of substances which have
attracted much attention in the past because of their
apparent flavour enhancing or modifying properties,
especially in the field of meat flavour since inosinic
acid was discovered amongst flavour precursors isolated
from raw meat (Batzer *et al.*, 1960; Batzer, Santoro and
Landmann, 1962).  Shimazono (1964) suggested that
inosine-5'-monophosphate (IMP) was probably the most
important single contributor to meat flavour but this
hypothesis has not found general support.  In fact, most
studies since have shown that IMP and other nucleotides
are capable of modifying existing flavours rather than
contributing intrinsic notes.

Nucleotides occur in raw muscle; Wismer-Pedersen (1966)
found 210 mg 5'-ribonucleotides per 100 g raw beef which
decreased to 160 mg after roasting; similarly for pigs, the
concentration decreased from 180 to 165 mg/100 g pork.  M
Naumann and Bailey (1970, Pt 3) detailed similar changes
for individual 5'-nucleotides, inosinic acid decreasing
in beef from 278 to 170 mg per 100 g dry tissue during
cooking to an internal temperature of 77°C, and from 209

to 33 mg in lamb cooked at 60°C for one hour. Cytidylic, uridylic and guanylic acids, present initially in very much smaller concentrations, changed little during cooking; adenylic acid increased by a factor of 2 or 3 in all three species. Overall, the nucleotides decreased as cooking and flavour development progressed.

IMP also decreases in meat during post-rigor storage, at the time when potential meat flavour is only developing (Caul, 1957), suggesting that IMP cannot be significantly involved in the production of cooked flavour. Also, Rhodes (1965) found no correlation between nucleotide degradation and flavour change, nor any detectable effect on the flavour of cooked minced meat resulting from the addition of IMP. In meat, therefore, it appears that nucleotides do not play a major part in the formation of final cooked flavour.

## SUMMARY

Meat flavour has not yet been fully defined. Over 100 compounds of at least 10 chemical classes have been identified, many having zero or only weak odour of a non-meaty nature when examined in isolation, thus apparently making no obvious contribution to the flavour. However, synergistic and antagonistic effects may prevail in *milieu* when these apparently ineffective substances may be important as flavour modifiers. Compounds which are most likely to be involved in the formation of basic meat flavour are those with low odour thresholds, for example, sulphur compounds, certain pyrazines and a few carbonyls. The relevance of some of the more complex nitrogen-, oxygen- and sulphur-containing substances has not yet been fully established.

Pre-slaughter factors can affect flavour: beef animals slaughtered at an early age have only weak (veal) flavour compared to mature 18 month beef. The composition of forage can cause flavour variation, especially in lambs, often of an adverse nature; beef animals, however, seem to be more resistant to flavour adulteration from feedstuffs. Tainted swill or the use of feed additives (e.g. hormones) may introduce unusual flavour notes into pig flesh. Overall lean-meat flavour is not determined by degree of fatness, although some fat appears to be essential to provide the species-specific character. Breed has no significant effect on flavour, nor has the sex of cattle or sheep; taint, however, may be detectable during the cooking of meat from old boars.

REFERENCES

BARBELLA, N.G., TANNOR, B. and JOHNSON, T.G. (1939).
*32nd Ann. Proc. Am. Soc. Anim. Prod.*, p.320
BATZER, O.F., SANTORO, A.T., TAN, M.C., LANDMANN, W.A.
and SCHWEIGERT, B.S. (1960). *J. agric. Fd Chem.*, 8,
498
BATZER, O.F., SANTORO, A.T. and LANDMANN, W.A. (1962).
*J. agric. Fd Chem.*, 10, 94
BRANAMAN, G.A., PEARSON, A.M., MAGEE, W.T., GRISWOLD, R.M.
and BROWN, G.A. (1962). *J. Anim. Sci.*, 21, 321
BREIDENSTEIN, B.B., COOPER, C.C., CASSENS, R.G., EVANS,
G. and BRAY, R.N. (1968). *J. Anim. Sci.*, 27, 1532
BRINKMAN, H.W., COPIER, H., de LEUW, J.J.M., and TJAN,
S.B. (1972). *J. agric. Fd Chem.*, 20, 177
BRYCE-JONES, K., HARRIES, J.M., ROBERTSON, J. and AKERS,
J.M. (1964). *J. Sci. Fd Agric.*, 15, 790
CAUL, F. (1957). *U.S. Quartermaster Food and Container
Inst., Surveys Prog. Military Subsistence Problems*,
Ser. 1, (9), p.152
CHANG, S.S., HIRAI, C., REDDY, B.R., HERZ, K.O., KATO, A.
and SIPMA, G. (1968). *Chem. and Ind.*, p.1639
CRYSTALL, B.B. (1971). *Proc. 17th European Meeting of
Meat Res. Workers*, pp.87-90, Bristol
DOTY, D.M. and PIERCE, J.C. (1961). *U.S. Dept. Agric.,
AMS, Tech. Bull.*, 1231
ELLIOTT, M.K. (1971). Thesis, University of Aberdeen
FAGERSON, I.S. (1969). *J. agric. Fd Chem.*, 17, 747
FLAMENT, I. and OHLOFF, G. (1971). *Helv. chim. Acta*, 54,
1911
FORSS, D.A. (1969). *J. agric. Fd Chem.*, 17, 681
HENRICKSEN, R.L. and MOORE, R.E. (1965). *Oklahoma St.
Univ. Tech. Bull.*, T-115
HERZ, K.O. (1968). Thesis, Rutgers State University,
USA
HERZ, K.O. and CHANG, S.S. (1970). *Adv. Fd Res.*, 18, 1
HIRAI, C., HERZ, K.O., POKORNY, J. and CHANG, S.S. (1973).
*J. Fd Sci.*, 38, 393
HODGE, J.E. (1953). *J. agric. Fd Chem.*, 1, 928
HODGE, J.E. (1967). 'Nonenzymatic Browning Reactions',
*Chemistry and Physiology of Flavours*, p.465, Ed. H.W.
Schultz, E.A. Day and L.M. Libbey. Publ. Avi Publ. Co.
Inc., Conn., USA
HORNSTEIN, I. and CROWE, P.F. (1960). *J. agric. Fd Chem.*
8, 494
HORNSTEIN, I. and CROWE, P.F. (1963). *J. agric. Fd Chem.*
11, 147

HORNSTEIN, I. and CROWE, P.F. (1964). *J. Gas Chromatog.*, 2, 128

HORNSTEIN, I., CROWE, P.F. and SULZBACHER, W. (1963). *Nature, Lond.*, 199, 1252

JEREMIAH, L.E., CARPENTER, Z.L., SMITH, G.C. and BUTLER, O.D. (1970). *Tech. report No. 22*, Texas A & M Univ. Coll. Stat., Texas

KRAMLICH, W.E. and PEARSON, A.M. (1958). *Food Res.*, 23, 567

LAW, H.M., YANG, S.P., MULLINS, A.M. and FIEDLER, M.M. (1967). *J. Fd Sci.*, 32, 637

LIEBICH, H.M., DOUGLAS, D.R., ZLATKIS, A., MUGGLER-CHUVAN, F. and DONZEL, A. (1972). *J. agric. Fd Chem.*, 20, 96

MABROUK, A.F., JARBOE, J.K. and O'CONNER, E.M. (1969). *J. agric. Fd Chem.*, 17, 5

MACY, R.L., NAUMANN, H.D. and BAILEY, M.E. (1964). *J. Fd Sci.*, 29, 142

MACY, R.L., Jr., NAUMANN, H.D. and BAILEY, M.E. (1970). *J. Fd Sci.*, 35, Pt 3, p.78; Pt 4, p.81; Pt 5, p.83

MAGA, J.A. and SIZER, C.E. (1973). *J. agric. Fd Chem.*, 21, 22

MAILLARD, L.C. (1912). *Compt. rend.*, 154, 66

MARSHALL, N. (1960). *J. Home Econ.*, 52, 31

MERRITT, C., Jr and ROBERTSON, D.H. (1967). *J. Gas Chrom.*, 5, 96

MUSSINAN, C.J., WILSON, R.A. and KATZ, I. (1973). *J. agric. Fd Chem.*, 21, 871

PARK, R.J., CORBETT, J.L. and FURNIVAL, E.P. (1972). *J. agric. Sci., Camb.*, 78, 47

PARK, R.J., SPURWAY, R.A. and WHEELER, J.L. (1972). *J. agric. Sci., Camb.*, 78, 53

PATTERSON, R.L.S. (1968). *J. Sci. Fd Agric.*, 19, 31

PATTERSON, R.L.S. (1972). *Vet. Ann.*, 13, 37

PATTERSON, R.L.S. (1972). Unpublished results

PEARSON, A.M., HARRINGTON, G., WEST, R.G. and SPOONER, M.E. (1962). *J. Fd Sci.*, 27, 177

PEARSON, A.M., WENHAM, L.M., CARSE, W.A., MCLEOD, K., DAVEY, C.L. and KIRTON, A.H. (1973). *J. Anim. Sci.*, 36, 511

PEPPER, F.H. and PEARSON, A.M. (1971). *J. agric. Fd Chem.*, 19, 964

PERRY, G.C., PATTERSON, R.L.S. and STINSON, C.G. (1972). *Proc. VIIth Int. Congr. Anim. Reprod. and Art. Insem.*, pp.395-399, Munich

PERSSON, T. and von SYDOW, E. (1972). *J. Fd Sci.*, 37, 234

PERSSON, T., von SYDOW, E. and ÅKESSON, C.A.J. (1973a). *J. Fd Sci.*, 38, 386

PERSSON, T., von SYDOW, E. and ÅKESSON, C.A.J. (1973b).
 *J. Fd Sci.*, 38, 682
REAGEN, J.O., CARPENTER, Z.L., SMITH, G.C. and KING, G.T.
 (1971). *J. Anim. Sci.*, 32, 641
REYNOLDS, T.M. (1963). *Advan. Fd Res.*, 12, 1
REYNOLDS, T.M. (1965). *Advan. Fd Res.*, 14, 167
RHODES, D.N. (1965). *J. Sci. Fd Agric.*, 16, 447
RHODES, D.N. (1969a). *Meat Res. Inst. Ann. Rep.*, *1968-69*,
 p.34. Agric. Res. Council, London
RHODES, D.N. (1969b). *Meat Production from Entire Male
 Animals*, pp.189-198. Ed. D.N. Rhodes, Churchill,
 London
RHODES, D.N. (1970). *Meat Res. Inst. Ann. Rep.*, *1969-70*,
 p.16. Agric. Res. Council, London
RHODES, D.N. (1971a). *Meat Res. Inst. Ann. Rep.*, *1970-71*,
 p.28. Agric. Res. Council, London
RHODES, D.N. (1971b). *J. Sci. Fd Agric.*, 22, 436
RHODES, D.N. (1971c). *J. Sci. Fd Agric.*, 22, 485
RHODES, D.N. (1971d). *Meat Res. Inst. Ann. Rep.*, *1970-71*,
 p.29. Agric. Res. Council, London
RHODES, D.N. (1971e). *J. Sci. Fd Agric.*, 22, 667
RHODES, D.N. (1972). *J. Sci. Fd Agric.*, 23, 1483
RHODES, D.N. (1973). *Meat Res. Inst. Mem.*, No. 15,
 Bristol
SANDERSON, A., PEARSON, A.M. and SCHWEIGERT, B.S. (1966).
 *J. agric. Fd Chem.*, 14, 245
SHIMAZONO, H. (1964). *Fd Technol.*, 18, 294
SHORLAND, F.B., CZOCHANSKA, Z., MOY, M., BARTON, R.A. and
 RAE, A.L. (1970). *J. Sci. Fd Agric.*, 21, 1
SIMONE, M., CARROLL, F. and CHICHESTER, C.O. (1959).
 *Fd Technol.*, 13, 337
SPARK, A.A. (1969). *J. Sci. Fd Agric.*, 20, 308
STEWART, T.F. (1969). BFMIRA Scientific and Technical
 Surveys, No. 61
TONSBEEK, C.H.Th., PLANCKEN, A.J. and van de WEERDHOF, T.
 (1968). *J. agric. Fd Chem.*, 16, 1016
TONSBEEK, C.H.Th., COPIER, H. and PLANCKEN, A.J. (1971).
 *J. agric. Fd Chem.*, 19, 1014
TUMA, H.J. (1963). *Dissertation Abstr.*, 24, 1563
TUMA, H.J., HENRICKSON, R.L., STEPHENS, D.F. and MOORE,
 R. (1962). *J. Anim. Sci.*, 21, 848
WANG, P-S. and ODELL, G.V. (1973). *J. agric. Fd Chem.*,
 21, 868
WASSERMAN, A.E. (1972). *J. agric. Fd Chem.*, 20, 737
WASSERMAN, A.E. and SPINELLI, A.M. (1970). *J. Fd Sci.*,
 35, 328
WASSERMAN, A.E. and SPINELLI, A.M. (1972). *J. agric. Fd
 Chem.*, 20, 171

WASSERMAN, A.E. and TALLEY, F. (1968). *J. Fd Sci.*,
 <u>33</u>, 219
WATANABE, K. and SATO, Y. (1968). *Agric. Biol. Chem.*,
 <u>32</u>, 191
WATANABE, K. and SATO, Y. (1969). *Agric. Biol. Chem.*,
 <u>33</u>, 242
WATANABE, K. and SATO, Y. (1971a). *J. agric. Fd Chem.*,
 <u>19</u>, 1017
WATANABE, K. and SATO, Y. (1971b). *Agric. Biol. Chem.*,
 <u>35</u>, 756
WATANABE, K. and SATO, Y. (1972). *J. agric. Fd Chem.*,
 <u>20</u>, 175
WILSON, R.A., MUSSINAN, C.J., KATZ, I. and SANDERSON, A.
 (1973). *J. agric. Fd Chem.*, <u>21</u>, 873
WISMER-PEDERSEN, J. (1966). *J. Fd Sci.*, <u>31</u>, 980
YAMATO, T., KURATA, T., KATO, H. and FUJIMAKI, M. (1970).
 *Agric. Biol. Chem.*, <u>34</u>, 88

# 19

## DEVELOPMENT OF A RADIO-IMMUNOASSAY FOR MEASURING BOAR-TAINT STEROID 5α-ANDROST-16-EN-3-ONE

### R. CLAUS

*Institut für Physiologie, Technische Universität, München-Weihenstephan, West Germany*

The substance which is mainly responsible for boar taint has been isolated and identified as the steroid 5α-androst-16-en-3-one (androstenone). This compound (*Figure 19.1*) is very similar to the androgens, though no androgenic activity for the boar-taint steroid was found with the chick-comb-assay.

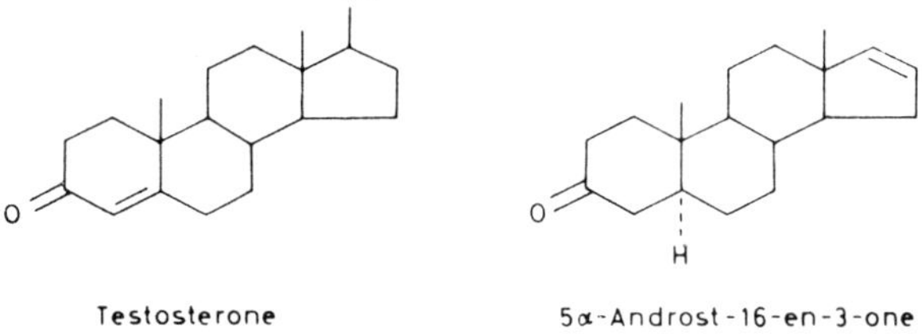

Figure 19.1 Testosterone and 5α-androst-16-en-3-one

Androstenone is detected by organoleptic examination
of fatty-tissue samples, although no quantification has
been attained, or by very time-consuming gas-liquid-
chromatographic (GLC) measurements.  Therefore we have
elaborated a rapid and sensitive radio-immunoassay (RIA)
for measuring the steroid in fatty tissue.  This test
enables the determination of up to 50 samples/week by one
person.

A radio-immunoassay depends on a reaction between
specific antibodies against the compound which should be
determined, and the compound itself.  The specificity of
the antibody allows the determination of the boar-taint
steroid in crude fatty-tissue extracts, which are only
subjected to a short solvent distribution to remove the
fat.

The method was first applied to fat samples from three
boars, one gilt and one barrow, in which the contents of
androstenone had been estimated previously by GLC
(*Table 19.1*).  A very good correlation ($r = 1.0$) was obtain

*Table 19.1*  5α-Androst-16-en-3-one in fatty tissue:
comparison of results obtained by RIA and GLC

| Number of animals | Sex | Age (days) | Weight (kg) | μg 5α-androst-16-en-3-one/g fatty tissue | |
|---|---|---|---|---|---|
| | | | | RIA | GLC |
| 39/1 | male | 175 | 98 | 1.13 | 1.03 |
| 23/19 | male | 217 | 122 | 6.94 | 7.49 |
| 15/50 | male | 229 | 138 | 1.78 | 1.74 |
| 15/58 | male castrated | 236 | 136 | 0.20 | 0.15 |
| 23/22 | female | 238 | 137 | 0.16 | 0.20 |

The test was also checked for reliability criteria.
The specificity, the reproducibility and the precision
were confirmed.  The sensitivity allows the determination
of less than 50 pg, which means that very small amounts
of fatty tissue, readily obtained by biopsy, are
sufficient for the determination.

In *Figure 19.2* one example for such determinations is
outlined.  The samples were obtained by biopsy.  It was
possible to demonstrate the increasing stores of androstenone in fatty tissue of 5 boars.  Strong individual
variations of boar-taint stores are obvious.  By RIA it
is also possible to measure androstenone in gilts and
barrows.  Perhaps those small amounts – which are not
regarded as 'unpleasant' – contribute to the typical
flavour of pork.

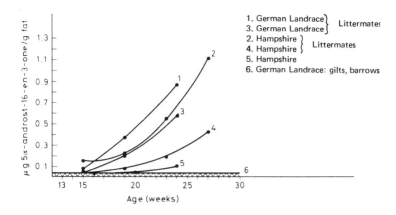

Figure 19.2  Development of 'taint' stores in fatty tissue with age

# 20

# MEAT COLOUR: THE IMPORTANCE OF HAEM CHEMISTRY

## C. L. WALTERS

*British Food Manufacturing Industries Research Association,
Leatherhead*

## INTRODUCTION

Haem proteins represent the chief colour constituents
of meat, the predominant components being myoglobin, of
molecular weight approximately 16,700, and haemoglobin,
with a molecular weight of about 67,000. Both of these
compounds contain one or more haem groups in which an
iron atom is placed centrally within a ring composed of
four pyrollic rings. Four of the co-ordination positions
of each iron atom within the plane of the tetrapyrollic
ring are satisfied by the nitrogen atoms of the five-
membered rings involved, the fifth is co-ordinated to an
imidazole residue contained within the protein structure
whilst the sixth co-ordination position is occupied by
either another imidazole residue or a water molecule or
a ligand such as oxygen or nitric oxide.

Both myoglobin and haemoglobin can exist in a reduced
deoxygenated form, purple red in appearance, which
combines reversibly with oxygen to form oxy-
derivatives, that have a brighter cherry-red tint
characterised by the colour rapidly acquired by freshly
cut meat. The capacity of haemoglobin to combine with
oxygen is dependent upon the oxygen tension, the pH over
a narrow range and its combination with carbon dioxide.
The last property facilitates the release of oxygen from

oxyhaemoglobin in the circulating blood, when it reaches
the tissues rich in carbon dioxide, and accentuates its
ability to combine with oxygen, when it reaches the lung
alveoli and releases its carbon dioxide. The increased
proton concentration resulting from a fall of pH weakens
the bond between oxygen and the iron atoms of haem groups
and facilitates the dissociation of the oxy- form.
Myoglobin, however, does not respond similarly to the
presence of carbon dioxide.

Controversy may exist concerning the valency state of
the iron atoms in oxymyoglobin and oxyhaemoglobin but
there is little doubt that its oxidation to the ferric
state as in metmyoglobin or methaemoglobin produces
forms which are brown or grey in appearance and which
are inactive in the physiological transport of oxygen.
This oxidation can readily be brought about by oxidising
agents such as ferricyanide or nitrite. It was noted by
Neill and Hastings (1925) that the rate of spontaneous
oxidation of haemoglobin was accelerated at low partial
pressures of oxygen. Later on, the reaction was more
fully investigated by Brooks (1935) and George and
Stratmann (1952) who found that the maximum rate of
oxidation of haemoglobin and myoglobin occurred at
partial pressures of oxygen in the range of 1-20 mm of
mercury, depending upon the pigment, pH and temperature.
In fact, the most convenient method for the oxidation of
oxymyoglobin without introducing oxidising agents is to
bubble through its solution for several hours a steady
stream of nitrogen or argon or some similar gas which
displaces the oxygen in solution and leads thereby to
reduced oxygen tensions.

The action of the curing agent, nitrite, on haem
pigments is interesting. Under aerobic conditions, it
rapidly oxidises oxyhaemoglobin to the met- form; the
action is a coupled one, with the oxidation of both
oxyhaemoglobin to methaemoglobin and of nitrite to
nitrate. Under anaerobic conditions, however, nitrite
can react with haemoglobin to form both methaemoglobin
and nitrosylhaemoglobin, the complex of nitric oxide
with haemoglobin in the ferrous form (Brooks, 1937);
the proportions of the two pigments formed are dependent
on the nitrite concentration employed (Marshall and
Marshall, 1945). This reaction has been extended in the
literature to myoglobin, without apparent justification
until recently. The only products observed in the
interaction of pig oxymyoglobin and nitrite, whether
under aerobic or anaerobic conditions, have been
metmyoglobin and, at high nitrite levels, its nitrite

salt, which differs somewhat in its absorption spectrum. In 1971, however, Koizumi and Duane Brown produced evidence of the formation of nitrosylmyoglobin during the anaerobic interaction of deoxygenated myoglobin and nitrite, although the spectral differentiation between nitrosyl- and oxymyoglobin was uncertain. Any difference between myoglobin and haemoglobin in this respect may result from the much greater tendency of the former to oxidise spontaneously in the absence of oxygen, a factor which is extremely difficult to control experimentally. Thus, it would appear that the formation of nitrosyl-myoglobin from oxymyoglobin and nitrite certainly requires the reduction of the latter to nitric oxide.

## THE INFLUENCE OF MEAT COLOUR ON CONSUMER SELECTION

Many if not most of the discolouration problems associated with fresh meat arise from the partial oxidation of oxymyoglobin and oxyhaemoglobin to their respective ferric met- forms. Within a survey of 141 Dublin housewives, for instance, 72% stated that they disliked 'darker red-brown beef', i.e. beef with a noticeable level of metmyoglobin (Riordan, 1973), since they considered such discoloured meat to be of deteriorating quality. It is extremely difficult, however, to dissociate completely preference based on colour from that associated with other qualities of meat. 24% of the housewives approached stated that they liked beef discoloured in this way because they expected it to be tender. However, a clear discrimination against round steak in which the pigment was subject to partial oxidation was revealed in a study conducted in two supermarkets (Hood and Riordan, 1973). A linear relationship was established between the level of discolouration in terms of metmyoglobin content over the range 5-33% and the proportion of total sales made up by the discoloured meat. When 20% metmyoglobin was present in the meat displayed for selection in a random manner amongst similar rounds of bright red meat of fresh appearance, the ratio of sales of discoloured to 'normal' beef was approximately 1:2. In order to provide a range of conversions of oxymyoglobin to metmyoglobin, the test steaks were held at room temperature for periods up to 16 h, whilst the control steaks were placed immediately and held in a refrigerator at 0°C.

So far as cured meat is concerned, consumer
discrimination can again be levelled at products in
which the haem pigment shows signs of oxidation, which
can, for instance, be brought about by the use of
excessive nitrite or by an impairment in the curing
process.   Furthermore, the cured-meat haem pigment is
photolabile probably because of the accelerated
dissociation of the nitric oxide ligand on irradiation
with light particularly in the presence of oxygen.   The
appeal of such products to the consumer can be strongly
influenced by an obvious variation in the haem pigment
contents of individual muscles in juxtaposition to each
other.

Thus, the valency of the iron atoms contained in the
myoglobin and haemoglobin of meat can play a vital role
in the colour and thence the attractiveness of meat
whether fresh or cured.   On a molecular basis, the
concentration of such haem pigments in pork is only of
the order of 0.02 mM so that very small amounts of a
deleterious reactant such as an oxidising agent are
sufficient to cause severe discolouration so far as the
housewife is concerned.


# FACTORS INVOLVED IN THE COLOUR OF MEAT

Muscle colour varies with species, beef being the
darkest, lamb intermediate, and pork the lightest;
muscle colour intensity also increases with advances in
the chronological age of the animal, as a result, in part,
of a rise in myoglobin concentration (Price and
Schweigert, 1971).   The colour of beef muscle varies from
a pale pink colour to an extremely dark colour of dark
cutting beef, which is discounted at the retail level
because it is not possible for the housewife to make a
distinction between it and beef from old animals.   A high
level of physical activity in a muscle is usually
accompanied by a high myoglobin concentration (Lawrie,
1964).

The colour of meat can be greatly influenced by the
morphology of the muscle (Price and Schweigert, 1971).
Dark-cutting beef, for instance, appears dark because
its surface will not scatter light to the same extent as
will the more 'open' surface of meat of lower ultimate
pH (Lawrie, 1964).   In the high pH muscle, the fibres are
swollen and tightly packed together (Bate-Smith, 1948),
presenting a barrier to the diffusion of oxygen; the

higher pH also favours respiratory activity. As a result, the surface layer of bright-red oxymyoglobin becomes much thinner, and the underlying purplish-red deoxygenated myoglobin becomes readily apparent. This type of muscle has not been found to have abnormal pigments or to have an unusual retention of haemoglobin (Lawrie, 1958).

At the other extreme, pale soft exudative (PSE) muscle involves a marked denaturation of sarcoplasmic and myofibrillar proteins resulting from a rapid post-mortem fall in pH while the muscle temperature remains high (Price and Schweigert, 1971). At the low ultimate pH, muscle myoglobin is readily oxidised to metmyoglobin, which does not contribute greatly to the depth of colour; furthermore, the structure of PSE muscle is 'open' and scatters light (MacDougall, 1970). One individual muscle may contain a gradual or even an abrupt transition from normal to PSE character (Briskey, 1964).

The striated or fibrous nature of muscle can lead to iridescence on the surface of sliced cured or cooked meats such as cold roast beef. The 'mother of pearl' effect results from the breaking up of incident white light in the manner of a diffraction grating; the phenomenon is not of any concern, but can cause unwarranted suspicions about the edibility of affected cooked meats.

A further contribution to the appearance of meat exposed for sale is that of the associated fat; for instance, the deficiency of fat in the carcasses of young bulls was stated (MacDougall and Rhodes, 1972) to give an unfinished appearance. White fat is considered to present evidence of high quality, since yellow fat is usually associated with meat obtained from older, dairy cows (Price and Schweigert, 1971). Pink or green discolourations are occasionally encountered in the fat of cured meat; these are probably due to the metabolic products of halophilic bacteria (Lawrie, 1964). The overall redness of a meat surface is increased significantly with increased marbling fat (Romans, Tuma and Tucker, 1965), owing presumably to greater contrast between lean and fat placed in juxtaposition.

Thus, it will be apparent that the colour presented by both fresh and cured meat is not only a property of the component haem pigments, their concentration and form, but also of the morphological structure and its effect upon the absorption and scattering of incident light and to its organisation in terms of proximity of lean and fat. This is not to mention completely unrelated factors such as the browning of pork resulting from Mailliardtype reactions between reducing sugars and amino groups produced by post-mortem enzyme action (Sharp, 1957 and 1958).

The death of the animal results in the cessation of the supply of circulating blood which maintains muscle myoglobin and haemoglobin in the oxygenated form. The respiration of the tissue continues, however, to scavenge residual oxygen with the production of carbon dioxide. On the surface of the meat, the access to oxygen is maintained and thus the tendency for the oxidation to metmyoglobin and methaemoglobin is minimal. In the deep interior of the muscle, no replenishment of oxygen is possible and there oxymyoglobin and oxyhaemoglobin are rapidly deoxygenated with the formation of the purple-red pigments, reduced myoglobin and haemoglobin. In between these two zones, the diffusion of oxygen from the exposed surface is able to maintain its partial pressure just below the surface at the low value most conducive to spontaneous oxidation to the met- pigments and thus a narrow brown shell can result behind the bright cherry-red of the surface. This effect is accentuated if the meat is wrapped in a film which limits the replenishment of oxygen from the surface. As the oxygen permeability of the film decreases, the partial pressure where oxygen utilisation by the tissues balances oxygen penetration at a pressure level which favours oxidation of the pigments approaches the surface more closely. Landrock and Wallace (1955) determined that a packaging film must have an oxygen penetration rate of at least five litres per square metre per day to prevent surface browning. The enzyme systems responsible for the continued respiration of excised muscles can also effect the reduction of metmyoglobin and methaemoglobin. Studie have been made of the capacities of ground beef to reduce metmyoglobin but in most instances an agent has been added, such as ferricyanide (Stewart *et al.*, 1965), capable of acting also as an electron carrier and therefore of distorting the issue. In general, the autoxidation of reduced myoglobin to metmyoglobin predominates at low oxygen tensions, since the ferric form is the more stable, unless the ferrous form is stabilised in some way as it is formed. For instance, the reduction of metmyoglobin by muscle respiratory enzymes proceeds only with great reluctance under an atmosphere of nitrogen but readily under carbon monoxide which combines with reduced myoglobin as soon as it is produced. In practical terms, a mixture of 2% carbon monoxide and 98% air has been used (El-Badawi *et al.*,

1964) to flush packaged beef before sealing the packages
and was found to be very effective in preserving and
stabilising the colour of fresh beef for 15 days. The
amount of carbon monoxide bound to the meat pigment was
most unlikely to provoke any toxicological response within
the blood of the consumer.

The primary action of nitrite in curing is the
oxidation of oxymyoglobin to metmyoglobin, without,
apparently, the production of an equivalent amount of
nitric oxide. Furthermore, Fox and Thomson (1963) found
that the formation of the nitric oxide required for the
cured meat pigment could not be accounted for by the
molecular dismutation of nitrous acid. It has been
postulated that certain elements of the muscle
respiratory enzyme system can bring about the reduction
of nitrite to nitric oxide in combination with metmyoglobin, the ferric form of the haem pigment (Walters,
Casselden and Taylor, 1967). Nitrosylmetmyoglobin is a
red pigment, with similar spectral characteristics to
nitrosyl-(ferro)-myoglobin; it is unstable in air,
decomposing into metmyoglobin and presumably nitrogen
dioxide. Whereas the ferric form, metmyoglobin, of
uncomplexed myoglobin is more stable than the ferrous,
reduced or deoxygenated myoglobin, the reverse applies
for the two complexes of the haem pigment with nitric
oxide. Thus, nitrosylmetmyoglobin is readily reduced by
muscle respiratory enzyme systems and other biological
components to nitrosyl-(ferro)-myoglobin; in fact, the
reduction of nitrosylmetmyoglobin to the ferrous complex
proceeds spontaneously in aqueous solution, in a similar
manner to nitrosylmethaemoglobin (Keilin and Hartree,
1937).

*Figure 20.1* demonstrates the changes in endogenous
haem pigment in model cures, analysed spectrophotometri-
cally in the Soret region of 400-450 nm wavelength,
during the anaerobic incubation at pH 6.0 of minces of a
pig *quadriceps femoris* muscle with 5-50 ppm (0.072-0.72
mM) sodium nitrite. Rapid oxidation of much endogenous
oxymyoglobin ($MbO_2$) and oxyhaemoglobin ($HbO_2$) to
metmyoglobin (MetMb) and methaemoglobin (MetHb) was
apparent throughout; the initial residues of $MbO_2 + HbO_2$
decreased with increasing nitrite concentration, falling
immediately to zero at a level of 50 ppm $NaNO_2$, and
remaining unaltered over two hours. At an initial
concentration of 5 ppm sodium nitrite, a gradual increase
of $MbO_2 + HbO_2$ was observed on analysis in air, probably
originating from reduced myoglobin and/or haemoglobin

formed during the anaerobic incubations simulating the interior of a block of meat, in spite of a residue of 0.7 ppm NaNO$_2$ or thereabouts. Continued incubation from zero time resulted in a steady production of nitrosylmyoglobin (NOMb) and nitrosylhaemoglobin (NOHb) at the expense of available MetMb + MetHb, the percentage of endogenous haem pigment converted into the nitrosyl form after two hours being virtually independent of nitrite concentration above 10 ppm.

The inclusion of 20% salt had little if any effect upon the relative proportions of the haem products found by the Soret procedure in the extracts of anaerobic incubations of minces of pig *quadriceps femoris* muscle

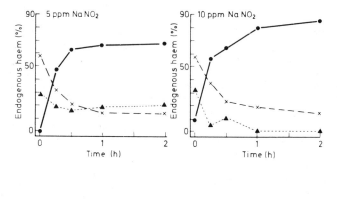

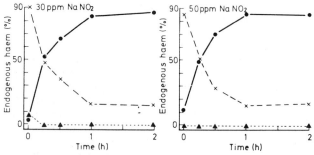

*Figure 20.1 Haem products of muscle mince with respect to nitrite concentration and period of incubation*
▲·····▲ *MbO$_2$ + HbO$_2$*   ×-----× *MetMb + MetHb*
●———● *NOMb + NOHb*

with nitrite for periods up to 45 h (*Figure 20.2*). The
formation of NOMb + NOHb equalled the concentration of
MetMb + MetHb from which they were formed after
approximately 10 h. Any nitrosylmetmyoglobin formed
during the simulated curing process would have reverted
to uncomplexed metmyoglobin during the aerobic
extraction procedure. However, evidence for its
intermediate formation has been obtained using skeletal
muscle mitochondria, the sub-cellular location of most
respiratory enzymes.

Ascorbic acid can promote the formation of the cured
meat pigment, nitrosylmyoglobin, from metmyoglobin and
nitrite in aqueous solution, with the intermediate
release of nitric oxide from a postulated ascorbic-acid –
nitrous-acid complex. In the same system, the vitamin
also acts in reducing nitrosylmetmyoglobin to the final
form of pigment, nitrosyl-(ferro)-myoglobin. In the

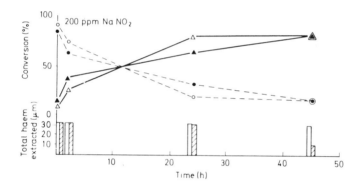

*Figure 20.2  Pigment conversion and total haem extracted
with (●▲ hatched area) and without (○△) 20% salt
----- MetMb + MetHb ——— NOMb + NOHb*

presence of the muscle constituent cytochrome-c,
ascorbic acid can simulate the actions of respiratory
enzyme systems in promoting the transfer of nitric
oxide to metmyoglobin and the subsequent reduction of the
product nitrosylmetmyoglobin to the cured meat pigment.

In the whole muscle situation, ascorbic acid is
effective in accelerating the conversion of metmyoglobin
into nitrosylmyoglobin during anaerobic incubation
(*Figure 20.3*) with nitrite. In addition, the equilibrium

percentage conversion of muscle haem pigment into the
cured form was increased.


# FACTORS INVOLVED IN THE DISCOLOURATION
# OF MEAT

The temperature of storage can exert an appreciable
effect on the discolouration of fresh meat through its
action on a number of properties of excised muscle.  This
was demonstrated by Snyder (1964) for beef rounds stored
at temperatures over the range of 6 to -2°C for up to
140 h, with all other conditions kept the same.  Higher
temperatures favour greater scavenging of oxygen by
residual respiratory enzymes and other oxygen-consuming
processes such as fat oxidation, leading thereby more
rapidly to the low tensions conducive to the autoxidation
of myoglobin and haemoglobin to their met- forms.
Conversely, lower temperatures promote increased
penetration of oxygen into the surface layer of the meat
as well as the amount of oxygen dissolved in tissue
fluids, both of which factors can assist in maintaining
myoglobin and haemoglobin in their oxygenated forms.
Furthermore, the dissociation of oxygen from its
reversible combination with myoglobin is enhanced on
raising the temperature, leading to an increased tendency
for autoxidation by the deoxygenated myoglobin produced.
  The development of bacteria in a meat product similarly
results in an enhanced depletion of the oxygen tension,

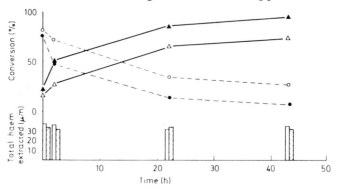

*Figure 20.3  Pigment conversion and total haem extracted
with (●▲ hatched area) and without (○△ ) ascorbate in
presence of nitrite and 130 ppm sodium nitrite.*
*----- MetMb + MetHb ——— NOMb + NOHb*

particularly when the access to air is restricted by a
plastic wrap. Kraft and Ayres (1952) noted that in many
instances the discolouration of packaged meats was
observed before disagreeable odours or other evidences of
spoilage became apparent at 4°C. The oxygen demand by
aerobic bacteria is particularly high during the
logarithmic growth phase, which coincided with a large
increase in the formation of metmyoglobin in prepackaged
beef steaks (Butler, Bratzler and Mallmann, 1953).
Further storage at 1°C of inoculated steaks resulted in
a reduction in metmyoglobin content below that of the
control samples, owing presumably to the ability of the
bacteria to convert the ferric form of the haem muscle
pigment back to the ferrous. Samples subsequently exposed
to oxygen developed a desirable red colour but rapid
deterioration to metmyoglobin occurred even on storage
at 0°C within an oxygen impermeable wrap owing to the
high bacterial populations.

Overall, bacterial growth reduced the saleable shelf-
life of all inoculated beef steaks. Changes caused or
accelerated by bacterial growth included not only
discolouration caused by the increased rate of metmyoglobin
formation but also the production of off odours and slime
formation; usually the changes were found to appear in the
order given. Not unexpectedly, the shelf-life was
prolonged by a lower initial level of bacterial contamination and by reducing the storage temperature. One method
of restricting the growth of putrefactive organisms has
been to package fresh meat in an atmosphere containing
carbon dioxide. Mixtures of carbon dioxide and oxygen
at high partial pressures were found to be particularly
successful (Taylor and MacDougall, 1973) since the layer
of oxygenated tissue maintained at the surface was
sufficient to conceal the development of underlying
metmyoglobin which causes discolouration in conventionally
pre-packaged meat. An initial atmospheric mixture of
60% oxygen and 40% carbon dioxide was found to maintain
the red colour of fresh beef for at least one week at
1°C but it was suggested that the best results in practice
would be obtained by using the maximum oxygen with the
minimum carbon dioxide level required to inhibit
microbial spoilage. Previously, it had been shown
(Rikert, Olin Ball and Stier, 1958), however, that the
rate of discolouration of fresh beef stored in air
enriched with carbon dioxide increased directly with the
partial pressure of carbon dioxide over the range from
0.2-72 cm of mercury, presumably as a result of the
corresponding depletion of oxygen pressures. On the

other hand, the discolouration of ham decreased with
increasing partial pressures of carbon dioxide, owing to
diminished oxidation of the cured meat pigment as the
oxygen tension was concurrently reduced.

A completely different type of bacterial discolouration
is that in which catalase negative organisms can produce
hydrogen peroxide and thus can cause defects in both
fresh and cured meat products - leading to the formation
of grey or green patches amongst the normal haem pigments.
Such phenomena result from either the oxidation of haem
pigments to a green form such as choleglobin or from
their degradation beyond porphyrins to bile pigments or
other similar products with little intrinsic pigmentation.
Tjaberg, Hangam and Nurmi (1969) have reported a
discolouration of this type in Norwegian salami sausage
promoted by lactobacilli used as starter cultures; the
colour defect appeared as a grey centre of the sliced
surface.

A number of metal ions has been reported to stimulate
the autoxidation of oxymyoglobin (Snyder and Skrdlant,
1966).  Of those tested in phosphate buffer at pH 5.7,
copper was by far the most active and iron, zinc and
aluminium far less so.  At a concentration of cupric
ions of 1.5 ppm equimolar to oxymyoglobin, a 25-fold
increase in autoxidation rate constant was apparent.
A more recent study, however, has demonstrated that the
rate of autoxidation of oxymyoglobin is not greatly
influenced by added cupric ions until a concentration of
200 molar equivalents is present; it was suggested that
previous results were invalidated by the presence of
trace residues of dithionite employed in the preparation
of oxymyoglobin.  In 1951, Coleman reported an increased
rate of autoxidation of oxymyoglobin and oxyhaemoglobin
by sodium chloride, possibly owing to concomitant changes
in the pH involved.  Part of the sodium chloride effect
was, however, found to be sensitive to the inclusion of
the chelating agent EDTA; thus it was concluded that
trace metal contaminants played a part in the enhanced
autoxidation but that salt at the 10% level did have a
stimulatory effect of its own.  Nevertheless, the
addition of EDTA to brines contaminated with copper did
not prevent the black discolouration of canned cured ox
tongues attributed to the metal    (Board and Ihsan-ul-
Haque, 1965).  Brines containing as little as 0.5 ppm
copper produced noticeable discolouration of canned
tongues whilst 5 ppm gave an intense black discolouration
One effective way of removing copper from a contaminated
brine was to add finely minced beef and to remove it by

filtration after stirring. Yellow-brown discolourations
appearing on the surface of canned cured pork have been
ascribed to the catalysis by iron, arising from damaged
tin coating, of the oxidation of the nitric oxide derived
from nitrite to nitrogen dioxide, leading to the formation
of metmyoglobin (Suvakov, Visacki and Marinkov, 1969).

Discolourations caused by exposure to light have become
a particularly serious problem because of modern methods
of merchandising which require exposure of retail cuts
in lighted display cases (Watts, 1954). Cured meats are
much more susceptible to light discolouration than fresh,
because it brings about the dissociation of nitric oxide
from the relevant pigment, nitrosylmyoglobin. Display
case lighting which does not significantly discolour
fresh meats over periods of three days can cause
noticeable fading of cured products within one hour in
the presence of oxygen. Kraft and Ayres (1954) reported
that soft white fluorescent light was unimportant in
causing discolouration, whereas ultraviolet light
promoted rapid discolouration even though it helped to
control bacterial growth on the surface of the beef
steaks studied. It was found that the irradiation of
cellophane-wrapped beef with ultraviolet light produced
marked dehydration of the meat; the resulting loss of
moisture appeared to be responsible not only for the
retardation of surface bacterial growth but also for
much of the darkening of the muscle pigment by
oxidation at the low oxygen tensions involved. Slight
dehydration was considered helpful for desirable fresh
meat colour by Urbin and Wilson (1958), but was
deleterious when it occurred to any great extent. In
general, light intensities that did not increase surface
temperature had no effect on discolouration of fresh
meat products.

The susceptibility of the cured meat pigment to
irradiation by fluorescent light is dependent upon pH,
nitrosylmyoglobin being more stable at pH 6.8 than at
pH 6.2 (Bailey, Frame and Naumann, 1964). In aqueous
solution, its stability towards fluorescent light is
enhanced by the addition of nitrogenous bases such as
nicotinic acid or nicotinamide, possibly through the
formation of a pigment complex more resistant to
irradiation.

The beneficial effects of ascorbate on the processes
leading to the formation of the characteristic haem
pigment during curing have already been mentioned. It is
particularly beneficial in terms of colour formation
where the nitrite level is low but less so where the

concentration of nitrite is adequate (Bauernfeind and Pinkert, 1970). Ascorbic acid also has a helpful effect on the stability of the cured meat colour once it has been formed. Several factors are probably involved in its protective action. Firstly, ascorbic acid can promote the replenishment of nitric oxide lost by dissociation from nitrosylmyoglobin through the action of light  etc. Secondly, it acts synergistically with tocopherols occurring naturally in meat in restricting the formation of fat peroxides detrimental to haem pigments. Finally, colour stability of cured products is better in the absence of oxygen and therefore improved by the oxygen scavenging action of ascorbate.

During the frying of bacon, a transition has been noted whereby the red uncooked cured meat pigment becomes converted to the red nitrosyl myochrome typical of the cooked product, with the intermediate formation of an uncharacterised grey pigment. The overall change requires a residual nitrite content of about 30 ppm (0.43 mM) at the time of cooking to give a good uniform cooked colour, which is quite unsatisfactory at levels of nitrite below 15 ppm (0.22 mM). To ensure a level of this order after long but otherwise satisfactory storage under both trade and domestic conditions, the residual nitrite content at the time of manufacture may need to be as high as 80 ppm (1.16 mM). The presence of ascorbate in the bacon can be detrimental in this context because it diminishes the amount of nitrite available at the time of cooking, since nitric oxide formed previously from nitrite consumed is not effective for this purpose. Thus, although the stability of uncooked nitrosylmyoglobin is improved by the action of ascorbate, the increase is not sufficient to offset the penalty incurred in cooking by the reduction in residual nitrite content. The requirement of additional nitrite in the formation of the cooked cured meat pigment may be bound up with the suggested content (Tarladgis, 1962) of two nitric oxide molecules co-ordinated to each iron atom.

## CONCLUSIONS

Many factors can contribute to the maintenance of the colour of fresh and cured meat products and to the promotion of discolouration so potentially damaging to the saleability on display to the consumer, and it has

by no means been possible to cover all practical
contingencies in this short summary.

Several general concepts apply, however, so far as the
maintenance of the pigment of fresh meat as oxymyoglobin
and that of cured meat as nitrosylmyoglobin is concerned.
The available evidence indicates that both haem pigments
oxidise far more readily after the dissociation of the
ligand, either oxygen or nitric oxide, attached to the
iron atom. The dissociation of oxymyoglobin increases
with decreasing oxygen tension. Thus, any process which
tends to lower gradually oxygen tensions towards the low
partial pressures, such as muscle respiration, bacterial
contamination, fat oxidation etc., promotes browning in
fresh meat products. A small amount of oxygen is
required for the oxidation of myoglobin to metmyoglobin,
so that its rapid and complete depletion from a meat
system will produce the deoxygenated form of myoglobin,
which is capable of re-combining with oxygen on exposure
to air.

In contrast, dissociation of nitrosylmyoglobin does not
increase at lower oxygen tensions, and the tendency for
the cured meat pigment to oxidise increases with the
oxygen supply. In this case, dissociation and thence
oxidation is markedly increased on exposure to light.

The formation of metmyoglobin from oxymyoglobin is
accelerated by conditions which cause denaturation of
the globin moiety of the haem protein, such as heat,
pH, salts and ultraviolet light. Porcine oxymyoglobin
is particularly susceptible to denaturation by acid
treatment (Satterlee and Zachariah, 1972) and by freezing
(Zachariah and Satterlee, 1973). Without its stabilising
bonds to native globin, the haem group loses its ability
to combine reversibly with oxygen to form the bright-red
oxy-form and is readily oxidised to the brown ferric
form. On the other hand, the red colour of nitrosylmyoglobin is not lost when the globin is denatured, as is
evident from the cooking of cured meat products (Watts,
1954).

REFERENCES

BAILEY, M.E., FRAME, R.W. and NAUMANN, H.D. (1964).
  *J. Agric. Fd Chem.*, <u>12</u>, 89
BATE-SMITH, E.C. (1948). *Advan. Food Res.*, <u>1</u>, 1
BAUERNFEIND, J.C. and PINKERT, D.M. (1970). *Advan. Food
  Res.*, <u>18</u>, 219

BEMBERS, M., ZACHARIAH, N.Y., SATTERLEE, L.D. and HILL, R.M. (1973). *J. Fd Sci.*, 38, 1122

BOARD, P.W. and IHSAN-UL-HAQUE (1965). *Fd Tech.*, 19, 1721

BRISKEY, E.J. (1964). *Advan. Fd Res.*, 13, 89

BROOKS, J. (1935). *Proc. Roy. Soc. Lond.*, B, 118, 560

BROOKS, J. (1937). *Proc. Roy. Soc. Lond.*, B, 123, 368

BUTLER, O.D., BRATZLER, L.J. and MALLMANN, W.L. (1953). *Fd Technol.*, 6, 8

COLEMAN, H.M. (1951). *Fd Res.*, 16, 222

EL-BADAWI, A.A., CAIN, R.F., SAMUELS, C.E. and ANGLEMEIER, A.F. (1964). *Fd Technol.*, 18, 159

FOX, J.B. and THOMSON, L.S. (1963). *Biochemistry*, 2, 465

GEORGE, P. and STRATMANN, C.J. (1952). *Biochem. J.*, 51, 418

HOOD, D.E. and RIORDAN, E.B. (1973). *J. Fd Technol.*, 8, 333

KEILIN, D. and HARTREE, E.F. (1937). *Nature, Lond.*, 139, 548

KOIZUMI, C. and DUANE BROWN, W. (1971). *J. Fd Sci.*, 36, 1105

KRAFT, A.A. and AYRES, J.C. (1952). *Fd Technol.*, 6, 8

KRAFT, A.A. and AYRES, J.C. (1954). *Fd Technol.*, 8, 290

LANDROCK, A.H. and WALLACE, G.A. (1955). *Fd Technol.*, 9, 194

LAWRIE, R.A. (1958). *J. Sci. Fd Agric.*, 11, 721

LAWRIE, R.A. (1966). *Meat Science*, p.272. Pergamon Press, Oxford

MACDOUGALL, D.B. (1970). *J. Sci. Fd Agric.*, 21, 568

MACDOUGALL, D.B. and RHODES, D.N. (1972). *J. Sci. Fd Agric.*, 23, 637

MARSHALL, W. and MARSHALL, C.R. (1945). *J. biol. Chem.*, 158, 187

NEILL, J.M. and HASTINGS, A.B. (1925). *J. biol. Chem.*, 63, 479

PRICE, J.F. and SCHWEIGERT, B.S. (1971). *Science of Meat and Meat Products*, p.371, W.H. Freeman and Co., San Francisco

RIKERT, J.A., OLIN BALL, C. and STIER, E.F. (1958). *Fd Technol.*, 12, 17

RIORDAN, E.B. (1973). *Beef to suit the consumer*. Report of a pilot study of consumers, The Agricultural Institute, Dublin

ROMANS, J.R., TUMA, H.J. and TUCKER, W.L. (1965). *J. Anim. Sci.*, 24, 686

SATTERLEE, L.D. and ZACHARIAH, N.Y. (1972). *J. Fd Sci.*, 37, 909

SHARP, J.G. (1957). *J. Sci. Fd Agric.*, 8, 14 and 21
SHARP, J.G. (1958). *Ann. Rep. Fd Invest. Bd., Lond.*,
  p.7
SNYDER, H.E. (1964). *J. Fd Sci.*, 29, 535
SNYDER, H.E. and SKRDLANT, H.B. (1966). *J. Fd Sci.*, 31,
  468
STEWART, M.R., HUTCHINS, B.K., ZIPSER, M.W. and WATTS,
  B.M. (1965). *J. Fd Sci.*, 30, 487
SUVAKOV, M., VISACKI, V. and MARINKOV, M. (1969). *15th
  European Meeting of Meat Research Workers, Helsinki*,
  p.325
TARLADGIS, B.G. (1962). *J. Sci. Fd Agric.*, 13, 485
TAYLOR, A.A. and MACDOUGALL, D.B. (1973). *J. Fd Technol.*,
  8, 453
TJABERG, T.B., HANGAM, M. and NURMI, E. (1969). *15th
  European Meeting of Meat Research Workers, Helsinki*,
  p.138
URBIN, M.C. and WILSON, G.D. (1958). *Proc. 10th Res.
  Conf., Amer. Meat Inst. Found.*, p.13
WALTERS, C.L., CASSELDEN, R.J. and TAYLOR, A.McM. (1967).
  *Biochim. Biophys. Acta*, 143, 310
WATTS, B.M. (1954). *Advan. Fd Res.*, 5, 1
ZACHARIAH, N.Y. and SATTERLEE, L.D. (1973). *J. Fd Sci.*,
  38, 418

# 21

# INFLUENCE OF HEATING METHODS

PAULINE C. PAUL

*School of Human Resources and Family Studies,
University of Illinois, Urbana, USA*

Some type of heat treatment of meat is of concern to
all of us, since we seldom eat meat raw. However, our
interest in this aspect is much more apt to be a personal
one rather than a scientific one, since we are strongly
influenced by our food patterns - a subject which is
surrounded by our personal likes and dislikes, family
customs, and all the other social and psychological
factors which determine our food habits.

  Much of our meat supply is heated during the processing
stages, as in the making of frankfurters, bologna or
pre-cooked hams, canned meat items, or precooked frozen
or freeze-dried products. These are usually reheated at
the time of consumption. Meat items purchased raw are
cooked in a variety of ways, in homes or in quantity
food service operations.

## COOKING MEAT FOR INDIVIDUAL OR GROUP
## CONSUMPTION

When one attempts to summarise meat cooking from the
teaching standpoint, one finds a very confusing picture.
There are probably about as many methods of cooking meat
as there are people doing the cooking. However, there
are some general methods and guidelines which can be used.
Most widely used meat cooking methods have a long history

of development through common practices. These may be
grouped by principal media of heat supply to the food,
as shown in *Table 21.1*.

*Table 21.1*  Meat cooking methods at time of consumption

| *Primary heat supply to meat* | *Cooking methods (US terminology)* |
|---|---|
| Heated air plus radiation | Roast, broil, grill barbecue |
| Heated metal | Pan broil |
| Heated oil or fat | Pan fry, deep fat fry |
| Steam and/or hot water | Braise, stew, cook in liquid, pressure cook |

An early study to establish a controlled basis for meat
cookery was that of Sprague and Grindley (1907), who used
thermometers to measure internal temperature of roasts as
a means of determining heating endpoint.  During the
1920s and 1930s an informal conference, known as the
National Cooperative Meat Investigations, made up of
representatives from the US Department of Agriculture and
twenty or more State Agricultural Experiment Stations,
worked on various problems related to production and
consumption of meat.  The Committee on Preparation
Factors of this Conference summarised their
studies in a publication entitled *Meat and Meat Cookery*
(Committee on Preparation Factors, 1942).  Their
conclusions supplied the basis for most of the meat
cookery methods recommended in the US since then.  These
methods are summarised in *Table 21.2*.  Some of their
findings have been modified by results of subsequent
studies.  The most recent standard recommendations
(AHEA Terminology Committee, 1971) are also listed in
*Table 21.2*.

Recent developments in cooking supplies and equipment
include temperature-controlled small electric appliances
such as hot plates or grills, frying pans, roasters, deep
fry kettles and 'slow' cookers, and aluminium foil and
oven-stable transparent films or bags.  There has been
some concern that the 'slow' cookers might present microbiological hazards if the rate of temperature rise was
very slow.  A very limited test with one brand (Paul,
1973), suggests that this problem probably is not serious
since the centre of the pot contents attained a temperatu
of 59°C within 2 h at the 'low' setting and 70°C within

2½ h at the 'high' setting. However, a good deal more
work would be needed to establish this definitely. Early
recommendations for roasting meat wrapped in aluminium
foil specified higher oven temperatures than customarily
used, to avoid lengthened cooking times. More recently,
Baity, Ellington and Woodburn (1969) investigated the
use of tight or loose foil wraps on roasting at 94°C or
232°C. At the lower oven temperature, the loose wrap had
an insulating effect owing to the air layer trapped
between the foil and the meat, but the tight wrap did
not increase, and in some cases decreased, the cooking
time required as compared to unwrapped meat. At 232°C,
the foil tended to increase the cooking time needed for
both the loose and tight wraps, the foil apparently
acting as a radiant heat shield. Transparent film wraps
and bags do not appear to act as radiant heat barriers,
since meat roasted in these does show considerable
surface browning. The films cannot be completely tight,
since tight wraps break with explosive force from
internal steam pressure. Another factor with films,
especially bags, is the necessity to add some kind of
finely divided, insoluble material to reduce the 'bumping'
tendency of the mixture of water-melted fat produced
during heating.

The development of microwave ovens has introduced a
new method of cooking meat, since the energy is supplied
to the meat as high-frequency electromagnetic waves. The
alternating electric field causes vibration of the polar
molecules in the food and the molecular motion heats the
food. This is a very rapid method of cooking. However,
the most successful uses to date have been for cooking
small uniform pieces of meat such as sliced bacon, for
reheating precooked frozen meat items or for use in
combination with another heating method in food
processing plants and large-quantity food service units.
A good example of the influence of the amount of food
placed in the microwave oven on the cooking time required
is found in the data of Culotta and Chen (1973), who
stated that while one chicken breast required approximately
3 minutes, four at a time took 11 minutes. Jiménez (1972)
studied restaurant packs of precooked frozen foods in
open-topped shallow aluminium containers and found that
a combination microwave and convection oven heated the
foods satisfactorily and in a shorter time than either
a conventional oven or a convection oven alone. He
cautioned that optimum heating times need to be
established for each product, to maximise quality and
efficiency. Penner and Bowers (1973) compared pork loin

Table 21.2 General recommendations for meat cookery

| Cooking method | Brief description | Species | Temperature (°C) | | | | Stage of doneness |
|---|---|---|---|---|---|---|---|
| | | | External | | Internal | | |
| | | | 1942 | 1971 | 1942 | 1971 | |
| Roast | Shallow, open pan, rack under meat if needed, in heated oven | Beef | 149 (small roasts 163) | 163 | 60 | 60 | rare |
| | | | | | 71 | 71 | medium |
| | | | | | 77 | 77 | well done |
| | | Pork | 177 | 163 | 85 | 77–85 | well done |
| | | Veal | 149 | 163 | 74–77 | 77 | well done |
| | | Lamb | 149 | 163 | 79–82 | 65 | medium |
| | | | | | | 82 | well done |
| Broil, grill | Meat on rack, either under or over heat source | Beef | 198 for cuts 2.5 cm thick 177 for cuts 5 cm thick | | 57–60 | | rare |
| | | | | | 71 | | medium |
| | | Lamb | | | 71–77 | | medium to well done |

| Method | Procedure | | | | | |
|---|---|---|---|---|---|---|
| Pan broil | Heated pan (may or may not be lightly greased), brown exterior of meat, reduce heat and cook to desired doneness. | | | | | |
| Pan fry | Heated pan with shallow layer of fat, cook as for pan broil. | | | | | |
| Deep fat fry | Immerse meat in heated fat | 149-163 | | | | |
| Braise | Brown meat in small amount of fat in heated pan, add small amount of liquid, cover and simmer until tender | | top of stove or oven at 163, to maintain 85-99 | 77-93 | 85-99 | well done |
| Stew, cook in liquid | Brown meat or not as desired. Cover with liquid, cover pan, simmer until tender. | 85-98 | 85-99 | 85-98 | 85-99 | well done |
| Pressure cook | Brown if desired, place on rack, add small amount of liquid, seal pan and cook under steam pressure. | 10 lb 12-15 min per lb meat | | | | |

roasts (a) frozen raw, thawed, roasted conventionally to 77°C in a 163°C oven; (b) roasted conventionally to 65°C, frozen, thawed, reheated to 55°C in a 163°C oven; and (c) roasted, frozen, thawed, reheated to 55°C by microwave. A brief summary of their results is shown in *Table 21.3*. Although cooking, then reheating by microwave, did increase the cooking losses over single cooking, the quality of the meat was satisfactory, and the reheating procedure might be more suitable for quantity service time schedules.

*Table 21.3* Summary of data on roasting pork loins (after Penner and Bowes, 1973)

|  | *Cooking method* | | |
|---|---|---|---|
|  | a[1] | b | c |
| Total moisture in meat when ready to serve (%) | 63.6 | 60.7 | 57.8 |
| Total cooking losses (%) | 28.5 | 30.9 | 35.5 |
| Total cooking and reheating time | shortest | longest | intermediate |
| Cooking time (a) or reheating time (b and c) (min) | 69 | 50 | 9 |
| TBA value | 3.31 | 4.97 | 4.01 |
| Flavour and aroma | very good | less sweet aroma, more metallic flavour | very good |
| Juiciness | juicy | drier than others | juicy |

[1] a - roasted conventionally at 163°C to an internal temperature of 77°C.
  b - roasted conventionally at 163°C to 65°C, frozen, thawed, reheated conventionally to 55°C.
  c - as in 'b' except reheated by microwave to 55°C.

## HEATING MEAT FOR EXPERIMENTAL STUDIES

Paul (1972) tabulated a wide variety of methods that have been used in experimental studies which involved heating meat. A survey of the studies published since that time indicates just as wide a variation. In 1973, the National Livestock and Meat Board sponsored a conference of a small group of research people involved in meat studies to discuss problems and methodology in preparing cooked meat samples for evaluation by objective and/or subjective methods. Two members of this group were asked to explore the possibility of preparing a publication which was tentatively entitled 'A Guide To Selected Methods Used In Meat Cooking Research'. The availability of such a compilation would assist research workers in selecting methods to be used for any particular problem. In the meantime, perhaps one way to look at the problem of method selection is to review the factors in meat which are altered by heating, and then look at some of the kinds of decisions which need to be made.

The characteristics of meat which are altered by heat treatment, and which are of most interest in relation to the quality of the product as food, are listed in *Table 21.4*. Those I have categorised as economic have had less consideration than many of the others. But these economic factors are increasingly important in food processing and in preparation for consumption both in the home and in quantity-food service establishments.

*Table 21.4* Characteristics of meat altered by heating

---

1. Economic factors
   (a) yield - cooking losses, volume changes
   (b) time and energy requirements
2. Palatability
   (a) appearance: external and internal colour, shape changes
   (b) flavour and aroma, pH change
   (c) juiciness, water and fat distribution
   (d) tenderness
3. Nutritive value
   (a) loss, retention or concentration of nutrients
   (b) digestibility
4. Sanitary quality

---

*Table 21.5* gives a brief summary of some of the kinds
of decisions which need to be made in either experimental
studies or procedures used in processing and in
preparation for consumption. To date, many people have
failed to recognise the considerable body of information
that is available as a basis for such decisions. Perhaps
some summary of these should be included in the projected
publication on methods in meat cooking research.

In considering the evidence concerning heat-induced
changes in muscle tissue for use as food, it is obvious
that the bulk of the experimental work has been done
with bovine muscle, with fewer studies on porcine or
ovine. However, it is generally found that the basic
pattern of changes is similar for all three species,
although they may differ in details. In selecting
studies to illustrate various points, I have drawn
principally on recently published studies, since there
have been a number of previous reviews which included
results from cooking studies, the most recent of these
being that of Laakkonen (1973). I have also included a
few studies on chicken, since these were especially good
illustrations of certain types of problem.

From the point of view of the research worker, the
first decision area (*Table 21.5*) of course, is the
objective or objectives of the particular study and the
particular situation or setting within which it will
occur. These will frequently determine many specification
such as the species, age, breed, grade of the meat, the

*Table 21.5* Decision areas for selection of heating
procedures for research

---

1. Objectives of research, facilities and supplies
   available
2. Evaluation methods needed
3. Cut to be used
4. Container for meat
5. Form in which energy is supplied, heat transfer medium
6. Heating rate and endpoint
7. Time restrictions
8. Suitability of products for analytical methods and
   types of data to be collected

---

experimental variables, as well as the equipment,
analytical methods, time, and funds available.  Once
these dimensions are defined, more detailed decisions can
be made.

The second major decision area would probably be the
methods of evaluation to be used.  These include a wide
variety of subjective and objective procedures, many of
which have been discussed in earlier chapters.  I would
like to consider a few of these in terms of heat effects
on meat.

Subjective evaluation usually means the participation
of people as evaluators of the products.  Since the
eventual purpose of any study of meat as food is a
product acceptable for consumption panel evaluations
provide the ultimate criteria.  There are many experimental meat heating procedures and problems which rule out
panel evaluation.  The meat may be unacceptable, or
potentially harmful, to anyone consuming it.  For example,
among some groups of people rare beef is unacceptable,
while other groups will not eat it if it is too well
done.  A few years ago, at a conference on collagen which
included consideration of the pathological changes in
collagen caused by consumption of plant materials
containing lathyritic compounds, as well as the role of
collagen in meat tenderness, a physician suggested that
we might explore the possibility of injection of these
compounds into meat animals to tenderise the collagen.
With such a study, possible retention of deleterious
materials in the cooked meat could endanger anyone
tasting the meat.  With some problems, time considerations
could make panel use difficult, as in a study which
produces samples at 10 or 15 minute intervals over a
period of several hours.

For trained panel evaluation in a laboratory setting,
the research personnel can use well specified and
carefully controlled heating procedures.  However, if one
purpose of the study is to ascertain consumer acceptance,
then a much larger group of people must be used, and often
one has to leave it to each consumer to cook their own
samples - does one leave the preparation wide open, or
does one try to give at least some general directions?

There are so many possible chemical and physical.
analyses used in meat studies that it would be too time
consuming to consider all of them.  However, a good
illustration of the need for careful selection of heating
methods in sample preparation can be found in some of the
recent studies on lipid fractions in meat.  Keller and

Kinsella (1973) determined oxidative changes in meat lipids caused by cooking ground beef in a teflon-coated skillet, in a metal frying pan, or by charcoal broiling. They reported changes in ratio of unsaturated to saturated fatty acids in the phospholipids, as well as changes in TBA values and total carbonyl present, indicating lipid oxidation during cooking. A summary of part of their data is given in *Table 21.6*.

*Table 21.6* Differences in TBA numbers and total carbonyls for the different cooking methods (after Keller and Kinsella, 1973)

| Average for | TBA Number (mg malonaldehyde/ kg meat) | Total carbonyls (μmoles/g meat) |
|---|---|---|
| Raw beef | 0.46 | 0.31 |
| Cooked in teflon skillet | 4.30 | 1.95 |
| Cooked in metal fry pan | 4.60 | 4.47 |
| Charcoal broiled | 4.21 | 4.64 |

Lee and Dawson (1973) studied the lipid changes occurrin when chicken parts were fried in deep fat, using pressurised frying equipment. Part of their data is given in *Table 21.7*. They found that this method of frying increased the unsaturated fatty acids in both skin and muscle, and previously heated corn oil tended to cause larger increases than did fresh corn oil.

While the type of meat animal may be determined earlier in the decision process, it is seldom that a laboratory has the facilities for studying an entire or half carcass of any of the large animals commonly used for food. So in choosing a cooking method, it is necessary to decide which part or parts are to be studied. Should the meat be cut by customary commercial methods (usually having more than one muscle included in a given cut), or separated by muscles? Should the individual samples be large and blocky for roasting, thinner for broiling, cut in small chunks as for a stew, or cut into pieces of pre-determined dimensions? The last type has the advantage of the most uniform rate of heat penetration, but may bear little resemblance to cuts as usually cooked for consumption.

A number of studies have shown marked differences in the characteristics of different muscles of the same carcass in their response to heat treatments. As an example,

Table 21.7 Changes in lipid fractions of chicken parts fried in deep fat and of the fat used for cooking (after Lee and Dawson, 1973)

| Fatty acids | Uncooked | | Cooked in fresh corn oil | | Cooked in corn oil previously heated 24 h | |
|---|---|---|---|---|---|---|
| | saturated | unsaturated | saturated | unsaturated | saturated | unsaturated |
| Total | | | | | | |
| chicken muscle | 29.8 | 70.6 | 28.4 | 71.6 | 27.4 | 72.8 |
| chicken skin | 35.4 | 64.7 | 25.2 | 74.9 | 25.2 | 74.8 |
| Phospholipids | | | | | | |
| chicken muscle | 43.9 | 56.2 | 37.5 | 62.2 | 33.1 | 66.6 |
| chicken skin | 29.9 | 70.2 | 26.3 | 73.9 | 20.8 | 79.3 |
| Neutral lipids | | | | | | |
| chicken muscle | 31.4 | 68.7 | 27.0 | 72.9 | 27.8 | 72.2 |
| chicken skin | 35.8 | 64.2 | 24.0 | 76.2 | 24.1 | 75.8 |
| Total | | | | | | |
| oil before cooking | | | 16.1 | 83.9 | 16.0 | 84.0 |
| oil after cooking chicken | | | 16.6 | 83.5 | 18.6 | 81.3 |

Hostetler *et al.* (1973) reported significant differences in panel scores for juiciness and 6 tenderness characteristics among 9 muscles of the bovine carcass, even though pieces uniform in size and shape were used, and all were heated the same amount at the same rate.

Meat loses water, water soluble compounds, and melted fat during the heating process. So meat is usually put into some kind of container to retain these extruded materials. Of course, the size and shape of the meat sample must be considered. Within these limitations, a shallow open pan in a dry atmosphere will maximise evaporation and surface changes in the sample, while a deep one will reduce these. If the container is covered this will further reduce evaporation unless the cooking time is much longer.

In an open container, the shape and material of the container may cause differences in results. *Table 21.8* shows differences in cooking time, internal temperature rise after removal from oven, and cooking losses, of ground beef loaves roasted on racks in shallow pans or on racks in 600 ml beakers. It is apparent that the use of a deep, small diameter container instead of a shallow wide one tended to decrease the time required to cook, increase the temperature rise after removal from heat source, decrease the total losses at 75 and 82°C end points, and resulted in more of the losses as drippings and less as evaporation.

Transparent oven film bags will permit surface browning of the meat, but minimise evaporation. Pouches for small meat samples, or shallow containers to catch drippings can be shaped from aluminium foil. Foil may act as a transmitter of heat, or as a baffle.

The shape and the material of which the container is made will influence heat transmission rate to the surface of the meat. Metals are good conductors of heat, but do not transmit radiant heat. Glass is not as good a conductor of heat, but will transmit radiant heat. A closed or covered container instead of an open one will usually increase the total cooking losses, and alter the distribution between volatile and drip losses. Also, in covered containers the heat transfer medium is steam rather than dry air, so heating will be more rapid once the system is heated enough for steam to form. In some cases, it is advisable to have a rack under the meat to keep the meat from contact with the container and the drippings. In our oven-broil method (Paul *et al.*, 1956), we use racks on 4-inch legs, to allow for heat circulation all around the meat so that it does not have to be turned during the cooking process.

*Table 21.8* Influence of shape of heating container on rate of heating, endpoint temperature, and cooking losses of ground beef loaves (Paul, class data)

| Container | Internal temperature (°C) | | Heating time (min) | Cooking loss (%) | | |
|---|---|---|---|---|---|---|
| | at heating endpoint | maximum reached | | Total | Evaporation | Drip |
| Shallow pan | 60 | 63.5 | 55 | 20.4 | 7.1 | 13.2 |
| | 68 | 70.5 | 63 | 25.6 | 10.3 | 15.3 |
| | 75 | 76.5 | 81 | 32.1 | 20.7 | 11.5 |
| | 82 | 83 | 94 | 36.4 | 26.0 | 10.4 |
| 600 ml beaker | 60 | 69 | 50 | 23.4 | 6.7 | 16.7 |
| | 68 | 76 | 52 | 25.1 | 7.4 | 17.7 |
| | 75 | 82 | 54 | 27.2 | 6.8 | 19.4 |
| | 82 | 83 | 72 | 31.7 | 13.9 | 17.8 |

The method of heat transfer utilised will produce marked differences in the results obtained.  Traditional methods use heat energy, transmitted either by conduction, convection or radiation, or some combination of these. The heat comes to the outer surface of the meat through air, steam, oil, water, and by any direct contact with the container.  Heat transfer is much faster through water, oil or steam than through dry air.  So in the latter case, the heating rate for the meat is slower than in the other three.  Since many of the changes which occur as meat is heated involve denaturation of proteins, the rate of heating is important, since the type and extent of heat denaturation of proteins are functions of both time and temperature.

Heat transfer within the meat itself is by conduction. Despite the high water content of raw meat (70-75%), heat transfer is slow owing to the structure of the muscle. Heat transfer through fat deposits is also slow, until the temperature rises sufficiently to melt the fat.  The liquid fat is a much more efficient conductor of heat than is solid fat (Irmiter, Alrich and Funk, 1967). Insertion of metal skewers or 'heating pins' will increase the rate of heat penetration within the meat, as the metal is more efficient as a heat conductor than is the muscle substance.  However, these may shorten the total heating time enough to decrease the amount of tenderisation which occurs.

As indicated previously, the microwave oven supplies energy as electromagnetic waves, the heat being generated in the meat itself.  With the electromagnetic waves as the only energy source, the air surrounding the meat stays cool, and the container heats only by transfer from the meat.  When using combinations such as microwave plus a broiling element, or microwave-convection ovens, the air and the container will be heated by the conventional heat source.

The rate and extent of heating of the meat will be determined not only by the method of heat transfer but also by the external temperature and by the internal temperature to which the meat is heated.  In the conventional oven, as the oven temperature is increased from about 100 to 260°C, the cooking losses increase, but when oven temperatures less than 100°C are used, losses increase again, since the time needed in the oven is so long (10 to 30 h) that evaporation losses are unusually large.  When cooking meat in deep fat, the temperature of the fat usually needs to be lowered as the thickness of the meat increases, tc avoid having the

exterior surface become dry and hard before the interior
is cooked. This also is true in broiling, although with
that method the heat reaching the surface of the meat can
be decreased by placing the samples farther from the heat
source.

Many of the studies on thermal conductivity of meat
itself have considered temperatures from below freezing
to around 20°C. Conductivity at temperatures used in
cooking is complex, being influenced by meat composition
and water content, distribution of fat, bone and lean,
direction of heat flow in relation to direction of the
muscle fibres, and the energy demands of increasing meat
temperature, evaporation of water, melting of fat, and
the chemical reactions involved in protein denaturation
and colour and flavour changes.

For any one cooking method, increasing the internal
temperature to which the meat is heated also increases
the cooking losses and decreases juiciness (for example,
Bayne *et al.*, 1973), as well as changing the surface and
interior colour. The internal temperature of larger
pieces of meat will vary considerably from place to place
within the cut, depending on the location of the sensor
in relation to the surface of the meat. Since meat
expands and contracts during heating, the sensing element
may be displaced from its original location so this must
be checked periodically.

Harries, Rhodes and Chrystall (1972) found that the
internal temperature of roasts needed to be brought to
74°C for uniform doneness, since the degree of doneness
varied from sample to sample at lower internal
temperatures. Lewis *et al.* (1973) found that different
muscles of the same porcine carcass were not at the same
stage of doneness at the same internal temperature. They
also stated that ante-mortem stress appeared to increase
the internal temperature required for a given degree of
doneness and suggested that protein solubility may be a
better estimate of stage of doneness than is internal
temperature.

When trying to cook meat to a specified internal
temperature, the research worker must allow for the
temperature rise which may occur after the meat is
removed from the heat source. For traditional cooking
methods, as a general rule, the higher the external
temperature and the lower the internal temperature to
which the meat is heated, the more the temperature will
rise after heating is stopped. However, the size of the
piece of meat is also important - pieces of around 500 g
or less often show no temperature rise after heating.

This subsequent temperature rise is especially marked in meat cooked by microwave heating. Decareau (1967) found that a seven-rib beef roast should be removed from the microwave oven at an internal temperature of about 27-30 °C. After standing at room temperature for 45 minutes, the internal temperature of the roast had risen to about 60°C.

A number of experimenters have used small, precisely shaped samples of muscle tissue, enclosed in metal, glass, foil or plastic, and heated in a water, steam or oil environment, to attain better external temperature control and minimise temperature variations within the sample. For studies such as those involved with detailed chemical and/or physical changes caused by heat, these methods are very appropriate, but the findings may need considerable interpretation or even modification before they can be applied to current methods of meat cooking at point of consumption.

Cooking time may be another factor influencing choice of methodology. Cooking time required varies with the size and composition of the sample, with the rate at which energy is supplied to the meat, and with the cooking endpoint required. Preparation time for the samples may vary from very little for retail cuts of meat to several hours for separation of muscles and preparation of samples cut to a specific size. A technique often used to minimise time problems when large quantities of meat must be handled is to freeze the meat, then thaw out a few pieces at a time for testing. However, this technique should be used with the recognition that freezing will alter the water-holding capacity of the meat, and that colour, flavour, tenderness, moisture content and oxidation level of the lipids present may change during frozen storage. For example, Locker and Daines (1973) found that one freeze-thaw cycle reduced the force required to shear but increased the cooking losses of bovine *sternomandibularis*. Results will also differ depending on whether the meat is heated from the frozen state or is thawed first. Jakobsson and Bengtsson (1973) compared slices of bovine *l.dorsi* cut 1.5 cm thick and cooked by pan frying. They found that while slices thawed first could be cooked at a higher temperature and in a shorter time, the samples cooked from the frozen state were juicier and had lower total thawing and cooking losses.

The experimenter must also consider the analytical methods to be used, the types of data to be collected, and the suitability of the samples for these. The palatability

characteristics produced will reflect the type and
extent of heating.  Shaffer, Harrison and Anderson (1973)
compared top round roasts roasted in open pans or in oven
film bags, and found marked differences in external and
internal appearance of the cuts, as well as in cooking
losses, temperature rise after removing the roasts from
the oven, moisture content and water-holding capacity of
the cooked lean.  Some of their results are summarised in
*Table 21.9.*

With broiled and braised steaks, Cover and Hostetler
(1960) observed that steaks broiled to 80°C internal were
the grey-brown of well-done beef, while those braised in
steam to 85°C were still slightly pink.  The contradictory
internal colour findings of Shaffer, Harrison and Anderson
(1973) and Cover and Hostetler (1960) could have been due
to different cooking times required, since the average
total cooking times for the two methods of roasting were
nearly the same, while the broiled steaks took four times
as long to come to 80°C as the braised steaks did to come
to 85°C.

Wassermann (1972) points out that the method of applying
heat influences flavour and aroma development produced by
changes in both the lean and lipid portions of meat.
Rhodes (1971) chose pan frying for evaluating consumer
acceptability of bacon, on the basis that any unusual
odour would be most obvious by this method of cooking.

Schock, Harrison and Anderson (1970) compared two
methods of cooking by dry heat (deep fat drying, oven
roasting) and two of cooking by moist heat (oven braising,
pressure braising at 10 lb psig).  They used bovine *semitendinosus* (ST), heated to 70°C internal.  Their results
showed that pressure braising gave the highest cooking
losses, lowest moisture, press fluid yield and water-
holding capacity in the cooked meat, and the lowest
overall acceptability by panel scores, while oven roasted
samples had the smallest cooking losses, highest moisture
content, press fluid yield and water-holding capacity,
and highest overall acceptability scores.  The other
methods gave results intermediate between these extremes.

Bennett *et al.* (1973) in a study of PSE, normal and
DFD pork, compared the results obtained when loin chops
were oven broiled or fried in deep fat.  They concluded
that oven-broiling was more effective as a cooking
method than deep fat frying to demonstrate differences
among the three types of carcasses, since of the ten
measures studied, eight showed highly significant
differences when oven broiling was used, but only two
did with deep fat frying.

Table 21.9 Comparison of roasts cooked in open pans or in oven film bags, to internal
temperatures of 60, 70 or 80°C (after Shaffer et al., 1973)

| Cooking method:<br>internal temperature (°C) | Open pan (dry heat) | | | Oven film bag (moist heat) | | |
|---|---|---|---|---|---|---|
| | 60 | 70 | 80 | 60 | 70 | 80 |
| CHARACTERISTICS MEASURED | | | | | | |
| Apparent degree of doneness | bright red in centre (5–7 cm), fading to grey-brown at edges, exudes red juice on standing | pink in centre (3–4 cm), fading to grey-brown about half-way to edge | slightly pink centre (2–3 cm) fading to grey-brown throughout the remainder | slightly pink centre (2–3 cm), remainder light grey-brown | uniformly grey-brown | uniformly grey-brown |
| Mean post-oven temperature rise (°C) | <1 | none | none | 7.8 | 5.6 | <1 |
| Surface colour of meat | --rich------dark------brown-- | | | --------grey-brown--------- | | |
| Surface texture | -----------fine grained----------- | | | ------------coarse------------ | | |
| % Cooking losses | | | | | | |
| Total | ---------- 27.69 ---------- | | | ---------- 34.89 ---- | | | p<0.001 |
| Drippings | ---------- 2.16 ---------- | | | ---------- 19.06 ---- | | | p<0.001 |
| Total moisture of cooked lean (%) | ---------- 67.03 ---------- | | | ---------- 63.63 ---- | | | p<0.001 |
| Water-holding capacity, No. | ---------- 0.80 ---------- | | | ---------- 0.76 ---- | | | p<0.01 |

HEAT-INDUCED CHANGES IN BOVINE MUSCLE

Histological studies on cooked meat have shown changes
in fibrous organisation of muscle fibres and collagenous
connective tissue and in staining characteristics of
collagen, often accompanied by increasing tenderness of
the meat. The recognition that hydroxyproline (hypro)
analysis could be used to identify collagen content
suggested that this might be a way of quantifying changes
in ease of extraction of collagen caused by heat. Also,
it seemed worthwhile to see if changes in collagen
correlated with tenderness measures such as depth of
penetration or force required to shear. Information was
also obtained on changes in composition of the muscle
samples caused by known heat treatments.

The experiments previously reported (Paul, McCrae and
Hofferber, 1973) have been extended to include *l.dorsi*
(LD) and *triceps brachii* (TR). Also, a series on strips
of all four muscles heated according to the braising
curve for ST steaks (McCrae and Paul, 1974) has been
completed. Endpoint temperatures used for the braised
samples were 70, 98, 98°C plus 30 min holding at
that temperature and 98°C plus 90 min holding. Both
heating curves are shown in *Figure 21.1*.

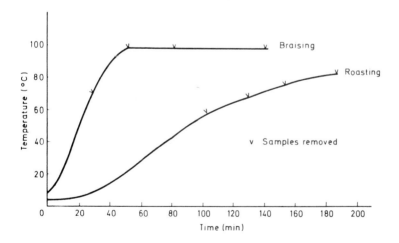

*Figure 21.1   Time-temperature curves for roasting and
for braising*

Data were collected on yield of solid meat and of
fluid expressed during heating, per cent moisture, non-
fat solids and fat content of heated meat and fluid,
per cent extractable collagen (as hypro) in the meat, per
cent hypro in the fluid, force required to shear and
depth of penetration. Methods described in Paul, McCrae
and Hofferber (1973) were used. *Table 21.10* lists some
of the averages and F-values from analysis of variance
for the samples heated according to the roasting curve,
and *Table 21.11* the same for the braising curve. Since
the initial studies on roasting curve heating of ST and
*biceps femoris* (BF) were done on muscles from two
different groups of animals, the data from these could
not be combined for statistical analysis.

*Cooked meat*

The amount of collagen (as hypro) which could be extracted
by water at 40°C increased with increasing heat treatment.
The yield of cooked meat decreased steadily from 58 to
98°C plus 30 min holding time, but did not change with an
additional hour at 98°C.

*Liquid expressed*

The yield of fluid from the heated meat increased with
increasing temperature and time, up to 98°C for 30 min,
but did not increase with the additional hour at 98°C.
The hypro in the juice represented primarily collagen
that had been solubilised by heat. With the roasting
curve, the major increase occurred between 58 and 67°C.
On the braising curve, ST-BF showed significant
increases with each time-temperature increment. While
the per cent hypro in juice of braised LD-TR increased
at each step, only the 98°C plus 30 min to 98°C plus
90 min increase was significant.

*Tenderness measures*

The percentage of the total collagen which was
solubilised by heat treatment increased gradually from
58°C to 98°C, with much larger increases during holding
at 98°C. The percentage converted by braising to 70°C
was smaller than by roasting to 67°C, but the time
required to 70°C by braising was less than half that
required to 67°C by roasting.

Table 21.10 Averages[1] and F- values for samples heated by roasting curve[2]

| | Internal temperature (°C) | | | | F-value |
|---|---|---|---|---|---|
| | 58 | 67 | 75 | 82 | |
| **LD-TR MUSCLE** | | | | | |
| Yield cooked meat (%) | 91.22 | 81.34 | 68.06 | 62.14 | 124.80** |
| Penetration depth (mm) | 9.86 | 8.04 | 6.11 | 5.14 | 59.62** |
| Shear (kg) | 4.50 | 3.25 | 3.49 | 3.48 | 3.47* |
| Yield juice from heating (%) | 9.29 | 19.20 | 32.03 | 38.64 | 112.09** |
| Collagen solubilised by heat (%) | 2.69 | 4.77 | 6.38 | 7.69 | 10.67** |
| Heating time (min) | 103.5 | 130 | 153.5 | 186.5 | |
| **ST MUSCLE** | | | | | |
| Yield cooked meat (%) | 86.22 | 72.92 | 62.85 | 58.18 | 641.24** |
| Penetration depth (mm) | 9.98 | 7.76 | 5.93 | 4.91 | 124.23** |
| Shear (kg) | 3.52 | 3.10 | 3.32 | 3.33 | < 1ns |
| Yield juice from heating (%) | 14.56 | 27.74 | 38.68 | 45.34 | 795.02** |
| Collagen solubilised by heat (%) | 4.25 | 6.04 | 7.95 | 11.03 | 21.76** |
| **BF MUSCLE** | | | | | |
| Yield cooked meat (%) | 89.30 | 75.62 | 64.46 | 58.48 | 296.09** |
| Penetration depth (mm) | 10.38 | 8.78 | 6.12 | 5.37 | 50.39** |
| Shear (kg) | 4.27 | 3.93 | 3.61 | 3.37 | 1.59ns |
| Yield juice from heating (%) | 11.62 | 25.20 | 36.57 | 42.30 | 186.50** |
| Collagen solubilised by heat (%) | 6.28 | 8.68 | 8.46 | 13.57 | 19.17** |

[1]Averages underlined at the same level are not significantly different by Duncan's multiple range test.

ns: non-significant; *: significant, $p<0.05$; **: highly significant, $p<0.01$

[2]Data on composition changes may be obtained in mimeographed form from the author.

*Table 21.11* Averages[1] and F- values for samples heated by braising curve

| | Internal temperature, $^{\circ}$C (+ min) | | | | F-value |
|---|---|---|---|---|---|
| | 70 | 98 | 98 (+30) | 98 (+90) | |
| **LD-TR MUSCLE** | | | | | |
| Yield cooked meat (%) | 81.85 | 61.06 | 58.53 | 58.68 | 331.76** |
| Penetration depth (mm) | 8.22 | 4.60 | 4.79 | 6.85 | 24.18** |
| Shear (kg) | 3.93 | 3.40 | 2.85 | 2.22 | 30.45** |
| Yield juice from heating (%) | 18.79 | 39.19 | 43.31 | 44.61 | 219.05** |
| Collagen solubilised by heat (%) | 3.66 | 11.25 | 21.50 | 44.91 | 173.38** |
| Heating time (min) | 29 | 52 | 82 | 142 | |
| **ST-BF MUSCLE** | | | | | |
| Yield cooked meat (%) | 80.40 | 57.84 | 54.90 | 55.14 | 869.34** |
| Penetration depth (mm) | 7.06 | 3.41 | 3.22 | 4.11 | 157.93** |
| Shear (kg) | 4.39 | 3.78 | 2.98 | 2.16 | 17.14** |
| Yield juice from heating (%) | 20.34 | 42.68 | 45.18 | 45.17 | 533.64** |
| Collagen solubilised by heat (%) | 4.80 | 10.23 | 23.33 | 52.02 | 392.40** |

[1]Averages underlined at the same level are not significantly different by Duncan's multiple range test.

**: highly significant, $p < 0.01$

The softness of the meat, as measured by depth of penetration, decreased with increasing internal temperature on the roasting curve, and from 70 to 98°C on the braising curve. However, holding for 30 min at 98°C did not produce any additional hardening, while additional holding appeared to soften the meat. The force required to shear did not change appreciably in meat heated by the roasting curve, but decreased with increasing time - temperature on the braising curve.

Correlation coefficients among these three measures are given in *Table 21.12*. The highly significant correlation coefficients suggest increasing tenderness with increasing collagen solubilisation in braising, but increasing hardness of the meat despite increasing collagen solubilisation in roasting. Overall, it appears that very large changes in collagen will increase tenderness as measured by shear force and by penetration, but that the smaller changes occurring during heating by the roasting curve and the first part of the braising curve are not particularly effective in this respect.

*Muscle differences*

Where the pairs of LD-TR and BF-ST muscles came from the same animal, it was possible to isolate statistically variations between the two muscles in each pair. These data are summarised in *Table 21.13*.

In general, the meat tended to maintain the same pattern of differences among muscles in both heating sequences. However, with a few measures, the muscles responded differently, depending on the heating curve used (*Table 21.14*).

When one looks at the parts of the carcass, rib and round, represented by the two sets of muscles, the major changes with method of heating were in percentage of extractable hypro in the meat and in penetration depth. The round muscles had higher percentage extractable hypro by roasting, but by braising TR and BF were higher than ST, with LD the lowest. The braising results may reflect partly the differences in collagen hypro of the raw meat, since TR was highest and LD lowest. For depth of penetration, the round muscles were somewhat more susceptible than the rib to the hardening effect of the braising time-temperature.

Table 21.12 Correlation coefficients among 'tenderness' measures between muscles

| | Roasting | | | Braising | |
| --- | --- | --- | --- | --- | --- |
| | LD-TR | ST | BF | LD-TR | ST-BF |
| % collagen (as hypro) solubilised by heat × shear | -0.44** | 0.09ns | -0.45 | -0.66** | -0.52** |
| % collagen (as hypro) solubilised by heat × penetration depth | -0.64** | -0.80** | -0.55** | -0.13* | -0.36* |
| Penetration depth × shear | 0.18ns | 0.06ns | 0.05ns | 0.30* | 0.26ns |

ns: not significant
*: $p < 0.05$
**: $p < 0.01$

Table 21.13 Averages and F- values showing variation between muscles

|  | Muscle | | | | | |
|---|---|---|---|---|---|---|
|  | LD | TR | F-value | ST | BF | |
| **ROASTING** | | | | | | |
| Yield cooked meat (%) | 75.39 | 75.99 | <1ns | 70.04 | 71.97 | |
| Penetration depth (mm) | 6.90 | 7.67 | 7.92** | 7.14 | 7.66 | |
| Shear (kg) | 2.63 | 4.73 | 48.79** | 3.32 | 3.80 | |
| Yield juice from heating (%) | 25.27 | 24.31 | <1ns | 31.58 | 29.00 | |
| Collagen solubilised by heat (%) | 6.21 | 4.56 | 6.23* | 7.32 | 9.25 | |

|  | Muscle | | | | | |
|---|---|---|---|---|---|---|
|  | LD | TR | F-value | ST | BF | F-value |
| **BRAISING** | | | | | | |
| Yield cooked meat (%) | 64.52 | 65.54 | 2.73ns | 61.60 | 62.54 | 5.08* |
| Penetration depth (mm) | 5.26 | 6.97 | 23.62** | 4.46 | 4.45 | <1ns |
| Shear (kg) | 2.20 | 4.00 | 184.47** | 3.86 | 2.79 | 20.77** |
| Yield juice from heating (%) | 36.52 | 36.42 | <1ns | 38.89 | 37.79 | 4.44* |
| Collagen solubilised by heat (%) | 25.14 | 15.52 | 49.93** | 24.55 | 20.64 | 13.45** |

ns: not significant
*: significant, $p < 0.05$
**: highly significant, $p < 0.01$

*Table 21.14*  **Effect of heating method on responses of
different muscles**

| Measure | Roasting curve | | Braising curve | |
|---|---|---|---|---|
| | Lowest | Highest | Lowest | Highest |
| Penetration depth | LD | TR | BF | TR |
| Percentage solubilised collagen | TR | BF | TR | LD |

*Heating curve*

The roasting curve gave a bigger yield of cooked meat,
greater depth of penetration and slightly higher force
required to shear than did the braising curve.  The
quantity of juice expressed during heating was lower by
roasting than by braising.  Braising solubilised about
three times as much collagen.


## CONCLUSION

Many of the heat-induced changes in meat that influence
its acceptability involve heat denaturation and
coagulation of proteins and surface pyrolysis of proteins
and lipids.  So the method chosen for heating the meat is
very important in determining the characteristics of the
product.  Selection of the appropriate heating method
should take into account previous research results that
show the importance of cut of meat, container, type of
energy supply, heat-transfer medium, heating time and
temperature, internal temperature endpoint, and total
time requirement.  Attention to such details will increase
the probability of obtaining results suitable for the
analytical methods and types of data to be collected in
fulfilling the objectives of the research.  Since most
studies have the eventual aim of producing an acceptable
product for consumption, the potential for usefulness of
the results in home or commercial preparations should
also be considered.

REFERENCES

AHEA Terminology Committee (1971). *Handbook of Food Preparation*, 6th edn. rev. Amer. Home Econ. Assoc., Washington, D.C.

BAITY, M.R., ELLINGTON, A.E. and WOODBURN, M. (1969). *J. Home Econ.*, **61**, 174

BAYNE, B.H., ALLEN, M.B., LARGE, N.F., MEYER, B.H. and GOERTZ, G.E. (1973). *Home Econ. J. Res.*, **2**, 29

BENNETT, M.E., BRAMBLETT, V.D., ABERLE, E.D. and HARRINGTON, R.B. (1973). *J. Fd Sci.*, **38**, 536

Committee on Preparation Factors, National Cooperative Meat Investigations (1942). *Meat and Meat Cookery*, National Livestock and Meat Board, Chicago

COVER, S. and HOSTETLER, R.L. (1960). *Texas Agri. Exper. Sta. Bull.*, No. 947

CULOTTA, J.T. and CHEN, T.C. (1973). *J. Fd Sci.*, **38**, 860

DECAREAU, R.V. (1967). *Proc. 20th Ann. Reciprocal Meat Conference*, p.216. National Livestock Meat Board, Chicago, Ill.

HARRIES, J.M., RHODES, D.N. and CHRYSTALL, B.B. (1972). *J. Texture Studies*, **3**, 101

HOSTETLER, R.L., LINK, B.A., LANDMANN, W.A. and FITZHUGH, H.A., Jr. (1973). *J. Fd Res.*, **38**, 264

IRMITER, T.F., ALRICH, P.J. and FUNK, K. (1967). *Fd Technol.*, *Champaign*, **21**, 779

JAKOBSSON, B. and BENGTSSON, N. (1973). *J. Fd Sci.*, **38**, 560

JIMÉNEZ, M.A. (1972). *Fd Technol.*, **26**, No. 9, 36

KELLER, J.D. and KINSELLA, J.E. (1973). *J. Fd Sci.*, **38**, 1200

LAAKKONEN, E. (1973). *Adv. Fd Res.*, **20**, 257

LEE, W.T. and DAWSON, L.E. (1973). *J. Fd Sci.*, **38**, 1232

LEWIS, P.K., Jr., CAMPBELL, K.R., YOUNGER, L., HECK, M.C. and BROWN, C.J. (1973). *Amer. Soc. Anim. Sci. Meeting*, July 30–August 2, 1973, Lincoln, Nebraska

LOCKER, R.H. and DAINES, G.J. (1973). *J. Sci. Fd Agri.*, **24**, 1273

MCCRAE, S.E. and PAUL, P.C. (1974). *J. Fd Sci.*, **39**, 18

PAUL, P.C. (1972). In *Food Theory and Applications*, pp.408–412. Ed. P.C. Paul and H.H. Palmer. John Wiley and Sons, Inc., New York

PAUL, P.C. (1973). Unpublished data

PAUL, P.C., MCCRAE, S.E. and HOFFERBER, L.M. (1973). *J. Fd Sci.*, **38**, 66

PENNER, K.K. and BOWERS, J.A. (1973). *J. Fd Sci.*, **38**, 553

RHODES, D.N. (1971). *J. Sci. Fd Agri.*, <u>22</u>, 485

SCHOCK, D.R., HARRISON, D.L. and ANDERSON, L.L. (1970). *J. Fd Sci.*, <u>35</u>, 195

SHAFFER, T.A., HARRISON, D.L. and ANDERSON, L.L. (1973), *J. Fd Sci.*, <u>38</u>, 1205

SPRAGUE, E.C. and GRINDLEY, H.S. (1907). *Univ. Illinois, Univ. Studies*, Vol. II, No. 4

WASSERMANN, A.E. (1972). *J. Agri. Fd Chem.*, <u>20</u>, 737

# V

## NUTRITIONAL ASPECTS

# 22

# NUTRITIONAL VALUE OF MEAT

A. E. BENDER

*Department of Nutrition, Queen Elizabeth College,
University of London*

## INTRODUCTION

For a variety of different reasons meat occupies a
particular place in meat-eating communities - it has
prestige value, its taste appears to be particularly
attractive and it is generally regarded as having high
nutritive value. A popular survey (Anon., 1969) placed
meat equal second with fruit and behind butter with a
91% popularity score; it was eaten by 99% of the people
questioned, the highest score of all foods. The popular
impressions of the nutritional virtues of meat are summed
up by the 69% who thought it 'body building in growing
children' (behind milk and eggs), 40% who thought it
'good for expectant and nursing mothers', for 'building
resistance to disease', for energy (second only to sugar)
and 'good for the blood' (second only to green vegetables)

In Great Britain meat accounts for 30% of all money
spent on food, a figure that has remained fairly constant
between 1956 and 1971 despite changing prices and incomes.
This is about three times as much as is spent on dairy
produce, or cereals or fruit or vegetables.

The amount of meat eaten falls as one passes down the
income groups but not to a very great extent. There is
a marked regional variation, Scots being the biggest beef
eaters and the smallest pork eaters, as well as the
smallest eaters of mutton and lamb. Londoners eat 43 oz
(1221 g) of meat each per week; the national average is

39 oz (1108 g) and the smallest eaters of meat are the
Scots and the inhabitants of the smaller towns with 36 oz
(1022 g) per week.


## PROTEIN

*Table 22.1* shows that in 1971 meats of all kinds
contributed 17% of the intake of energy, 28% of the
protein and 30% of the fat of the average UK diet.  At
the same time meat contributed 28% of the iron, 20% of
vitamins $B_1$ and $B_2$ and 37% of the niacin (National Food
Survey, 1973).

Meat is popularly regarded as being nutritionally
superior to plant foods but so far as protein quality is
concerned the difference is not very great.  The Biological
Values (BV) of individual plant proteins are generally
lower than those of individual animal proteins but even
this is not always true, e.g. BV meat 75, soya 70 and
gelatin 0.  When we consider whole diets as distinct
from single foods, the difference is less marked since
the BV of a mixture of plant proteins can be as high as
that of animal foods if there is adequate complementation
between the amino acids of the proteins.  The average
quality of Western diets, based largely on two animal and
one vegetable source, namely meat, milk and wheat, is 80,
while that of the worst-fed areas in the world based on
plant proteins is 70; 'situations may exist where quality
may be as low as 60' (WHO, 1973).

One reason why diets are examined for their content of
animal and plant proteins is that the amount of animal
food serves as an overall index of quality since it
includes vitamins and minerals that are not found to such
an extent in plant foods.  A second reason is that
animal foods are generally more concentrated sources of
protein so that it is easier (and more attractive) to
increase the amount of protein in the diet with animal
foods.  Protein quantity compensates for quality so that
60 g of a diet BV 60 is equivalent to 45 g of a diet
BV 80.  Unfortunately, areas where protein is of the lower
quality also tend to have only limited amounts of protein-
rich foods available.

There is a remarkable similarity in amino-acid
composition, i.e. protein quality, of the muscle of
animals of different species and also between the different
tissues.  Crawford (1968) showed that despite differences

Table 22.1 Contribution to the average British diet (1971) by meat of various types – per cent of total intake (National Food Survey Committee, 1973)

| | Intake per person per week (oz) | (g) | Energy | Protein | Fat | Iron | Vitamin $B_1$ | Vitamin $B_2$ | Niacin |
|---|---|---|---|---|---|---|---|---|---|
| Beef and veal | 8.0 | (227) | 3 | 6.8 | 5.2 | 9.2 | 0.9 | 3.5 | 9.5 |
| Mutton and lamb | 5.4 | (153) | 2.3 | 3.8 | 4.3 | 2.9 | 1.4 | 2.7 | 5.8 |
| Pork | 3.0 | (85) | 1.7 | 1.8 | 3.5 | 0.8 | 5.2 | 1.2 | 3.2 |
| Bacon (uncooked) | 5.1 | (145) | 3.2 | 2.8 | 6.8 | 1.5 | 4.8 | 1.6 | 2.0 |
| Liver | 0.8 | (23) | 0.2 | 0.8 | 0.2 | 3.3 | 0.6 | 5.5 | 2.6 |
| Poultry (uncooked) | 4.7 | (134) | 0.8 | 3.7 | 0.8 | 1.6 | 0.4 | 1.2 | 4.8 |
| Sausages | 3.7 | (105) | 1.9 | 2.1 | 3.2 | 1.5 | – | 0.7 | 1.7 |
| Other meat | 8.2 | (234) | 3.6 | 6.5 | 5.7 | 7.6 | 5.5 | 4.3 | 7.0 |
| Total meat | 39.0 | (1107) | 16.7 | 28.1 | 29.8 | 28.2 | 18.8 | 20.6 | 36.6 |

| | Intake per person per week | |
|---|---|---|
| | (oz) | (g) |
| [1]OTHER MEAT: | | |
| Offals other than liver | 0.5 | (14) |
| Bacon, ham (cooked and canned) | 0.9 | (26) |
| Chicken, cooked | 0.2 | (6) |
| Corned meat | 0.4 | (12) |
| Other canned and cooked meat | 2.5 | (70) |
| Rabbit, game | 0.1 | (3) |
| Meat and sausage pies | 0.7 | (20) |
| Other meat products | 2.8 | (80) |

in species and in diet, ranging from grass to acacia bush
for herbivores and from zebra to warthog for carnivores,
the amino-acid composition of leopard, hyena, man,
warthog, kob, buffalo, topi, hartebeeste, eland, giraffe,
elephant, domestic ox, pig and sheep were remarkably
constant.  The composition of the protein tissues is
under genetic control and diet has no influence.
Fattening bulls fed on diets in which urea replaced half
of the conventional protein showed no change in chemical
composition or BV of muscle (Patjas, Sommu and Palanska,
1972).

*Table 22.2* shows the lysine and sulphur amino-acid
content of a variety of meats; lysine is constant between
0.51 and 0.57% and methionine plus cystine between 0.21
and 0.26%.  *Table 22.3* shows the constancy of these
amino acids in different parts of the body, even
including such diverse tissues as brains and giblets.
Lysine varies only between 0.42 and 0.52% and the sulphur
amino acids between 0.21 and 0.23% (excluding pancreas
for which cystine values were not given).

Consequently it is generally stated that all types of
meat, including the cheapest as well as the more
expensive cuts, have the same nutritive value (Sinclair
and Hollingsworth, 1969).  One report, however, Dvorak
and Vognarova (1969) provides theoretical and some
experimental evidence that this is not so.

These authors point out that the amount of connective
tissue differs in different cuts of meat and since there
is very little cystine and methionine in collagen and
elastin, this should reduce the nutritive value of meat
rich in connective tissue; collagen is also low in
lysine (*Table 22.3*).  They show that this is true for
total essential amino acids, 28.6% for beef shank and
42.7% for beef fillet; 29.2% for veal shank and 39.1%
for veal fillet.  Methionine is also shown to be much
lower in the 'inferior' cuts (the beef figures being in
g/16 g N): shank 1.45, flank 1.85, shoulder 2.13 and
fillet 2.73.  Similar figures for veal are shank 1.28,
flank 1.60, round 1.92 and fillet 2.56.

Dvorak and Vognarova also show that when the hydroxyproline (an index of connective tissue) is less than
1.2 g/16 g N,  methionine is the limiting amino acid, but
this changes to phenylalanine when the amount of hydroxyproline is above this level.  Unfortunately cystine and
tyrosine were not measured so that it is not possible to
calculate a chemical score, and no biological work was
carried out to verify this calculation.

*Table 22.2* Lysine and sulphur amino acids in muscle of various species (Orr and Watts, 1957)

| Meat | Lysine | Methionine | Cystine |
|------|--------|------------|---------|
| | | (g/g total N) | |
| Beef | 0.55 | 0.16 | 0.08 |
| Veal | 0.52 | 0.14 | 0.07 |
| Pork | 0.51 | 0.15 | 0.07 |
| Mutton | 0.51 | 0.15 | 0.08 |
| Rabbit | 0.54 | 0.16 | – |
| Chicken | 0.55 | 0.16 | 0.08 |
| Duck | 0.54 | 0.16 | – |
| Turkey | 0.57 | 0.17 | 0.09 |

*Table 22.3* Lysine and sulphur amino acids in different tissues (Orr and Watts, 1957)

| Tissue | Lysine | Methionine | Cystine |
|--------|--------|------------|---------|
| | | (g/g total N) | |
| Brain | 0.46 | 0.13 | 0.09 |
| Giblets | 0.49 | 0.14 | 0.08 |
| Heart | 0.51 | 0.15 | 0.06 |
| Kidney | 0.45 | 0.13 | 0.08 |
| Liver | 0.47 | 0.15 | 0.08 |
| Pancreas | 0.46 | 0.11 | – |
| Tongue | 0.52 | 0.14 | 0.08 |
| Muscle | 0.55 | 0.16 | 0.08 |
| Collagen | 0.27 | 0.05 | 0.005 |

## IRON

Anaemia is a common nutritional problem. An estimate from the World Health Organization puts the world-wide prevalence at 20% and even in Great Britain it affects 2-8% of men and 10-20% of women.

One remedy, that of enriching a staple food with iron salts, presents severe problems. The iron added to the bread in Great Britain for the past 25 years is poorly absorbed (HMSO, 1968) and if enough is added to provide an effective dose, about 10 mg per day (Anon, 1973a) it affects baking properties and induces rancidity on storage. Moreover, sufferers from the rare disease of haemochromatosis would be at serious risk from extra iron supplementation of staple foods.

One of the nutritional virtues of meat is its iron content and availability. Much of the dietary iron is poorly absorbed but the iron in meat products is not only relatively well absorbed itself but appears to assist the absorption of iron from other foods.

The availability of iron from various foods is not only low but variable; absorption is enhanced by ascorbic acid and protein and decreased by phytate and phosphates, enhanced by the simultaneous consumption of meat and markedly decreased by the simultaneous consumption of egg (HMSO, 1968). Absorption varies with the chemical form of the iron. Against this background meat occupies a special position in the diet.

The importance of meat in this respect is illustrated by an experiment on 200 subjects using maize, wheat and veal labelled with radioactive iron (Martinez-Torres and Layrisse, 1971; Layrisse *et al.*, 1972). When iron salts were added to vegetable foods the absorption was very low, 0.3 mg from an intake of 60 mg. Meat differed markedly, 5 mg of iron in a veal hamburger resulted in the absorption of 0.85 mg.

The experiment revealed the enormous variation between individuals. An average absorption of 23% of the iron from veal covered a range from 10 to 48%. The average absorption of 30% of iron from veal supplemented with ferric chloride covered a range from 6 to 77%.

Maize appeared to reduce the absorption of iron from ferric chloride even when fed with veal, but did not affect the absorption of iron from veal alone.

The most dramatic figures were obtained with the standard diet of the region, a mixture of maize, beans and rice. Only 0.09 mg of iron was absorbed by normal

subjects and 0.3 mg by iron-deficient subjects. When
veal was added to make one-quarter of the meal 0.46 mg
was absorbed by normal subjects and 0.93 mg by deficient
ones. Meat doubled the absorption of iron from the
vegetable foods as well as contributing its own iron.
Eighty-six per cent of the iron was present in the veal
as haem, and 15-20% as ferritin. The iron content of
meat can vary considerably (Jacobs and Greenman, 1969).
The iron in six samples of cooked steak ranged between
2.9 and 4.65 mg/100 g; in the raw form the figures ranged
between 2.10 and 3.25; cooked sausages ranged between
1.40 and 3.90; corned beef showed the biggest variation,
in five samples the range was 2.5 to 9.0 mg. While some
of the iron in the cooked meat, the sausages and the
corned beef may have come from processing, the figures
for raw steak must presumably be genuine variability of
the tissue.


## COPPER AND OTHER MINERALS

Copper has the distinction of having only a tenfold range
between requirements and harmful level. The requirements
are estimated at 2 mg/day and the average daily intake
is estimated at 2-5 mg/day; the maximum acceptable daily
load is 0.5 mg/kg body weight, i.e. 30 mg for an adult
(FAO/WHO, 1971). Copper is fed to pigs as a growth
stimulant at levels of 150 ppm of diet and the amount
found in livers can be as high as 100 mg/kg.

Trace minerals are present in many animal tissues and
it is not always clear whether they play an essential
role in that tissue or whether they are there only because
the animal consumed them with the diet. Tamate and
Ohtaka (1972) found aluminium, silicon, zinc, lead and
manganese in all pig samples examined and, in some
samples only, silver, chromium, molybdenum, barium and
tin.


## POLYUNSATURATED FATTY ACIDS

Their relative lack of polyunsaturated fatty acids (PUFA)
has somewhat detracted from the nutritional virtues of meat
fats in recent years. Not only are PUFA of benefit to
a variety of disorders ranging from schizophrenia to
ischaemic heart disease, but saturated animal fats may

be harmful. In attempts to rectify this situation
vegetable oils rich in PUFA have been fed to chickens and
pigs with a resultant increase in the PUFA content of
their products. For example pigs normally have 2-4% PUFA
in their tissues (except for fat from the heart with 20%)
but after feeding 20% soya-bean oil, the fat from the
various cuts of meat rose to 40-47% linoleic acid (Brooks,
1971) with 4-8% arachidonic acid in the liver fat. This
cannot be practised in ruminants since the unsaturated
fats are hydrogenated in the rumen. However, recent
technology has led to a stabilised form of the polyunsaturates which escape this fate when consumed by the
animal (Cook *et al.*, 1970 and 1972). A mixture of equal
parts of safflower oil and casein is treated with
formalin which reduces the digestibility of the protein
and protects the fat so that it can proceed to the
abomasum where it is absorbed unchanged. When fed to
lambs at 20% level there was an increased amount of
linoleic acid in the carcass fat; the amount in the
mesenteric fat rose from 2% to 11% in 3 weeks and to 16%
after 6 weeks. Similar results were obtained in cattle
and the milk from a cow fed 1.5 kg of the safflower oil-
protein mixture showed an increase in the linoleic acid
of the milk from 2% of the total fatty acids to 35%,
with corresponding falls in the saturated and mono-
unsaturated fatty acids (Plowman *et al.*, 1972).


## OTHER FUNCTIONAL EFFECTS


Meat extract and therefore, by extrapolation, cooked
meat contains a factor that stimulates the secretion of
gastric juice (Wood, Adams and Bender, 1962). There is
no evidence that this plays any essential role in
digestion but it may be one of the reasons why meat is
regarded so highly.


## ENVIRONMENT, FEED AND BREED


The constancy of the amino-acid composition of muscle,
despite differences of species and of diet, has already
been discussed. Vitamin content of tissues depends to
a large extent upon diet but false conclusions have
sometimes been reached by comparing animals fed under

controlled and free conditions. Under free conditions
an animal may consume excessive amounts of a nutrient,
for example far more vitamin A than it needs for its own
metabolism, and store the surplus in the tissues. Under
controlled feeding the supply of the nutrient may be
limited to the animal's own needs with little surplus
stored in the tissues. So it is possible to find eggs
from free-range chickens with considerably more vitamin A
than battery-fed ones and livers of grass-eating cattle
with more vitamin A than stall-fed animals. This has
sometimes been interpreted as a reflection of the
conditions of husbandry but is simply a reflection of
the diet. If the farmer wants to produce eggs or livers
rich in vitamin A he can simply feed more to the animals.
Vitamin $B_2$ has been fed to pigs as a growth stimulant
and higher levels accumulate in the meat (Antipova, 1972).

Species, breed, feed and environment all affect both
the amount and type of fat. Animals living in woodland
in Uganda and Tanzania have 2% lipid in their meat
containing 30% polyunsaturated fatty acids; those
grazing on grassland have 3% lipids with 15% PUFA. Lean
domesticated cattle have 5% lipids with 8% PUFA; the
average for all domesticated cattle that were examined
was 9% lipids, 3% PUFA (Crawford *et al.*, 1970).

Observations of this kind have led to suggestions that
it is more 'natural' to consume animal fats with the
higher PUFA content and that domestication of animals has
led to an impairment of human health for this reason.
However, the comparisons cannot be valid since they are
made between entirely different species and breeds, and
while the domesticated animals are well-fed and cared
for, animals living under natural conditions may be
under-nourished. There is certainly no reason to believe
that in the wild, where animals have to hunt their food
or search the fields, their body composition is optimal,
either for their own survival or as food for man.


## SEASONAL VARIATIONS


Nutritional values of animal tissues vary seasonally,
partly for physiological reasons but probably more
because of changes in diet. Vitamin A in cattle liver
in one series of measurements (Antila, Varesmaa and
Niinivaara, 1968) was seven times higher in autumn than
early summer, while that of pig liver was twice as high
in winter as in summer; vitamin A content of the liver

is a direct reflection of the diet.  In the same series
of observations no seasonal change was observed in
vitamin $B_1$ content (0.21-0.39 mg/100 g cattle liver,
0.28-0.36 mg/100 g pig liver); iron varied little with
the season (12 mg in cattle and 20 mg in pig); copper
was also constant (5 mg in cattle and 2 mg in pig).
Linoleic acid content of cattle liver was lower than
that of pig liver but arachidonic acid was higher, with
little seasonal variation.


## PROCESSING AND COOKING

Nutritional losses during processing and cooking depend
on the temperature, time, heat penetration, size of
sample, composition and method of applying heat.
Consequently the mass of data accumulated in the
literature cannot be expected to do other than indicate
general effects.

So far as proteins are concerned the changes that take
place affect BV only when the limiting sulphur amino
acids are reduced; lysine is in surplus and so losses of
lysine can occur without any fall in BV.  There could be
a fall in the nutritive value of the complete diet in
which the affected meat is included but only when, as
happens rarely, lysine is the limiting amino acid in the
diet.

The BV of fresh, uncooked meat is 75.  One type of
damage that occurs in proteins on heating and more slowly
on storage is a linkage of the amino acids in a form that
cannot be hydrolysed during digestion so that they are
biologically unavailable.  Since chemical analysis is
always preceded by acid hydrolysis this is revealed only
by biological assay.

The most thoroughly investigated of the reactions of
this type is the Maillard reaction in which lysine is
linked and rendered unavailable.  Microbiological methods
have shown that other amino acids can also be rendered
unavailable although little is known of the chemistry of
these reactions.

Lysine was lost, according to one report, at as low a
temperature as 70°C after 3 h; at 121°C the loss was 20%,
at 140°C, 40% and at 160°C, 50%   (Dvorak and Vognarova,
1965).  Bognar (1971) reported that only methionine and
cystine were lost on cooking, the former by oxidation
and the latter as $H_2S$, with small amounts of both being
leached into the cooking water.  Losses amounted to 6-8%

of the essential amino acids when cooked at 100°C for
180 min and 8-12% when cooked at 120°C for 50 min.
Chemical score indicated a fall of 7-14%. Beuk, Chornock
and Rice (1949) showed that the availability of several
amino acids was reduced when pork was autoclaved. After
24 h at 112°C, 44% of the cystine was destroyed and 70%
rendered unavailable. More detailed investigation
(Donoso *et al.*, 1962) showed that both sulphur amino
acids and lysine were reduced when pork was heated. The
sample was cooked in water at 110°C for 24 h then dried
at 100°C for 6 h. NPU (i.e. BV x Digestibility) fell
from 76 to 41, available lysine fell by 34% and 20% was
destroyed; 16% of the methionine and 44% of the cystine
were destroyed. Addition of methionine partly restored
the NPU to 60, indicating that lysine loss was
responsible for the rest of the damage.

It is the temperature reached by the meat rather than
the cooking temperature that is important; there are
reports of no damage when meat is roasted in an open pan
(Mayfield and Hendrick, 1949) but the internal temperature
reached only 80°C.

The Maillard compounds formed during heating confer an
attractive flavour on the meat even at the expense of a
small reduction in nutritional value. A dehydrated beef-
pork mixture prepared at the Ministry of Food Experimental
Factory showed no loss in nutritive value during
processing and only a small loss, namely a fall from
BV 76 to 69 after 15 months storage (Bender, 1962).
There must have been a considerable reduction in lysine
if the BV fell and evidence for the formation of Maillard
compounds is suggested by the observation that
palatability improved during the first 3 months storage.

Mayfield and Hendrick (1949) reported no loss on
corning of beef but Bender (1962) reported a 30% fall
in NPU to 55, on corning, but no further loss after 9
years storage.

Two extremely old samples of canned meat, obtained
from the British Food Manufacturing Industries Research
Association, indicated the type of change that eventually
takes place. These were the oldest samples of canned food
recorded, one was veal canned in 1823 and the other mutton
canned in 1849; they were opened and examined in 1959.
The cans were sterile, the fat had decomposed and appeared
as a surface scum but the meat appeared to have retained
its texture. However, on drying by stirring with acetone
preparatory to incorporating into animal diets, the
fibrous material fell to a powder, indicating that the
structure had broken down. The NPU of the veal was 29

and that of the mutton 27. Column chromatography of the
amino acids after an acid hydrolysis showed that all the
amino acids were present to the same extent as in fresh
meat, so it must be deduced that the fall in nutritive
value was due to unavailability of the amino acids
(Bender, 1962).


## VITAMINS


A general figure of the vitamin losses in cooking meat is
that one-third of $B_1$, $B_6$, $B_{12}$ and pantothenic acid is
lost and less than one-tenth of the $B_2$ and niacin. After
freezing 10% of the water-soluble ingredients may be lost
in the drip on thawing.  In canning up to 10% of the $B_2$
and niacin, 20% biotin, 20-30% pantothenic acid and
20-40% $B_1$ may be lost (Harris and von Loesecke, 1960).
For vitamin $B_1$ the same source gives values of 15-40% on
broiling, 40-50% on frying, 30-60% on roasting, and
50-75% on canning.  A recent publication (Rognerud, 1973)
reports retention of vitamin $B_1$ on cooking broiler
chickens; 58% retention on boiling, 65% on frying, 76%
on broiling and 65% in a commercial barbecue machine.
  Some general figures for losses of vitamins in
experimentally cooked meats are given in *Tables 22.4* and
*22.5*.  The standard values for some meat products are
given in *Tables 22.6, 22.7, 22.8* and *22.9*.

*Table 22.4* Retention of vitamin $B_1$ in meat cooked at
three different temperatures in the conventional oven
compared with electronic cooking (Lushbough et al., 1962)

| Beef round | | Retention (%) |
|---|---|---|
| Oven cooked | | |
| Temperature 93°C | Inner part | 88 |
| | Outer part | 77 |
| Temperature 149°C | Inner part | 88 |
| | Outer part | 102[1] |
| Temperature 204°C | Inner part | 67 |
| | Outer part | 60 |
| Electronic | Inner part | 86 |
| | Outer part | 67 |

[1]Result possibly due to flow of tissue fluids from the
centre to the dried outer parts of the meat with local
destruction and concentration of water-soluble substances
including vitamin $B_1$.

Table 22.5 Retention of vitamins $B_1$ and $B_2$ in different cuts of braised meat (Noble, 1965)

| | Weight (lb) | Time of cooking (min/lb) | Vitamin $B_1$ (% retention) | Vitamin $B_2$ (% retention) |
|---|---|---|---|---|
| BEEF | | | | |
| Short ribs | 4.5 | 30 | 25 | 58 |
| Chuck | 6 | 35 | 23 | 74 |
| Flank steak | 1.75 | 28 | 30 | 72 |
| Round (roast) | | 27 | 40 | 73 |
| Round (steak) | | 18 | 40 | 65 |
| VEAL | | | | |
| Chops | 1.75 | 28 | 38 | 73 |
| Round steak | 1.75 | 27 | 48 | 76 |
| PORK | | Total time of cooking | | |
| Chops | | 50 min | 44 | 64 |
| Spare ribs | | 2 h | 26 | 72 |
| Tenderloin | | 40 min | .57 | 83 |

*Table 22.6* Standard values for nutrient composition of meat (per 100 g) (McCance and Widdowson, 1960)

| | Protein (g) | Fat (g) | Energy (kcal) | B₁ | B₂ | Niacin |
|---|---|---|---|---|---|---|
| | | | | $B_1$ | $B_2$ | Niacin |
| | | | | | Vitamins (mg) | |
| Beef raw | 20.3 | 7.3 | 150 | 0.07 | 0.20 | 5 |
| roast | 21.3 | 32.1 | 385 | 0.05 | 0.22 | 5 |
| Chicken roast, breast | 29.6 | 7.3 | 190 | 0.05 | 0.06 | 8 |
| leg | | | | | 0.20 | 4 |
| Mutton, leg roast | 25.0 | 20.4 | 290 | 0.10 | 0.25 | 4.5 |
| Pork, roast | 24.6 | 23.2 | 320 | 0.80 | 0.20 | 5 |

*Table 22.7*  Vitamins in cooked meats (Minoccheri and Cantoni, 1972)

| | $B_1$ (µg/100 g) | $B_2$ (µg/100 g) | $B_{12}$ (µg/100 g) |
|---|---|---|---|
| Italian hams (6 samples) | 485–535 | 149–172 | 4.5–4.8 |
| Liver sausage (2 samples) | 215–325 | 682–1290 | 59–107 |
| Mortadella (23 samples) | 87–151 | 92–195 | 3.1–4.4 |

*Table 22.8*  Nutrients in blood from slaughterhouses (per 100 ml) (Frentz and Perron, 1972)

| | |
|---|---|
| Protein | 18.6–20.8% |
| Vitamin C | 2 mg |
| Carotenoids | 7.5 µg |
| Retinol | 25 µg |
| Folic acid | 0.66 µg |
| Niacin | 0.47 µg |
| Pantothenic acid | 33.5 µg |
| $B_2$ | 3 µg |
| $B_{12}$ | 1 ng |
| D | 0.25–0.3 µg |

*Table 22.9*  Nutritional value of meat protein (FAO, 1968)

| | |
|---|---|
| Beef and veal | BV 74 (62–78) |
| Beef heart | NPU 67 |
| Beef liver | NPU 65 |
| Chicken | NPU 74 (72–78) |
| Pork | NPU 74 |

(NPU = BV × Digestibility)

## IS MEAT ESSENTIAL?

Vegetarians can develop anaemia through lack of iron and vitamin $B_{12}$ (Anon., 1973b) and although a selection of vegetables will provide an adequate supply of iron the problem of its absorption remains. There are no good non-meat sources of vitamin $B_{12}$ and the advice offered in the *British Medical Journal* was to take one or more of four proprietary products, one made from yeast, some varieties of which do contain vitamin $B_{12}$, and the others were preparations containing the $B_{12}$ from fermentation sources. However, the advice given concludes that even with these products difficulty may still be experienced in obtaining the recommended daily intake of 3-4 µg vitamin $B_{12}$. Such statements suggest that meat is an essential article of diet but it must be borne in mind that about half the world is vegetarian, from necessity if not from choice, and still manages to survive.

Anon. (1969). *Food Facts & Fallacies*. Margarine & Shortening Manuf., London

Anon. (1973a). *Lancet*, ii, 189

Anon. (1973b). *Br. Med. J.*, 2, 402

ANTILA, P., VARESMAA, E. and NIINIVAARA, F.P. (1968). *J. Sci. Agric. Soc. (Finland)*, 40, 19-21 cited in *Nutr. Abst. Rev.* (1969), 39, Abstr. 4534

ANTIPOVA, N.I. (1972). *Nutr. Abst. Rev.*, 42, Abstr. 4697

BENDER, A.E. (1962). *Proc. 1st Internat. Cong. Fd Sci. and Tech.*, 3, 449

BEUK, J.F., CHORNOCK, F.W. and RICE, E.E. (1949). *J. Biol. Chem.*, 180, 1243

BOGNAR, A. (1971). *Ernähr.-Umsch.*, 18, 200

BROOKS, C.C. (1971). *J. Anim. Sci.*, 33, 1224

COOK, L.J., SCOTT, T.W., FAICHNEY, G.J. and DAVIES, H.L. (1972). *Lipids*, 7, 83

COOK, L.J., SCOTT, J.W., FERGUSON, K.A. and MCDONALD, I.W. (1970). *Nature, Lond.*, 228, 178

CRAWFORD, M.A. (1968). *Proc. Nutr. Soc.*, 27, 163

CRAWFORD, M.A., GALE, M.M., WOODFORD, M.H. and CASPED, N.M. (1970). *Internat. J. Biochem.*, 1, 295

DONOSO, G., LEWIS, O.A.M., MILLER, D.S. and PAYNE, P.R. (1962). *J. Sci. Fd Agric.*, 13, 192

DVORAK, Z. and VOGNAROVA, I. (1965). *J. Sci. Fd Agric.*, 16, 305

DVORAK, Z. and VOGNAROVA, I. (1969). *J. Sci. Fd Agric.*, 20, 146

FAO (1968). *Amino acid content of foods and biological data on proteins,* Rome

FAO/WHO (1971). *Toxicological evaluation of some extraction solvents and certain other substances. Rpt. Series No. 48*

FRENTZ, J.C. and PERRON, P. (1972). *Fd Sci. and Tech. Abstr.,* 4, Abstr. 10 S 1252,

HARRIS, R.S. and VON LOESECKE, H. (1960). *Nutritional evaluation of food processing.* Wiley, New York

HMSO (1968). *Ministry of Health Reports on Public Health and Medical Subjects No. 117,* 'Iron in Flour'

JACOBS, A. and GREENMAN, D.A. (1969). *Br. Med. J.,* 1, 673

LAYRISSE, M. MARTINEZ-TORRES, C. and WALTER, R. (1972). *Amer. J. Clin. Nutr.,* 25, 401

LUSHBOUGH, C.H., HELLER, B.S., WEIR, E. and SCHWEIGERT, B.S. (1962). *J. Amer. Dietet. Assoc.,* 40, 35

MCCANCE, R.A. and WIDDOWSON, E.M. (1960). *The Composition of Foods.* MRC Rpt. No. 297

MARTINEZ-TORRES, C. and LAYRISSE, M. (1971). *Amer. J. Clin. Nutr.,* 24, 531

MAYFIELD, H.L. and HENDRICK, M.T. (1949). *J. Nutr.,* 37, 487

MINOCCHERI, F. and CANTONI, C. (1972). *Ind. Aliment.,* 11, 81

NATIONAL FOOD SURVEY (1973). HMSO, London

NESTEL, P.J., HAVENSTEIN, N., WHYTE, H.M., SCOTT, T.J. and COOK, L.J. (1973). *New Engl. J. Med.,* 379, 1973

NOBLE, I. (1965). *J. Amer. Dietet. Assoc.,* 47, 205

ORR, M.L. and WATTS, B.K. (1957). 'Amino acid content of foods', *Home Econ. Res. Rep. No. 4,* USA.

PATJAS, M., SOMMU, A. and PALANSKA, O. (1972). *Archiv. Tierernähr.,* 22, 149

PLOWMAN, R.D., BITMAN, J., GORDON, C.H., DRYDEN, L.P., GOERING, H.K., WRENN, T.R., EDMONDSON, L.F., YONCOSKIE, R.A. and DOUGLAS, W. (1972). *J. Dairy Sci.,* 55, 204

ROGNERUD, G. (1973). *Nutr. Abstr. Rev.,* 43, Abstr. 102

SINCLAIR, H.M. and HOLLINGSWORTH, D.F. (1969). In *Hutchinson's Food and the Principles of Nutrition,* p.390. Edward Arnold, London

TAMATE, R. and OHTAKA, F. (1972). *Jap. J. Zootech. Sci.,* 43, 251

WHO (1973). *Energy and Protein Requirements,* WHO Tech. Rept. Series No. 522, Geneva

WOOD, T., ADAMS, E.P. and BENDER, A.E. (1962). *Nature, Lond.,* 195, 1207

# 23

# MEAT AS A SOURCE OF LIPIDS

M. A. CRAWFORD

*Department of Biochemistry,*
*Nuffield Institute of Comparative Medicine*
*Zoological Society of London*

## INTRODUCTION

The direct use of plant proteins or proteins from waste
for human food is approximately five to ten times more
efficient than conversion through animals. On this
premise it is often suggested that animal production
should be replaced by plant agriculture and factory
manufacture of protein (e.g. *The Times,* 11th February,
1974).

This approach assumes protein to be the only nutrient
that we have to worry about. To reduce food to protein,
vitamins and minerals, is to claim that human nutrition
is fully understood, which is clearly not true. If we
knew all that there was to know, we would not still be
wrestling with such problems as obesity, diverticular
disease, cancer of the large gut (Burkitt and Painter,
1971), multiple sclerosis (Bernsohn and Stephanides,
1967) and atherosclerosis (Keys, 1970; Crawford and
Crawford, 1973) which are thought to be linked with
modern developments in food production.

The simplification of human nutrition to 'protein'
intake is dependent entirely on the speculation that
protein is the only major body-building nutrient relevant
to man. Yet whilst plants may be more efficient at
producing protein, animals are more efficient sources
of other nutrients. Future reliance on cereal crops to
the exclusion of animal husbandry on the assumption that

protein only is needed for maintaining future human development is questionable and could have long-term consequences.

Meat as a food is very much more than protein and no matter how good the quality of a biosynthetic protein, it will not be a substitute for meat. Indeed, the real value of animal products may not be in their protein content, but as Professor Bender indicated earlier, in other nutrients such as vitamin $B_{12}$ and iron. In addition, there is now evidence that lipids may be important and this paper is principally concerned with the lipid value of meat, although the general principles apply to other animal products.

## PROTEIN AND LIPIDS: THE COMPARATIVE APPROACH

In attempting to evaluate food one often has resource to so-called 'natural foods', but today there is perhaps only one food which can still make some valid claim to be 'natural' and that is breast milk. Some consider the human species to have reached a degree of evolutionary excellence measurably superior to that of other species, yet if we compare the chemistry of milks we find there is less protein in human milk than in any other species. The rat has 9 g/100 ml, the dog 8 g, the hippopotamus 7 g, the pig 6 g, the ox 4 g and man 1.5 g. During suckling the mean growth rate of the pig is 295 g/day, of the calf 580 g/day and of the human infant only 25 g/day.

Such data do not suggest the human species to be an example of outstanding quantitative protein nourishment! It is interesting that in contrast cows' milk has about three times more protein than human milk, whilst human milk has four times more of the essential fatty acids.

## PROTEIN AND GROWTH: MILK COMPOSITION - THE LABORATORY AND COMPARATIVE EVIDENCE

To compare one protein with another, experimentalists often use the growth curve in rats. Improvement of growth performance has been generally considered to imply nutritional excellence and reduction in growth rate nutritional poverty.

More recently, Miller and Payne (1968) showed that animals fed a high protein intake had a shorter life span. We found that histidine supplementation to a normal guinea-pig diet improved growth rate and reproductive performance (Gale and Crawford, 1969), but the long-term follow-up showed that while the supplemented animals may have grown faster, they also died faster.

The comparative data tell us that the remarkable growth rate of large mammals like the cow and the rhinoceros is accompanied by an accumulation of protein- and mineral-rich tissues, but little development of the lipid-rich brain with its peripheral developments of nervous tissue. It is in the slow-growing primates that we see the developments of the lipid-rich nervous system and it is interesting that in related 'secondates' we find it is the carnivores which exhibit a relatively high degree of development in the central and the peripheral nervous system such as is evidenced by the articulated claws and the night vision of the large cats.

THE HERBIVORE AND CARNIVORE

Until recently it has been assumed that the difference between the food structure of the carnivore and the herbivore was the high protein diet eaten by the carnivores, or that the herbivores existed on a diet poor in protein. The brain in these large herbivorous mammals is very small; but the brain, unlike other tissues in the body, is not protein-rich, but lipid-rich. There can be little doubt that a characteristic of the herbivore is a rapid rate of growth in terms of mineral-rich tissues like bone, and protein-rich tissues such as muscle. Now if growth rate is an important criterion of protein quality, then the comparative evidence that the herbivores have the fastest rate of growth does not support the idea that the herbivore food is poor in protein. Indeed the opposite seems to be the case.

It is now known that the ruminant has solved the problem of protein supply in a unique way by employing a spectrum of micro-organisms in the rumen to convert the vegetation and cellulose into single-cell systems. Mammals cannot digest plant cellulose but the micro-organisms can. The fermentation process in the rumen provides protein-rich protozoa and bacteria which the

ruminant then eats (Hungate, 1966).  There is no oxygen
supply to the rumen contents and the end-product of
energy metabolism by the micro-organisms is hydrogen
instead of water, hence highly unsaturated molecules may
act as hydrogen acceptors, but in so doing become
saturated.  Therefore, much of the polyunsaturated lipid
nutrients in the food is saturated in the rumen (Wilde
and Dawson, 1966).  The ruminants gain protein from
their micro-organisms and also vitamin $B_{12}$.  On the
other hand, this is at the expense of a loss in polyunsaturated fats ultimately used in brain growth.  Thus,
the ruminant is an extreme adaptation to a high-quality
protein and low-quality lipid food structure.

The ruminant herbivore is capable of producing
substantial amounts of protein-rich muscle tissue, but
the features associated with the central and peripheral
nervous system, which are dependent on lipid considerations, are poorly developed; the brain to body weight
ratio is small.

By comparison the carnivorous system involves a far
higher degree of development in the central and
peripheral nervous system.  In other words, the carnivore
does not illustrate development in the protein-rich
tissues, but the lipid-rich; the primates reach an even
greater peak of development.

The primates, in contrast to the secondates, have a
slower rate of growth, long gestation and lactation
periods and exhibit substantially less muscle development, but a considerable advance in the brain.  Again,
within the large apes we can see the contrast between
the massive, strictly vegetarian gorilla, by comparison
with the smaller but larger-brained, omnivorous ape,
man.

In fact, the chemistry of the human species is
remarkable, not from its protein content, but its lipid
characteristics.  The structural lipids employed in cell
construction are quantitatively the most important
structural group in the brain and the second most
important in all other soft tissues in the body.  Hence,
in species like the rhinoceros, the buffalo and gorilla,
the preponderance of muscle tissue and bone emphasises
the protein and mineral aspects of animal biology,
whereas the development of central, peripheral nervous
and vascular systems in the human primate emphasises the
lipid aspects.

HUMAN MALNUTRITION - A PRIORITY OF PROTEIN OR FOOD?

The above discussion suggests 'protein' to be an overvalued currency on the basis of comparative and biochemical evidence. However, the need to revalue ideas on protein and food value is now receiving impressive support from the epidemiology of human malnutrition. Studies by Sukhatme (1970; 1974) in regions in which malnutrition, referred to as 'protein malnutrition', tends to be endemic, have failed to identify diets short of protein on a percentage calorie (or energy) basis, but have identified a shortage of energy, that is, a shortage of food, not protein.

Probably, more harm has been done to nutrition by the introduction of the word protein in food descriptions than any other single innovation in this somewhat inexact subject. To define the problems of Maharastra as 'protein malnutrition' dictates a solution of protein supplementation. Yet despite some 20 years of protein supplementation, McLaren and Pellet (1970) claim there is not one major success. The reason is not hard to find. Protein fed when food is short is burnt for energy. If you define the problem as a food shortage, then food works when 'protein' or supplementation does not. Protein fed to infants produced essential fatty acid deficiency (Hansen *et al.*, 1963). The pellagra of the depression years in the Southern States of America was not abolished by supplementation programmes, but by a broad development of agriculture (Davies, 1964).

In summary, the view expressed here is that our knowledge of human nutrition is still primitive. Previously, the simplification applied to food quality was protein. I am suggesting this is inadequate; 'energy' may soon be recognised to be as important as protein and whilst protein may be relevant to body growth, lipids are important to brain growth. As the most outstanding biological development in man is his brain, it may be that lipids are especially important to the human species.

## THE ESSENTIAL FATTY ACIDS OF STRUCTURAL LIPIDS

### THE DISTINCTION BETWEEN PLANT AND ANIMAL LIPIDS

The relationship of essential lipids to brain growth has been discussed in detail elsewhere (Crawford and Sinclair, 1972; Crawford and Crawford, 1973); this discussion will only be summarised here.  All cells use both protein and lipid for their construction and as with protein where we have essential and non-essential amino acids, we also have in the lipids both essential and non-essential fatty acids.  But the relationships between the fatty acids are more complicated.  Because of the significance of the lipids in the nervous system I would like to set this discussion in the context of the source of the lipids that are found in the brain.  Two essential fatty acids occur in the vegetation:

1.  Linoleic acid (C18:2,$n$-6) which has two double bonds and is found mainly in seeds.
2.  Linolenic acid (C18:3,$n$-3) which has a third double bond in a different position from that of linoleic acid and hence it cannot be made from linoleic acid by animals.  Linolenic acid occurs mainly in leaf material, i.e. dark green vegetation.

Generally speaking, plant and animal systems require different physical properties with respect to 'fluidity' (Chapman, 1972).  In the lipids fluidity increases with the degree of unsaturation, i.e. the number of double bonds.  In the lipids of plant and animal cells we find the animal structural lipids have a much higher degree of unsaturation and longer chain length.  The explanation for this difference is relatively simple.  Although animal systems cannot insert the first groups of double bonds in linoleic and linolenic acids, once they are presented with these double bonds they can add more.  Hence, when animals eat the C18 vegetable acids these are metabolised by the liver to produce two families of the long-chain polyunsaturated acids (LCP) which are generally specific to animal products and it is especially interesting that the brain cells contain these LCP and not the short-chain polyunsaturates (SCP), i.e. the parent vegetable acids *(Table 23.1)*.

Table 23.1 Metabolism and chain lengthening of essential fatty acids

| Family | Source | | |
|---|---|---|---|
| | SCP vegetable | | LCP animal (all tissues) |
| Linoleic | C18:2n-6 linoleic | metabolism (liver) → C20:3n-6 | → C20:4n-6 → C22:4n-6 arachidonic [brain] |
| Linolenic | C18:3n-3 linolenic | metabolism (liver) → C20:5n-3 | → C22:5n-3 → C22:6n-3 docosahexaenoic |

Not only is the fluidity of the membrane fat markedly increased by chain elongation and desaturation, but also the number of individual molecules and the diversity of molecular fluidity is broadened to meet the different requirements of the animal systems, which walk about, as opposed to plants, which stand still. Furthermore, the chain elongation and desaturation products of linoleic acid give rise to the group of hormone-like substances, the prostaglandins.

It is sometimes thought that given the parent C18 vegetable acids, the rest can be made by animals. This may be true in small species with a rapid metabolism like the rodents, but is certainly not true for larger herbivorous mammals where the chain elongation process is not completed in the sense that $C22:6,n-3$ is not found in significant quantities in liver structural lipids but only the precursor $C22:5,n-3$ (Crawford and Sinclair, 1972). This finding suggests that although the mechanism is available, the rate at which the process can occur is limited.

As opposed to the large herbivores the food structure of the carnivorous animals commences not with the parent SCP, but with a spectrum of both parent SCP and the LCP derivates accumulated and developed by the herbivore. Consequently, there is a progression of the structural fats in the food chain from the vegetation to the herbivore to the carnivore (Figure 23.1) and it is of special interest that this progression brings the spectrum of lipids in the carnivore closer to that which is used in brain construction (Figure 23.2); the brain contains only the LCP. In a sense the herbivore seems

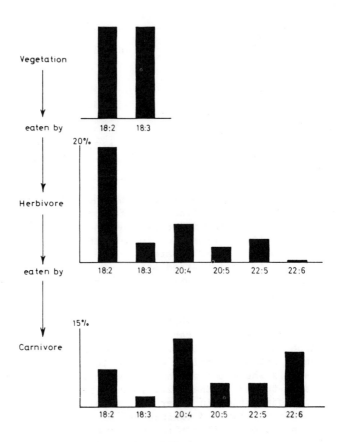

Figure 23.1 Diagrammatic representation of polyunsaturated fatty acids in the food chain to illustrate the progression or 'biomagnification' of the long-chain (C20-C22) polyunsaturated acids. It is the long-chain and not the short-chain (C18) polyunsaturated fatty acids which are used in brain cell structures (Crawford and Sinclair, 1972)

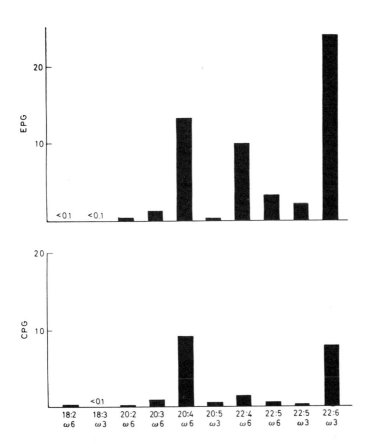

*Figure 23.2 Distribution of polyunsaturated fatty acids in ethanolamine and choline phosphoglycerides (EPG and CPG, respectively) from human grey matter as a percentage of total fatty acids and aldehydes (average results obtained from motor cortex of 7 adult humans)*

to provide prefabricated construction units for the
brain. This idea suggests that the LCP of animal
products may be qualitatively different from the SCP of
vegetable origin, an idea which is further supported by
our finding that these long-chain polyunsaturated acids
are incorporated some 20-60 times faster into rat liver
(Brenner and José, 1965) and the developing rat brain by
comparison with the shorter chain parent acids of
vegetable origin (Sinclair and Crawford, 1972).

Man is a large mammal and a qualitative requirement
for LCP by man could mean a nutrient role, previously
not suspected for animal products, such as milk, meat,
eggs and fish, which is concerned with lipid nutrients
rather than protein. This suggestion is consistent
with the comparative evidence discussed previously. For
example, the horse satisfies all its protein requirements
from a primary food resource like grass and in one year
lays down more protein in its tissues than we do in
twenty years of growth. But the horse does not lay down
a large lipid-rich brain.


SCP VERSUS LCP

Obviously it would be very valuable if instead of
eating animals we could simply eat the vegetation.
Vegetarians are often quoted as supporting evidence for
such an argument and Professor Bender reminded us that
one half of the world exists mainly on vegetables, if
not by choice, by force of necessity. At the same time,
it is these communities which are affected by acute
nutritional problems and which have a relatively short
average life span.

In Britain, vegetarians mostly eat eggs, drink milk
and some will even eat fish. In other parts of the world
people may eat frogs, snails, small birds, reptiles,
insects and small mammals. In our adulation of meat we
tend to overlook the value of such foods which, in some
respects, can be considerably richer than conventional
meat with respect to lipid nutrients. The same applies
to offal; liver, for example, has about 5-7% (fresh
weight) as structural lipid by comparison with the
1-2% of meat. Furthermore, liver is a rich source of
B vitamins, including vitamin $B_{12}$ and the oil soluble
vitamins D and A which are not found in significant
quantities in meat. Communities which eat less meat
tend to eat more offal and more of the other animal
products.

Another point frequently overlooked is that those communities which are forced to exist today mainly on vegetables are the communities which practise breast feeding for the longest – frequently up to two years. Human milk does contain LCP; at the same period in the infant's life, our children are fed on high-protein cereals and cows' milk which is not as rich as human milk with respect to essential fatty acids.

Comparison between vegetarians and omnivores is also complicated by the theoretical balance between the two families of essential fatty acids.

Animal tissues provide a balance of the two distinct families of essential fatty acids. Leaving aside the question of LCP, this balance could be achieved in terms of SCP by eating a diversity of vegetable products, including green leaves. Today, green foods are available only in small quantities, as plant agriculture has primarily developed the grains and cereals with the one exception of grass pasture. Grass is a green leaf and a main source of linolenic acid. In this way the herbivores which we eat make available to us an essential fatty acid from grass which would not be readily available if this food supply was removed.

This discussion is presented to illustrate the complexities in considering lipids within the food chain. The difficulties are increased when it is realised that lipid deficiencies take a long time to exert their effect and marginal deficiencies may only produce an observable effect, for example, on brain development in subsequent generations (Sinclair and Crawford, 1972)

With regard specifically to the question of SCP and LCP, it is clear that many animal species do not eat LCP. It is also clear that absence of LCP in the diet will not stop protein or lipid biosynthesis. However, the message which emerges is that in terms of LCP and the associated nutrients we would be concerned with qualitative factors. The sheep and the dog both survive, but they have different qualities of life.

The points which emerge from this discussion could be summarised as follows. Firstly, it would be wrong to assume that we could dispense with animal agriculture in favour of protein-rich plant crops. If there is as I suggest a qualitative role for animal products, then there is a strong case for the extension of animal husbandry. Although I have discussed this aspect in terms of LCP it should be appreciated that this parameter may only be an easily recognisable index of broader issues in nutrition which we have yet to understand.

The current enthusiasm for calculating the efficiency of land-use in terms of protein production is, in my view, a simplification not justified by any good scientific evidence and calculations based on arachidonic acid or vitamin $B_{12}$ would give totally different answers. The comparative biology suggests that protein is about the easiest nutrient to obtain from plant produce and structural lipids are the most difficult; animal products may be of more value for their structural lipids than for their protein. Finally, it seems to me that the human species is an optimum expression of development of the nervous and vascular systems, and one's objective should be guided by a positive search for the optimum conditions, not a subsistence on the minimum.

## MANAGEMENT AND LIPID NUTRIENTS

### TWO TYPES OF FAT IN THE ANIMAL BODY

The discussion on essential lipids in relation to animal metabolism has so far defined a distinction between plant and animal products in terms of the short- and long-chain polyunsaturated fatty acids of the structural lipids However, in the animal itself there are two main types of fat creating a further distinction; these are:

1.  Storage fats - adipose tissue, triglycerides and mainly non-essential.
2.  Structural lipids - mainly phospholipid and rich in long-chain essential fatty acids.

The fatty-acid composition of the storage fats is very dependent on diet, both qualitatively and quantitatively. Diets rich in essential fatty acids or ruminants fed essential fats protected from the hydrogenating influence of the rumen will raise the essential fatty-acid content of the storage fats. A high-energy dietary intake will dramatically increase the deposition of body fat and body weight. However, studies on sheep have revealed that this increase in body fat and body weight is not accompanied by an increased production of wool or the carcass nutrient value (Blaxter, 1967).

The energy balance of a production system is obviously relevant to carcass fat and although seldom considered, is also relevant to the nutrient value of the animal products which we ourselves eat.

The subdivision of fat as 'structural' or 'storage' enables us to re-examine the end-product of contemporary animal production in the light of its contribution to human nutrition. That this re-examination is important is illustrated by Western communities who have pioneered intensive animal production systems and are today affected by severe and chronic degenerative diseases of the nervous and vascular system. Although there is still discussion regarding the precise cause of, for example, obesity, cardiovascular disease and multiple sclerosis, there is little doubt that they are specific to technically advanced communities where heart disease is the major cause of premature death. Also, there may be diverse opinions regarding the exact cause that precipitates the heart attack, ranging from lack of exercise and smoking, to stress. There is no dispute that the underlying process of arterial disease is a matter of lipids.

In the context of lipids it is usual to blame too much saturated fat or cholesterol for atherosclerosis, or sugar for everything; however, we have suggested that in a chronic degenerative disease one should surely be concerned with the building bricks required for tissue construction, repair and maintenance (Crawford, 1968; Crawford and Crawford, 1973). If the study of nutrition is meaningful, then it is the balance of nutrients and energy that is important; what is interesting is that the animal provides us with both the non-essential saturated constituents in the storage fats and the polyunsaturated building bricks for cell construction and maintenance in the structural lipids. On the one hand, the elements of 'destruction' and on the other, 'construction'!

MANAGEMENT AND LIPID BALANCE IN MEAT

The evidence from meat-eating tribes like the Hadza, Eskimos, Bushmen, Masai and El Molo, is that they have no hypertension or evidence of heart disease and remarkably low blood cholesterols; this is interpreted to mean that the saturated fat aspect of animal production can have nothing to do with heart disease (Yudkin, 1972). However, on looking at the meat which the consumer buys in Europe today, one is impressed not by the amount of meat, but the amount of visible or

*Table 23.2* Constituents of meat from some free-living animals

| Species | Composition of meat | | | |
|---------|-----|-----|-----|-----|
| | Ash | Fat | Solid nutrient[1] | Calories |
| | (g/100 g fresh weight) | | | |
| Eland | 1.1 | 1.9 | 23 | 125 |
| Hartebeest | 1.2 | 2.1 | 24 | 130 |
| Topi | 1.3 | 2.3 | 23 | 126 |
| Giraffe | 1.4 | 2.2 | 22 | 123 |
| Buffalo | 1.1 | 1.8 | 22 | 120 |
| Warthog | 1.2 | 2.3 | 25 | 132 |
| Free range cattle | 1.1 | 2.0 | 22 | 120 |
| Intensive fat stock | 0.9 | 15.0 | 18 | 230 |

[1]Protein, structural lipids, etc.

*Note:* In the rainy seasons free-living animals eat large quantities of fresh young grass. The fat content of meat from these species varies, therefore, according to the season.

storage fat. This fat is not only around the meat, but infiltrated between the muscle cells and this is illustrated by the high calorific content of the meat itself (*Table 23.2*). The technical name of 'marbling' has been coined to describe this infiltration, yet pathologists would describe a fatty infiltration as pathological.

The end-product of animal production today is itself an end-product of a development over a number of years of intensification, high-energy feeds and restriction, leading to the abolition of exercise. If all these individual factors are said to be wrong for human health, then can they be correct for animal health? However, regardless of whether these management principles are appropriate to the interests of pigs, sheep and cattle, it is especially important to know if their products are appropriate for human health.

In seeking to answer this question, the difficulty lies in finding some form of standard from which to operate. To simplify the discussion we can examine the balance of storage fat to the solid nutrient value of meat, which includes the structural lipids, protein,

vitamins and minerals. As there is no standard for
'extensively' reared animals prior to 'intensification'
we can take as our standard the general performance of
many free-living animal species including nomadic cattle,
sheep and goats, buffalo, venison, wild pig and the like,
living under 'extensive' or free-living conditions. On
the basis of this standard, free-living animals used for
food by man provide a 5% carcass fat and 75% lean
(Ledger, 1968), whereas intensively reared beef and
sheep can provide a 25-30% carcass fat and 50% lean.

Lean meat is itself 80% water and if we subtract this,
we are left with 5% storage fat to 15% solid nutrients
from the lean of free-living animals. By contrast,
intensively reared animals provide 30% storage fat and
10% solid nutrients, i.e. the solid nutrient value of
the free-living animal is three times that of its
storage fats; in the intensively reared beef the storage
fats are three times greater in amount than the solid
nutrients (*Table 23.3*). Details of fatty-acid analyses
and the balance of lipids from these two extremes are
given in *Tables 23.4, 23.5* and *23.6*.

The bush people, the El Molo and Hadza are eating free-
living animals which provide more nutrients, including
the polyunsaturated structural lipids, than saturated-type
fats, and these people have low blood cholesterols and
triglycerides (Crawford, Crawford and Hansen, 1971). We
eat animals high in saturated storage fats and low in
essential structural lipids and we have high blood
cholesterol and triglyceride values and a high risk to
atherosclerosis.

In the case of pig intensification the ratio of
storage fat to nutrient value can be even higher, but
in hill sheep or lamb the balance can be closer to the
balance found in nature. Welsh hill lamb has been
slaughtered at a 10% carcass fat and 14% solid nutrient
value (Crawford, unpublished observations).

It seems to me that the epidemiology backed by the
experimental induction of atherosclerotic-like changes
in the artery by feeding excess saturated fats, does
question the validity of producing animals grossly
overweight with surplus fat for us to eat. The one
feature about which there seems to be universal agreement both in the realms of life insurance statistics and
in medicine is that obesity is ill advised.

In summary, it is clear that the carcass of land
animals in a free-living environment provides more solid
nutrients than saturated-type fats; intensively reared
domestic animals provide more saturated-type fats than

*Table 23.3* Simplified balance of carcass fat to solid
nutrient in free-living and intensively-reared ruminants

| System | Carcass composition (%) | | | Fat/solid nutrient |
|---|---|---|---|---|
| | Fat | Lean | Lean without water | |
| Extensive, e.g. buffalo, eland, wild pig, venison | 5% | 75% | 15% | 1:3 |
| Intensive: beef, pork | 30% | 50% | 10% | 3:1 |

*Note:* In most instances a 5% carcass fat is high for
free-living mammals and is achieved in the spring or
rainy seasons. A 30% carcass fat for beef is also a
high average, but in intensive systems the infiltrated
triglyceride fat can be as much as 15% for fatty beef
and 7% for very lean beef. It is seldom more than 2.5%
of the fresh weight in free-living situations.

solid nutrients. In strictly lipid terminology, the
profile of the lipids extractable from the meat of
free-living animals is predominantly polyunsaturated and
phospholipid or structural in character, whilst that
extracted from intensively reared beef and pork is
predominantly of the triglyceride, saturated-type fats.
It also seems clear that the lipid value of meat is
dependent directly on the management system. This means
that the situation is reversible; that one could produce
animal products for the community to eat in which the
balance favoured the structural lipids and solid
nutrients rather than an excess of storage fat. We do
know that an excess of storage fats is biologically
harmful in general and harmful to man in particular. We
now also know that the structural lipids are quantitative
the most important group in the brain and nervous system,
and the second most important in all other soft tissues
in the body. The storage fats are a valuable energy
source, but in excess are known to be harmful. In my
submission, because of the potential importance of the
positive role of the structural lipids in meat, there
is as yet no case to replace animal products by protein
analogues, but there is a case for re-evaluating the

Table 23.4  Polyunsaturated acids in the ethanolamine phosphoglycerides from animal and plant sources

| | Liver | | | | Muscle | | | | Brain | |
|---|---|---|---|---|---|---|---|---|---|---|
| | Oxcalf | Eland | Kob | Warthog | Oxcalf | Eland | Kob | Warthog | Kob | Man |
| 18:2, *n*-6 | 4.0 | 8.5 | 10.2 | 22.0 | 1.7 | 9.2 | 10.4 | 20.0 | 0.8 | 0.5 |
| 18:3, *n*-3 | 1.0 | 2.8 | 4.6 | 3.5 | 0.3 | 3.8 | 4.1 | 7.4 | 0.3 | 0.1 |
| 20:2, *n*-6 | 0.9 | 0.2 | 0.5 | 0.1 | 2.9 | 0.3 | 0.6 | 0.3 | 0.5 | 0.8 |
| 20:3, *n*-6 | 1.2 | 0.3 | 0.5 | 0.6 | 3.4 | 0.8 | 2.1 | 0.8 | 1.3 | 1.1 |
| 20:4, *n*-6 | 16.0 | 14.0 | 14.4 | 12.0 | 13.3 | 12.4 | 10.4 | 18.5 | 13.0 | 12.3 |
| 20:5, *n*-3 | 3.1 | 1.7 | 4.6 | 3.1 | 3.9 | 3.6 | 4.8 | 5.1 | 1.6 | 0.8 |
| 22:4, *n*-6 | 1.1 | 2.4 | 1.8 | 1.1 | 2.2 | 0.7 | 0.5 | 0.95 | 8.3 | 7.4 |
| 22:5, *n*-6 | 0.3 | 0.2 | 0.2 | 0.6 | 0.5 | 0.3 | 0.1 | 0.14 | 0.9 | 1.6 |
| 22:5, *n*-3 | 11.0 | 7.8 | 11.0 | 5.0 | 6.3 | 11.5 | 11.3 | 3.7 | 1.4 | 0.7 |
| 22:6, *n*-3 | 2.3 | 6.4 | 2.2 | 1.6 | 2.1 | 1.9 | 1.3 | 2.2 | 25.0 | 26.0 |

*Note:* The ethanolamine phosphoglyceride is the phospholipid fraction which is richest in the C20 and C22 polyenoic acids. C22:6,*n*-3 is low in relation to C22:5,*n*-3, but in the brain the C22:6,*n*-3 is the principal component; this is true for all species we have studied (Crawford and Sinclair, 1972).

Table 23.6 A comparison of the fatty acids in muscle total lipids, triglycerides and ethanolamine phosphoglycerides[1]

| Fatty acid designation | Bos taurus[2] (Beef) | | | Taurotragus oryx (Eland) | | Syncerus caffer (Buffalo) | | Bos indicus[3] (African beef) | | |
|---|---|---|---|---|---|---|---|---|---|---|
| | Total | Triglyceride | EPG | Total | EPG | Total | EPG | Total | Triglyceride | EPG |
| 16:0 | 29.0 | 33.0 | 12.0 | 19.0 | 9.0 | 16.0 | 11.0 | 17.0 | 29.0 | 15.0 |
| 18:0 | 18.0 | 16.0 | 36.0 | 20.0 | 33.0 | 22.0 | 32.0 | 18.0 | 16.0 | 33.0 |
| 18:1 | 47.0 | 45.0 | 16.0 | 17.0 | 9.0 | 23.0 | 14.0 | 28.0 | 48.0 | 18.0 |
| 18:2,n-6 | 1.2 | 1.0 | 1.6 | 22.0 | 8.5 | 16.0 | 4.5 | 14.0 | 3.6 | 3.7 |
| 18:3,n-3 | 0.9 | 0.5 | 0.8 | 4.1 | 3.8 | 5.0 | 3.0 | 3.5 | 2.1 | 2.0 |
| 20:3,n-6 | 0.3 | 0.1 | 3.0 | 0.8 | 0.4 | 0.3 | 0.9 | 1.1 | 0.1 | 1.1 |
| 20:4,n-6 | 1.2 | 0.2 | 12.0 | 6.4 | 15.0 | 7.1 | 12.0 | 7.0 | - | 13.0 |
| 20:5,n-3 | 0.2 | - | 2.9 | 1.0 | 1.9 | 1.1 | 2.5 | 1.8 | - | 1.9 |
| 22:5,n-3 | 0.5 | - | 8.0 | 3.8 | 9.3 | 3.0 | 9.0 | 4.0 | - | 11.0 |
| 22:6,n-3 | 0.2 | - | 1.4 | 0.8 | 3.4 | 0.2 | 0.9 | 0.8 | - | 1.2 |

[1]The results are expressed as the percentage (by weight) of the total fatty acids. These results illustrate that whole tissue fatty-acid analyses need not reflect the fatty-acid composition of cell structural lipids.
[2]Intensive reared for the London market.
[3]Non-intensive obtained from the Siroti market, Uganda.

parameters of animal production based on the balance
between the structural and storage lipids or what I
have somewhat loosely referred to as the 'constructive'
and 'destructive' fats.

*Table 23.5* Ratio of non-essential and essential fats
in meat from beef and free-living eland

| Lipid | Intensive beef | Nomadic beef | Eland |
|---|---|---|---|
| NON-ESSENTIAL/ESSENTIAL FATTY ACIDS IN: | | | |
| Total extractable fat | 20.9 | 2.0 | 1.4 |
| Ethanolamine phospho-glyceride | 0.62 | 0.71 | 1.2 |
| Polyunsaturated/saturated P/S ratio total | 0.09 | 0.92 | 1.02 |
| Ethanolamine phospho-glyceride | 0.62 | 0.71 | 1.01 |

REFERENCES

BERNSOHN, J. and STEPHANIDES, L.M. (1967). *Nature,
Lond.*, 215, 821
BRENNER, R.R. and JOSÉ, P. (1965). *J. Nutr.*, 85, 196
BURKITT, D. and PAINTER, N.F. (1971). *Br. Med. J.*,
2, 450
BLAXTER, K.L. (1967). *The Energy Metabolism of
Ruminants*. Hutchinson, London
CHAPMAN, D. (1972). In *Lipids, Malnutrition and the
Developing Brain*, p.31. Ed. K. Elliott and J. Knight.
Associated Scientific Publishers, Amsterdam
CRAWFORD, M.A. (1968). *Lancet, i*, 1419
CRAWFORD, M.A. and CRAWFORD, S.M. (1973). In *What we
eat today*, Neville Spearman, London
CRAWFORD, M.A., CRAWFORD, S.M. and HANSEN, A.E. (1971).
*Biochem. J.*, 122, 11-12p
CRAWFORD, M.A., GALE, M.M., SOMERS, K. and HANSEN, I.L.
(1970). *Br. J. Nutr.*, 24, 393
CRAWFORD, M.A., GALE, M.M., WOODFORD, M.H. and CASPERD,
N.M. (1970). *Int. J. Biochem.*, 1, 295

CRAWFORD, M.A. and SINCLAIR, A.J. (1972). In *Lipids, Malnutrition and the Developing Brain,* p.267. Eds. K. Elliott and J. Knight. Associated Scientific Publishers, Amsterdam

DAVIES, J.N.P. (1964). *Lancet,* ii, 195

GALE, M.M. and CRAWFORD, M.A. (1969). *Trans. Roy. Soc. Trop. Med. Hyg.,* 63, 826

HUNGATE, F.P. (1966). *The Rumen and Its Microbes.* Academic Press, London

HANSEN, A.E., WIESE, H.F., BOELSCHE, A.N., HAGGARD, N.E., ADAM, D.J. and DAVIS, H. (1963). *Pediatrics,* 31, (Suppl. 1 pt. 2), 171

KEYS, A. (1970). 'Coronary Heart Disease in Seven Countries', in *Circulation,* 41, Suppl. 1

LEDGER, H.P. (1968). In *Symposium No. 21 Zool. Soc. Lond.,* p.289. Academic Press, London

MCLAREN, D.S. and PELLET, P.B. (1970). *World Rev. Nutr. Diet.,* 12, 43

MILLER, D.S. and PAYNE, P.R. (1968). *Exper. Geront.,* 3, 231

SINCLAIR, A.J. and CRAWFORD, M.A. (1972). *Febs Letters,* 26, 127

SUKHATME, P.V. (1970). *Br. J. Nutr.,* 24, 477

SUKHATME, P.V. (1974). In *The Man/Food Equation.* Ed. A. Bourne. In the press

WILDE, P.F. and DAWSON, R.M.C. (1966). *Biochem. J.,* 98, 469

YUDKIN, J. (1971). *Br. J. Hosp. Med.,* 5, 666

# 24

# MEAT IN THE ADAPTATION OF THE HUMAN ALIMENTARY TRACT

J. L. PATERSON

*University of Strathclyde*

In prevailing world conditions, the shortage of dietary protein tends to focus attention on human nutritional requirements and to distract attention from associated problems of the 'mechanics' of alimentation and digestion.  It is useful perhaps in considering the role of meat in the diet to survey the extent of carnivorism as a human character and the adaptations of the alimentary tract that it involves.  Such a survey leads to the conclusion that the alimentary tract in man is of a relatively unspecialised nature, i.e. it is basically omnivorous in function and highly adaptable with regard to nutrition.

Human dentition is of an all purpose variety.  The incisors are cutting in character.  The canines are reduced in parallel with the development of the hand as a grasping and retaining tool.  The premolars are intermediate between grinding and shearing (carnassial) forms.  The molars retain a grinding function.  In the alimentary tract itself the omnivorous propensity is typified by the digestive function of the saliva which is lost in true carnivores, the replication of digestive facility for natural protein and lipid and the loss of microbiotal cellulose digestion.  It can be concluded from this, coupled with evidence of regressive change in large bowel capacity, that human archetypes were probably originally primarily herbivorous but were at an early stage in our evolution tending towards carnivorism possibly initially by way of insectivorism.  Such a gradual transition to the intermediate omnivorous state offers considerable evolutionary advantage.

The inclusion of carnivorism in the dietary habits reduces the total time spent on ingestion of food to achieve nutritional adequacy and assures survival in a wide range of habitats as well as in changing conditions in a single habitat. Similar adaptive flexibility may be found in other anthropoids. In a natural habitat, gorillas are predominantly herbivorous but in captivity will readily adapt to and even prefer meat in their diet.

Human ethnic groups show extremes of dietary preference between carnivorism and herbivorism and achieve nutritional adequacy. Palaeolithic man leaves evidence of a high level of carnivorism which appears to have been modified with the emergence of agriculture in Neolithic times. This coincided with a transition from a nomadic existence to the establishment of social settlements and forage by hunting. The development of methods of animal husbandry reversed the herbivorous trend and achieved restoration of a broadly omnivorous habitus.

Subsequent changes in dietary preference have been largely ethical in character depending to a much greater extent upon social custom and taboo than upon physiological needs. The time scale of such ethical change is much shorter than that involved in species evolution and produces problems of physiological adaptation. Such problems arise less from meat eating than from sophistication of other dietary constituents though the availability of nutritional protein needs from meat has probably played a major role in the need for, and therefore the selection of, other foods. Intensive animal husbandry has been accompanied by a progressive refinement of carbohydrate and fat intake including a marked and rapid rise in consumption of sucrose and dairy produce in prosperous countries, in marked contradistinction to areas of protein deficiency where the dietary alternatives are less refined and include quantities of raw vegetable matter.

Since all human beings have essentially the same nutritional requirements, these considerable differences in dietary intake make widely varied demands upon the alimentary system. An important aspect of alimentary change induced by diet is disturbance of enteromicrobial ecology. Apart from the nutritional roles of intestinal micro-organisms in synthesising vitamins and amino acids and as competitors with the host for available nutrients, these organisms provide a morphogenetic stimulus which influences intestinal development.

   Investigation of the alimentary tract development in
gnotobiotic animal strains shows marked anatomical and
histological deviations which revert to a more normal
state when the alimentary tract is infected with its
usual organisms such as *Lactobacillus, Bacteroides,
Escherichia,* faecal streptococci etc.   The evidence
suggests that the level of alimentary epithelial growth
and function is modified by the intestinal flora which
is in turn modified by the nature of the diet.   Thus
antibiotic sterilisation of the alimentary tract is
followed by a very slow return to normal flora in
omnivorous species fed a very high protein diet by
comparison with those fed a mixed diet.   The implications
for human digestive function are not clear but there is
a suggestion that a high protein diet (i.e. one in which
protein is a major constituent in gross excess of the
amount required to prevent amino-acid imbalance) leads
to the development of an intestinal mucosa of abnormal
'delicacy'.   This may in part account for the increase
in intestinal disease which appears to accompany
increasing meat-eating in affluent ethnic groups.   It
is not, however, the sole aetiological factor and must
be considered together with the overall refinement of
diet which leads to the consumption of low-residue foods
with a marked effect on intestinal transit time.   Current
interest in the incidence of cancer of the lower bowel
highlights possible dietary influences like the foregoing.   In Britain and the United States, cancer of the
colon and rectum is second only to cancer of the lung
as a cause of mortality.   In general it is much commoner
in 'Western' countries than in Afro-Asian situations.
The possible dietary correlation is emphasised by its
relationship to country and even district of residence
rather than racial origin.   Thus there is a higher
incidence amongst Japanese immigrants in the USA than in
native Japanese; a higher incidence amongst American
negroes than in Africans.   It has been shown that
indigenous Japanese who have adopted a Western diet have
a higher incidence of bowel carcinoma than those adhering
to a traditional diet.   While it is impossible to determine any particular constituent of Western diet which
might directly contribute to the disease, the increased
consumption of meat *with the attendant alterations in
other dietary constituents* and increased gut transit
time appears to play a role.   Importantly there may be
an involvement of dietary fat and particularly of
saturated fats which are associated with meat-eating

communities. This correlates with observations on the relationship between dietary fat and atherosclerotic vascular disease since the latter and bowel cancer coincide in their distributions. It has been noted that faecal levels of steroids derived from cholesterol and bile salts are higher in those areas where cancer of the colon is common than in predominantly vegetarian regions where its incidence is low. Recorded differences in distribution of intestinal organisms show that in areas of high bowel cancer incidence, the intestinal flora shows a reduction in aerobic species and an increase in anaerobic forms like *Bacteroides,* usually associated with high protein intake. These anaerobes produce 7α-dehydroxylase which can convert cholic acid to deoxycholate; this has been shown to be carcinogenic in experimental animals (cf. structure of 20-methyl cholanthrene which is extremely carcinogenic).

Faecal concentrations of deoxycholic acid have been correlated with colonic cancer incidence. A similar correlation is established between the disease and excretion rates of other breakdown products of cholesterol, e.g. coprostanol and coprostanone which are again associated with Western dietary habits.

These observations support the suggestion of Burkitt (1971) that a diet high in refined carbohydrate, fat and protein, and of low residue, is an important aetiological factor in bowel malignancy. The low stool weight would tend to concentrate any faecal carcinogen present while the longer intestinal transit time would prolong contact with an intestinal mucose perhaps already rendered hypersensitive by altered ecology of the intestinal flora. Recent findings show that addition of fibre in the form of bran to the diet reduces intestinal degradation of bile salts probably by adsorbent activity. It appears likely that the correlation of bowel cancer with animal protein is the result of the indirect influence on gut flora and transit time rather than a direct carcinogenic effect of meat constituents.

A similar relationship may be found between meat consumption and other bowel dysfunctions, e.g. ulcerative colitis and constipation. It may be that the low residue diet is indirectly responsible for such conditions by encouraging the consumption of laxative drugs which have irritant actions. Similar irritation may arise from the use of spices and flavourings which may damage the intestinal mucosa. Evidence for the importance of intestinal transit time and the presence of adsorbent fibre in the gut lumen is available not only from

interracial studies but also from comparisons of dietary
habits in different regions of the UK.  Thus, in Scotland
and the Northern Counties of England where consumption of
fresh vegetables and fruit falls well below the national
average, the incidence of bowel disorders is proportion-
ally higher than in the Southern Counties where
consumption exceeds considerably the national average.
This difference is unrelated to the meat consumption
figures but is paralleled by increased consumption of
bakery produce and sugar which appear to substitute in
the North for Southern vegetable consumption.

Such figures as are obtainable for dietary habits in
the UK over the past 100 years indicate a steady change
in the direction of increased protein intake.

Since 1880 dietary increases (based on increased
consumption per capita) have included:

|  |  |  |
|---|---|---|
| Liquid milk | + 53% | |
| Sugar | + 72% | |
| Butter | + 67% | (including margarine + 150%) |
| Fish | + 17% | |
| Meat | + 48% | |

These increases have been accompanied by reductions in
consumption of vegetable products, for example:

|  |  |
|---|---|
| Potatoes | − 28% |
| Wheat flour | − 45% |

and these changes in dietary habit which typify the
Western pattern closely parallel increased incidence of
bowel disease.  It is notable that the overall reduction
in wheat flour consumption has been accompanied by
increasing refinement of flour with a disproportionate
reduction in fibre content.

The distribution of bowel disorders is similar to that
for disorders of lipid metabolism and the two taken
together suggest that increased meat consumption may be
symptomatic of a more generalised phenomenon of overnutrition among inhabitants of affluent countries.

Obesity is probably the commonest nutritional disease
in the Western world and is a serious aetiological
factor in the induction of metabolic disorders which lead
to degenerative tissue changes and the attendant
reduction in life expectancy.  While dietary fats are
principally incriminated there is evidence that protein
may play an indirect role when available to excess.

Comparison of breast-fed babies with those fed on higher protein cow's milk has shown that despite the more rapid growth of the latter their resistance to infection is reduced and that they are more prone to gastro-intestinal disorders. The role of protein excess in adults is more difficult to define.

It may be that the ready digestibility of meat and prepared animal fats fails to assuage natural hunger responses and encourages the more frequent ingestion of food including excess sugar and other carbohydrate and that these, in a situation where a substantial part of the calorific needs are provided by protein, represent a considerable nutritional excess. It is notable that true carnivores maintain health on less frequent meals than human populations which are predominantly carnivorous. It is also the case that modern meat-eating habits are made possible by the development of technologies which are energy – rather than labour – intensive so that along with intensive husbandry methods has developed a considerable reduction in the overall energy requirements of the populations. This makes overnutrition a probability for a majority of inhabitants of affluent societies. At the same time the emphasis on alternative energy sources changes with increasing meat consumption so that a higher proportion is derived from fats with a relative increase in saturated fats, e.g. in largely vegetarian countries such as Japan about 12% of calories in the diet are of fat origin (chiefly unsaturated); in industrial Western societies fats contribute about 40% of the energy intake (about half as saturated fat). This represents a substantial difference in diet and implies an ability for considerable alimentary and metabolic adaptation on the part of our species. Evolutionary and historical evidence indicates that the retention of truly omnivorous characters endows us with such adaptive capacity.

The changes in population stature and maturity in Western countries over the past few centuries are the results of improved nutrition and social conditions rather than of genetic change. Much of this improvement has been brought about by advances in methods of meat and dairy produce production, the vigorous pursuit of which is essential if projected global increases in population are to be fed. However, it may be that on a localised scale the achievement of a greater than adequate nutritional availability has, in terms of time-scale, outstripped the capacity for physiological adaptation to rapid dietary change. Certainly the

increase in life expectancy is as much the result of
ability to control infections as of the increased
nutritional standards achieved. There is as yet no
evidence for a belief that the basic processes of human
senescence have been slowed though there is an indication
that the elimination of malnutrition of the deficiency
variety may be followed by the malnutrition of plethora.
It is such plethora which throughout history has led to
the sublimation of eating from an act of nourishment to
a social activity indulged in for pleasure. This has
resulted in demands for more acceptable rather than more
suitable foods and has led to highly localised species
preferences and to complex methods of preparation which
may render food less suitable for the system it is to
nourish. It may well be that the alimentary tract which
has ensured the survival and propagation of our species
in a wide range of environmental situations with a wide
range of natural foods by virtue of its omnivorous and
adaptive capacity, is less able to provide for our
entertainment. We remain omnivorous but it is a highly
selective omnivorism that we practice, with the criteria
of selection more firmly based on psychological and
sociological considerations than upon nutritional needs.

The considerable acceleration in the rate of change of
dietary habits in the last two centuries appears to have
exceeded the capacity for physiological adaptation which
suggests that the satisfactory and satisfying fulfilment
of the nutritional requirements of increasing population
must involve not only increased production capacity. It
will require also increased re-education and redistribution: re-education is required to establish again more
representative omnivorism for which the alimentary tract
is equipped; redistribution seems necessary as a short-
term solution for populations in areas of deficiency.

While the present level of knowledge provides only
putative evidence of the involvement of the meat-eating
habitus as an indirect factor in the aetiology of
increasing 'disorders of civilisation', and since the
availability of adequate experimental control groups is
virtually non-existent, there can be no indictment of
animal protein as a causal agent in disease. It does
however represent a consistent factor in the multi-
factorial system associated with the stress-induced
disorders of industrialised and urbanised Western
societies and merits consideration with other identifiable
factors. In such considerations, statistically viable
evidence is dependent upon the validation of major
problems whose solutions are both costly and time-

consuming.  Prophylaxis by modest and timely re-education in selected areas of activity is in such circumstances both reasonable and rational.

There is little doubt that *given adequate time* the adaptive abilities of our digestive and metabolic systems will enable us to adapt to future developments in food production and to meet our nutritional needs by exploitation of a wider range of resources.

FURTHER READING

*Dietary Trends*

GREAVES, J.P. and HOLLINGSWORTH, D. (1966).  National Food Survey Committee - Annual Reports 1960-1963

*Mortality Statistics*

DOLL, R. (1969).  *Br. J. Cancer*, 23, 1
DOLL, R., MUIR, C. and WATERHOUSE, J. (1970).  *Cancer Incidence in Five Continents*

*Colonic Cancer and Low Residue Diets*

BURKITT, D.P. (1971).  *Cancer*, 28, 3
BURKITT, D.P., WALKER, A.R.P. and PAINTER, N.S. (1972).  'Effect of Dietary Fibre on Stools and Transit Times and Its Role in the Causation of Disease'.  *Lancet*, 2, 1408
DRASAR, B.S. and IRVING, D. (1973).  *Br. J. Cancer*, 27, 167
GREGOR, O., TOMAN, R. and PRASOVA, F. (1969).  *Gut*, 10, 1031
HILL, M.J., CROWTHER, J.S., DRASAR, B.S., HAWKSWORTH, G., ARIES, V. and WILLIAMS, R.E.O. (1971).  'Bacteria and Aetiology of Cancer of the Large Bowel'.  *Lancet*, 1, 95
PAINTER, N.S. and BURKITT, D.P. (1971).  'Diverticular Disease of the Colon: A Deficiency Disease of Western Civilization'.  *Br. Med. J.*, 2, 450
REDDY, B.S. and WYNDER, E.L. (1973).  *J. natn. Cancer Inst.*, 50, 1437
TROWELL, H. (1972).  *Am. J. clin. Nutr.*, 25, 926
WYNDER, E.L., KAJITANI, T., ISHIKAWA, S., DODO, H. and TAKANO, A. (1969).  *Cancer*, 23, 1210

*General Texts*

CLEAVE, T.L., CAMPBELL, G.D. and PAINTER, N.S. (1969).
  *Diabetes, Coronary Thrombosis and the Saccharine
  Disease.*  John Wright and Sons Ltd., Bristol
MELVYN HOWE, G. (1972).  *Man, Environment and Disease in
  Britain.*  Barnes and Noble Books, New York

# VI

## FUTURE PROSPECTS

# 25

# THE POTENTIAL FOR CONVENTIONAL MEAT ANIMALS

C. R. W. SPEDDING and A. M. HOXEY

*Grassland Research Institute, Hurley*

## INTRODUCTION

Meat animals are kept primarily for the production of lean meat but they often yield additional products of sufficient value to influence the economics of the total process and frequently fulfil other important roles as well. 'Conventional' meat animals are considered here to be those normally used for the production of meat, in appreciable numbers; it has to be recognised, however, that what is conventional in one part of the world may be highly unconventional in another.

The major conventional meat-producing species, looked at from a world point of view, are listed in *Table 25.1*, together with the other main products derived from them.

Before discussing the potential for such animals in the future, and the nature of the factors that are likely to determine this, it is as well to consider their production characteristics. *Table 25.2*, therefore, is a compilation of 'normal' performance levels, mature liveweights and data relating to reproduction.

Clearly, if for example buffaloes and goats are to be discussed comparatively, it is necessary to know that buffaloes are bigger than goats, and that they eat more as well as produce more. It is then necessary to know whether they produce more or less per unit of food, or body weight, or indeed per unit of any of the important resources used in the production process.

Table 25.1 Major conventional meat-producing species of the world and their main products other than meat (a selection from several sources, including Turner, H.N., 1971)

| Species | Products | | | | |
|---|---|---|---|---|---|
| | Milk | Skin | Fibre | Feathers/down | Faeces |
| Cattle | + | + | + | | + |
| Buffalo | + | + | | | + |
| Musk oxen | | | + | | |
| Yak | + | + | + | | |
| Sheep | + | + | + | | + |
| Goats | + | + | + | | + |
| Deer (red) | | + | | | |
| Horses | + | + | + | | + |
| Camels | + | + | + | | + |
| Alpaca | | | + | | |
| Llama | | | + | | |
| Rabbits | | + | + | | + |
| Guinea pigs | | | + | | |
| Capybara | | + | | | |
| Pigs | | + | + | | + |
| Dogs | | | | | |
| Hens | | | | + | + |
| Ducks | | | | + | + |
| Geese | | | | + | + |
| Turkeys | | | | + | + |
| Game animals and birds | | | | + | |

Table 25.2 'Normal' performance levels and attributes of animals in Table 25.1

| Species | Mature size (kg liveweight) Male | Female | Reproductive Rate (No. of young/ year) | Ratio of males to females for breeding | Yield/progeny (kg carcass weight) |
|---|---|---|---|---|---|
| Cattle | 700-800 | 450-700 | 0.9 | 1:30-50 | 200-300 |
| Buffalo | 665-718 | 509-548 | | | 144-279 |
| Musk oxen | 365 | | 0.5-1.0 | | |
| Yak | 230-360 | 180-320 | | | |
| Sheep | 30-150 | 20-100 | 1-2+ | 1:30-40 | 18-24 |
| Goats | 48-58 | 45-54 | 1-3 | 1:40 | 4.3-8.4 |
| Deer (red) | 124 | 75-82 | 1 | 1:1.6-6.6 | 20-64 |
| Horses | 1000 | 700-900 | 1 | 1:70-100 | 360 |
| Camels | 450-840 | 595 | 0.5 | 1:10-70 | 210-250 |
| Alpaca | 80 | | | | |
| Llama | 80-110 | | | | |
| Rabbits | 4.0-7.2 | 4.5-7.6 | 30-50 | 1:15-20 | 1-2 |
| Guinea pigs | | 0.6-1.0 | 20-30 | | 0.7 |
| Capybara | 60 | 45 | 8.7 | | 15.3 |
| Pigs | 350 | 220 | 20 | 1:20 | 45-67 |
| Dogs | 12-15 | | | | |
| Hens | 4 | 3 | 108 | 1:10 | 1.45 |
| Ducks | 4.5 | 4.0 | 110-175 | 1:5-8 | 2 |
| Geese | 5-10 | 4.5-9 | 25-50 | 1:2-6 | 4-5 |
| Turkeys | 13-23 | 8-12 | 40-100 | 1:10-15 | 3-9 |

It is not possible to arrive at 'average' performance figures, but it is possible to state 'normal' levels (or ranges) of performance that are adequate for species comparisons. The potential use of these species cannot necessarily be judged from their normal performance. An attempt needs to be made to assess what performance levels might be achieved, from evidence available in the literature, where improved nutrition, management or disease control have allowed exceptionally high performance, well above that regarded as normal.

Efficiency may be of the greatest importance and can be expressed in many different ways. It is used here to denote a ratio of an important output per unit of one or more of the important resources used (Spedding, 1973), recognising that any particular ratio may be the most relevant or useful for a particular purpose. In addition, both inputs and outputs can be expressed in a variety of ways (such as weight, energy content or monetary value) which must be regarded as more or less appropriate to the specified purpose of the calculation.

Only some of the more important ratios are illustrated in the tables that follow and it has not been possible to achieve a complete or comprehensive statement for all the main species.

Data of these kinds are the essential descriptions of the animals considered, in order that their future usage may be assessed. However, there are also many other factors involved in determining this.


## DETERMINANTS OF THE FUTURE OF CONVENTIONAL MEAT ANIMALS

The most important determinant is bound to be an economic one. Clearly, the numbers of meat-producing animals will depend chiefly on the total demand for their products at the prices for which they can be profitably produced. However desirable, they will not be produced unless they enable a profit to be made by those engaged in the production process and this, in turn, requires that enough people will buy the products at economic prices. So the first determinant could be considered to be the cost of production, relative to the value placed on meat by the consumer. There are many other factors that influence demand, beside the ability to pay for the product. There is, for example, the desire to eat meat of any particular kind.

On a world scale, it is often argued that meat production is too wasteful and inefficient a process for it to be the main basis of food production for the majority of the population. However, Blaxter (1970) has pointed out that as living standards rise, so does the total consumption of meat. Where this applies, therefore, there should be no shortage of demand or ability to pay.

This does not mean that a particular kind of meat will continue to be wanted, or that the desire for meat might not be channelled into forms of food that fulfil similar roles but are based on non-animal sources (*see* Chapter 27).

A further determinant of demand may be the tolerance of the community for the *methods* of meat production. Whether or not the community's standards of what is tolerable make sense to the animal production operator, they may still greatly influence demand, especially if attractive alternative foods are available.

In addition to the demand for meat, there also has to be a capacity to supply it, quite apart from questions of profitability. Such capacity has to be related to the availability of all the resources required. This is not really a different argument from the economic one, since scarce resources would naturally have a high cost, but it is possible to consider the alternative uses of such resources long before it is possible to estimate their probable future costs.

The most useful approach to an assessment of the potential for conventional meat animals would therefore seem to be to calculate the relative efficiency with which they use the more important resources. Any list of these resources may change with time, but is likely to include the following:

> (a) Feed.
> (b) Land.
> (c) Labour.
> (d) Capital.
> (e) Energy
> (f) Water.

The relative efficiencies of meat-producing animals, in the use of these resources, will be considered in turn.

(a)   FEED CONVERSION EFFICIENCY

*Table 25.3* shows the energetic efficiencies with which
'normal' populations of each species convert the feed
on which they commonly produce.  These efficiency values
might be increased with 'better' breeds or varieties,
by improved nutrition, in terms of level of feeding or
quality of the diet, and by more efficient populations,
in terms of higher reproductive rates or lower male/
female ratios.  Values might be lowered by adverse
environments, by greater disease incidence or by reduced
longevity of breeding animals.

In general, improvement in all these factors will
raise population efficiency (Dickerson, 1970; Large,
1970; Spedding, 1971a) towards that of the individual
meat-producing animal, but it cannot generally exceed
this value.  Individual values may be used, therefore,
as 'ceiling indicators' and are shown in *Table 25.4*.

The implication is that some species are basically
more efficient than others in the use of their preferred
food.  The extent to which they can perform, or even
survive, on each other's foods is very variable.

There is not a great deal of information on the
efficiency with which the main animal feeds are converted
by different species, but some work has been done on
efficiency in relation to the proportions of different
components in the diet.  *Table 25.5* illustrates the
effect of the proportion of dried grass in a dried-grass/
barley diet on the efficiency of protein utilisation.


(b)   LAND USE

Since the efficiency of feed conversion varies with the
type and quality of feed used, so will the efficiency of
land use.  It follows that the type of land must
influence efficiency by virtue of its ability to grow
different crops for animal feed, in addition to
associated effects of land-type and climate.  As with
feed, therefore, there are several ways of assessing the
efficiency of land use.

One method is to compare the relative efficiencies
with which different animal species produce meat from a
unit area of land capable of producing all the kinds of
feed required (*Table 25.6*).  Another is to take con-
trasting types of land and consider the efficiency with
which each can be used by a range of meat-producing
species.  Unfortunately, the data available are hardly
adequate for this purpose.

*Table 25.3* Range of efficiency values ($E^1$) for populations[2]

|  |  | $E^1$ |
|---|---|---|
| Suckler cows and calves[3] |  | 3.2 |
| Sheep[4] with | Singles | 2.4 |
|  | Twins | 3.4 |
|  | Triplets | 4.2 |
| Rabbits[5] |  | 8.0 |
| Pigs[6] | Pork | 23 |
|  | Bacon | 27 |
| Hens[7] |  | 14.6 |
| Geese[8] |  | 10 |

[1] $E = \dfrac{\text{Total energy in carcass produced}}{\text{Gross energy in feed for the progeny and proportion of parents}} \times 100.$

[2] No allowance has been made for replacements.

[3] Feed intake of the cow for one year plus feed intake of the calf to slaughter. No allowance for sire as AI could be used. Carcass production of 250 kg at 18 months. Sources include Baker and Barker (1971).

[4] Feed intakes of the ewe for one year plus feed intake of the lamb(s) to slaughter. Animals fed on dried lucerne nuts under housed conditions. Feed intake of the ram = 657 kg DM/yr and 1 ram to 35 ewes. Sources: Large (1965) and Bradfield (1967).

[5] 1 buck to 15 does, each producing 40 progeny per year. Sources include Walsingham (1972b).

[6] 1 boar to 20 sows, each producing 20 progeny per year.

[7] 1 cock to 10 hens each producing 108 progeny per year. Sources: Morris (1971) and Kivimäe *et al.* (1971).

[8] 1 gander to 3 geese each producing 18 progeny per year. Sources: Spedding (1973) and Kivimäe *et al.* (1971).

*Table 25.4* Range of efficiency values ($E^1$) for independent (e.g. weaned) individuals

|  |  | $E^1$ |
|---|---|---|
| Beef cattle[2] |  | 5.2-7.8 |
| Sheep[3] |  | 11-14.6 |
| Rabbit |  | 12.5-17.5 |
| Pigs | Pork | 35 |
|  | Bacon | 35 |
| Hens |  | 16 |
| Geese |  | 13.4 |

$$^1E = \frac{\text{Total energy in the carcass}}{\text{Gross energy in feed from independence to slaughter}} \times 100 \,.$$

[2]Cattle fed dried grass. Source: Tayler, (1970). Cattle fed barley. Source: J.B.P.C. (1968).

[3]Artificially-reared lambs fed mainly on (a) dried grass or (b) barley, diets from weaning to slaughter under housed conditions. Source: Treacher and Hoxey (1969).

Other animals as for *Table 25.3*.

*Table 25.5* Efficiency of meat production by bulls fed on different proportions of barley and grass (after Homb and Joshi, 1973)

| Percentage of dried grass in the dry ration | Efficiency of protein utilisation (%) | Area needed per 240 kg bull carcass (ha) | Carcass weight produced (kg/ha) | Edible carcass protein (kg/ha) |
|---|---|---|---|---|
| 21 | 20 | 0.53 | 450 | 65 |
| 39 | 18 | 0.49 | 490 | 71 |
| 55 | 16 | 0.45 | 530 | 77 |
| 69 | 15 | 0.41 | 580 | 84 |

*Table 25.6* The efficiency of land use for meat production (after Walsingham, 1972a)

| Species | $E = \dfrac{kg\ carcass}{ha\ land}$ |
|---|---|
| Cattle | 598 |
| Sheep | 429 |
| Poultry (broiler hens) | 852 |
| Rabbits | 932 |
| Pigs (pork) | 812 |

*Table 25.7* Labour demands of different meat-producing enterprises in the UK (Nix, 1972)

| Enterprise | Labour (h/month)[1] | |
|---|---|---|
| | Average | Premium |
| Beef fattening (per bullock fattened) | | |
| Yarded | 2.5 | 1.4 (1.0 self-feeding) |
| Summer grazed | 0.25 | 0.15 |
| Barley beef | 0.9 | 0.7 |
| Suckler herds (per cow) | | |
| Single suckling (average whole year) | 1.0 | 0.75 |
| Multiple suckling (average whole year) | 3.25 | 2.75 |
| Veal calves | | |
| 120–150 per man (at a time) | 2.0 | 1.5 |
| Sheep (per ewe) average whole year | 0.3 | 0.25 |
| Pigs | | |
| Breeding and fattening for pork/sow | 5.8 | 3.7 |
| Breeding and fattening for bacon and heavy hogs/sow | 6.2 | 4.0 |
| Poultry per 100 produced | 0.75 | 0.5 |

[1]Average and premium refer to number of animals looked after per man i.e. Pork production: average 40 sows and progeny/man, Premium 65. Bacon production: average 35–40 sows progeny/man, Premium 60. Table birds: average 30,000 birds/year/worker, Premium 50,000. Sheep: average 400 ewes and lambs/man, Premium 600.

*Table 25.8* Costs involved in different animal enterprises (data from MLC (1971 and 1972), Norman and Coote (1971), Nix (1972) and Young (1973)

| Animal | Costs[1]/ animal (£) | Costs/ha (£) | Costs/kg carcass (£) |
|---|---|---|---|
| Ewe and 1.75 lambs/year (11 ewes/ha) | 12.37 | 136 | 0.44 |
| Single suckled beef (18 mth) all-grassland system (2.2 cows/ha) | 64 | 140 | 0.40 |
| Beef-calves from dairy herds (18 mth) intensive grassland (3 beasts/ha) | 72 | 214 | 0.38 |
| Pork 4 animals/year | 49 | - | 0.27 |
| Bacon 2.6 animals/year | 41 | - | 0.25 |
| Table poultry 6 birds/year | 2.47 | - | 0.27 |

[1]Costs include fixed and working capital.

(c) USE OF LABOUR

There is considerable variation in the labour demands of different meat-producing enterprises; this is illustrated in *Table 25.7* and in *Figure 25.1* for sheep and suckler cows, but the requirements shown in the figure are characteristic of particular enterprises rather than species.

(d) CAPITAL REQUIREMENTS

The capital required obviously depends on the size and nature of the enterprise. The relative costs involved in different animal enterprises are illustrated in *Table 25.8*, where they are expressed per animal, per unit of land and per unit of output.

Table 25.9 Regional variation in maximum crop growth rate (after Cooper, 1970)

| Climate and country | Crop | Crop growth rate (g/m$^2$/day) | Total radiation input (cal/cm$^2$/day) | Conversion of light energy (%) |
|---|---|---|---|---|
| **TEMPERATE** | | | | |
| UK | *Lolium perenne* | 16.6 | 290 | 5.4 |
| UK | Barley | 23 | 484 | 4.0 |
| UK | Beet | 31 | 294 | 9.5 |
| New Zealand | *Lolium perenne* | 17.5 | 480 | 3.4 |
| USA | *Zea mays* | 52 | 500 | 9.8 |
| **SUBTROPICAL** | | | | |
| USA | *Zea mays* | 52 | 736 | 6.4 |
| Australia | *Digitaria decumbens* | 19.3 | 580 | 3.1 |
| **TROPICAL** | | | | |
| Puerto Rico | *Panicum maximum* | 16.8 | 480 | 3.3 |
| Australia | *Pennisetum typhoides* | 54 | 510 | 9.5 |

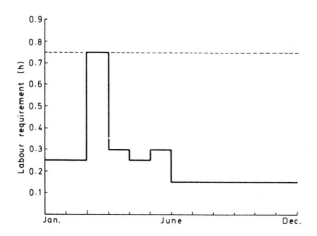

*Figure 25.1 Pattern of labour demand of different meat-producing enterprises in the UK.* ————— *Labour requirement per ewe, assuming mainly March lambing. 600 ewes with lambs per man (not employed full time).* --------- *Labour requirement per cow for single suckling herds. (After Nix, 1972)*

*Table 25.10 Variation in crop production from experimental plots (after Cooper, 1970)*

| Climate and country | Crop | DM production (t/ha/yr) | Estimated conversion of light energy (%) |
|---|---|---|---|
| **TEMPERATE** | | | |
| UK | *Lolium perenne* | 16.7 | 2.0 |
| New Zealand | *Lolium perenne* | 26.6 | 2.2 |
| **SUBTROPICAL** | | | |
| USA | *Cynodon dactylon* | 27.2 | 1.6 |
| Australia | *Paspalum plicatulum* | 31.9 | 1.8 |
| Australia | *Chloris gayana* | 23.5 | 1.3 |
| **TROPICAL** | | | |
| Puerto Rico | *Panicum maximum* | 48.8 | 2.9 |
| El Salvador | *Pennisetum purpureum* | 85.2 | 5.4 |

(e)   ENERGY USE

The efficiency with which energy is used has already
been dealt with as a method of expressing feed conversion
efficiency.  Land-use efficiency may be similarly
expressed, in order to take account of the different
amounts of solar energy received.  *Tables 25.9* and *25.10*
show the kind of variation to be expected in different
regions of the world, where the solar radiation receipt
differs markedly, in both maximum crop growth rates and
efficiencies of conversion of light energy.

In addition, there is an increasing need to take into
account the supplementary or 'support' energy (Black,
1971; Lawton, 1973) required by different enterprises.
This is largely derived from fossil fuels, although that
may not always be so and there are interesting examples
of the use of current solar energy other than by
photosynthesis (Hammond, 1972).  Electricity generated
by water power is an example of the transfer of energy
into a system without depleting the useful energy store
elsewhere.

In general, however, the consequences of importing
energy into systems of production have serious
implications and, without necessarily judging the
importance or urgency of the issue, clearly they should
be considered.

Black (1971) pointed out that modern methods of
agriculture that appeared vastly more productive than
primitive methods, might easily be very similar in terms
of the efficiency with which the total energy resources
are used.

There are, of course, many difficulties in trying to
take account of 'support' energy, notably in deciding
where to draw the line and in obtaining the necessary
information.  In *Table 25.11*, the additional consumption
of energy as fuel has been included and some attempt
made, where possible, to take account of the energy costs
of constructing the equipment used.  Clearly, some
enterprises are much greater consumers of 'support'
energy than others.

Table 25.11 Support energy usage in agricultural production (compiled by Jean M. Walsingham)

| | | Lamb[1] | Eggs[2] |
|---|---|---|---|
| SUPPORT ENERGY INPUTS | Upstream[3] (Mcal) | 71815 | 12695 |
| | On the farm (Mcal) | 30183 | 115527 |
| | Downstream[4] (Mcal) | 6816 | * |
| | Sub-total (Mcal) | 108814 | 128222 |
| LABOUR INPUTS | Upstream (h) | 38 | 102 |
| | On the farm (h) | 1745 | 1889 |
| | Downstream (h) | 530 | * |
| | Total | 2313 | 1991 |
| Energy content of the food required to support these hours of work (Mcal) | | 934 | 804 |
| TOTAL SUPPORT ENERGY INPUTS TO THE PRODUCTION PROCESS (Mcal) | | 109748 | 129026 |
| (MJ) | | 458747 | 539329 |
| ENERGY OUTPUT | Main product (Mcal) | 23026 (Meat) | 16863 (Eggs) |
| | By-product (Mcal) | 17040 (Wool) | 3480 (Cull carcasses) |
| | Output total (MJ) | 167476 | 85034 |
| RATIO | Energy output in products / Support energy input | 0.37 | 0.16 |

[1] Based on a flock of 300 ewes on arable farms.
[2] Based on a 1000 bird battery unit on arable farms.
[3] Used in the manufacture or production of inputs to farm enterprises.
[4] Used in the utilisation of farm products (processing, packaging and delivery).
*Data needed but not available when table was constructed.

(f)   WATER USE

Similar arguments can be applied to other scarce
resources, such as water, with one major difference,
that all but energy can be re-cycled.  Water cycles
naturally and is already often used and re-used many
times between precipitation and arriving at the major
water masses (seas, oceans and lakes).  All this involves
a cost, including an energy cost, but there is also some
limit to the extent that re-cycling can solve a problem
of shortage.

There is some relevance to future potential, therefore,
in looking at the water requirement of different animal
production processes (*Table 25.12*).  In general, however,
the water requirements of meat-production processes
appear to be small, provided that the water needed for
crop production is not included.

Animals that are kept under housed conditions may also
need water to dispose of their excreta (*Table 25.13*).

## FUTURE PRODUCTION SYSTEMS

Having regard to the foregoing, and bearing in mind the
difficulty of the relative weighting to be given to each
consideration, it is possible to estimate what kind of
meat-production systems would seem most appropriate to
particular circumstances.  It is accepted that economic
factors will determine what actually happens but that
the issues discussed here will ultimately influence these
same economic factors; to precisely what extent, it is
hardly possible to predict.

The relevant circumstances could be grouped in many
different ways, the following will be used in this
chapter:

1.   Resources of direct use to Man.
2.   Resources with alternative uses (including
     indirect use for Man).
3.   Resources with one (or a very few) limited use(s).
4.   Integrated use of resources for agricultural,
     amenity and recreational purposes.

Table 25.12 Estimated water requirements for animal production

| Species | Liveweight (kg) | Environment | Water (1/100 kg liveweight/ day) | Source |
|---|---|---|---|---|
| Beef cattle | 400 | 15–21°C | 7 | 1 |
| *Bos Taurus* | | Arid subtropics (summer) | 16.1 | 2 |
| *Bos Indicus* | | Equatorial arid (July) | 7.5 | 2 |
| Musk oxen | 365[1] | Arctic (April) | 3.5 | 2 |
| Sheep | 40 | 15–20°C | 9 | 1 |
| Haired sheep | 18–25 | Simulated desert | 3.8 | 3 |
| Sheep (Ogaden fat tail) | | Arid subtropics (summer) | 10.2 | 2 |
| Sheep (Corriedale) | | Arctic (April) | 6.2 | 2 |
| Goats | | Arctic (April) | 5.2 | 2 |
| Goats | 15–20 | Simulated desert | 4.2 | 3 |
| Pigs | 15 | Normal housing | 7 | 4 |
| | 90 | | 6 | |
| Horses | 600 | – | 4.1–4.6 | 5,6 |
| Reindeer | | Arctic (April) | 12.8 | 2 |
| Camel | | Arid subtropics (summer) | 8.2 | 2 |
| Rabbits | 2.5 | Housed: dry-diet | 6 | 7 |
| Hens | 2–3 | Housed: dry-diet | 7–35 | 8 |

[1]Inserted from data in *Table 25.2.*

Sources: 
1. ARC (1965)
2. Macfarlane et al. (1971)
3. Maloiy and Taylor (1971)
4. ARC (1966)
5. Rossier (1973)
6. Olsson (1969)
7. Kennaway (1943)
8. Tyler (1958)

*Table 25.13* A comparison of the $O_2$ demand of the excreta produced by agricultural animals. The quantities of water required to meet the demand are also included (after Owens, 1973)

| Source | Daily $O_2$ demand of the excreta (g) | Daily flow of fully oxygenated water required ($m^3$/day) |
|---|---|---|
| Cow | 500 | 50 |
| Pig | 140 | 14 |
| Sheep | 100 | 10 |
| Hen | 10 | 1 |

## 1. RESOURCES OF DIRECT USE TO MAN

The main possibilities in this category are the use of land and associated resources for (a) the direct production of human food as crops, and (b) their use in meat production.

There is little doubt that where these alternatives exist in relatively uncomplicated situations, more food energy and protein (and almost every other food constituent) can be produced by crops, compared with meat-producing animals, per unit of most, if not all resources (*see Table 25.14* for comparisons of energy production per unit of land). Thus, for example, cereals will be sensibly fed to animals only where (i) there are overriding economic considerations, (ii) where the resources are not regarded as sufficiently scarce, and (iii) where they are used as a small supplement, on which is based a level or efficiency of meat-production that leads to a greater *total* efficiency of resource use.

## 2. RESOURCES WITH ALTERNATIVE USES

Where resources cannot be used for the production of crops suitable for direct consumption by Man, there are alternative meat-producing processes to choose from. Apart from overriding economic considerations and, for simplicity, ignoring mixed enterprises such as mixed grazing, clearly the choice will be determined to a large extent by the efficiency with which the most costly resources are used in the production of meat. However, the most costly resources will differ between animal

*Table 25.14* Relative efficiency of production per unit of land by crops and animals (after Spedding, 1973). The data are shown to indicate orders of magnitude: methods of calculation vary and the sources given should be consulted before making detailed comparisons

| Product harvested | Protein (kg/ha/yr) | Energy (MJ/ha/yr) | Source |
|---|---|---|---|
| Dried grass | 700–2200 | 92000–218000 | 1 |
| Leaf protein | 2000 | – | 2 |
| Cabbage (edible) | 1100 | 33500 | 3 |
| Maize grain (North America) | 430 | 83700 | 4,5 |
| Potato (edible tuber) | 420 | 100400 | 3 |
| Barley grain (UK) | 370 | 62800 | 4,5 |
| Wheat (edible grain) | 350 | 58600 | 3 |
| Rice grain (Europe) | 320 | 87900 | 4,5 |
| Cassava roots (Malawi) | 246[1] | 133800[1] | 4,6 |
| Rabbit carcass | 180 | 7400 | 7 |
| Chicken (edible broiler) | 92 | 4600 | 3 |
| Lamb (edible meat, New Zealand) | 62 | 7500 | 8 |
| Beef (edible meat, mainly barley-fed) | 57 | 4600 | 9 |
| Pig (edible meat) | 50 | 7900 | 3 |
| Lamb (edible meat, UK and Eire) | 23–43 | 2100–5400 | 3,10 |
| Beef (edible meat, mainly grass-fed) | 27 | 3100 | 3 |

[1]kg/ha/crop (not year)

Source: 1. Castle and Holmes (1960)
2. Pirie (1971)
3. Holmes (1970)
4. FAO (1970a)
5. FAO (1970b)
6. Pynaert (1951)
7. Walsingham (1972a)
8. Campbell (1968)
9. Duckham and Lloyd (1966)
10. Conway (1968)

species and within the range of production methods
available for each species.

Thus, the relative efficiency of feed conversion by
sheep is unlikely to equal that of beef production from
dairy calves and if feed was the dominant resource,
sheep would not be kept on land where the choice existed.
However, if capital were limiting, or where return on
capital was most important, the answer might be quite
different.

It would be misleading to represent this issue as
being simpler than it is, and there appear to be few
generalisations that would not involve oversimplification
at this time. However, the major constituent of this
category is likely to be grassland and the scale of this
type of land resource across the world suggests a
continued major role for the ruminants.


3.   RESOURCES WITH LIMITED USE

The major cases in this category concern land that can
grow only a limited range of crops (economically), not
including those of direct use to Man. The reasons for
this may be extremely important because they may also
determine what kind of animals may be kept, or at what
level of productivity.

The main difference in this category, is that it is
relatively obvious that some animal production
enterprises would be impossible or uneconomic in
particular conditions; if the conditions are sufficiently
extreme, there may be only one possibility. Thus,
reindeer will presumably remain the preferred method of
producing meat from Arctic tundra and few calculations
relating to pigs are required. However, where the
resources are not characteristic of the land and climate
but of the enterprises already involved and already
producing them, the situation might be quite different.

Arable by-products, including straw, might have only
limited use but there may be nothing about the
environment that imposes further restrictions. The
question can then be posed of such by-products, as to
which meat-producing animal will make the most efficient
use of them, raw or after some processing. Clearly,
however efficiently a species might use a by-product,
there will be many other considerations, including the
efficiency with which all the other resources are
employed. Nevertheless, where large quantities of an
agricultural by-product, or an industrial waste or a

rapidly-used product such as newspaper, are available,
this is one factor that may influence the future of those
meat-producing animals that can utilise them.

This argument is most frequently applied to ruminants
but other animal species are also relevant.  Fish, such
as the herbivorous Chinese Grass Carp (*Ctenopharyngodon
idella*), have been used extensively in China and Rumania,
for example, and might be better utilisers of highly
seasonal food supplies.  Indeed, poikilotherms might
generally have advantages in this respect (Février,
1971; Spedding, 1972a).

One major agricultural by-product (or waste, according
to one's point of view) consists of faeces, especially
from large concentrations of housed animals.  Much effort
is currently being directed to methods of disposing of
such faeces, disposal being regarded as a major problem.
However, there have been some successful attempts
(Fosgate and Babb, 1972) to utilise faeces from dairy
cattle to produce saleable potting soil and earthworms
for anglers.  Furthermore, the worms produced can be
used as the basis for further meat production, after
processing, or for the production of pet food.  The
volume of pet food is now so great in some countries
that it may be very important to ensure that no
directly-usable meat is involved.  These possibilities
hardly count as conventional methods but they may be
applied to the feeding of conventional meat animals.

A full economic assessment is not possible but it may
be argued with some force that any animal that can
produce acceptable meat from wastes that themselves
constitute a disposal problem, must have considerable
future potential.


## 4.  INTEGRATED USE OF RESOURCES

The probable growth of demand for land for recreational,
amenity and field-sport purposes suggests that
agriculture may need to become more intensive on a
smaller total area of land.  However, the potential for
agricultural usage of amenity land should not be
ignored.

First, visual amenity may positively require land
to be farmed, although it might favour different
patterns of farming from those in current use or those
resulting from narrower economic considerations alone.

Secondly, large areas of grassland for amenity
purposes would require controlled animal populations to

keep the grass short and to prevent regeneration of scrub and woodland (Spedding, 1971b). Such animal populations would need management, including control of their numbers; this suggests the need for experienced animal managers and the possibility of a substantial meat output in the form of culled animals. These animals could be of any age, including animals at birth (or close to it) removed for rearing elsewhere as a quite separate enterprise.

There are several special attributes that would be required of such animals: they would need to be hardy and resistant to disease and they would have to require little attention and be able to defend themselves or escape from the unwelcome attentions of dogs and people. At the Grassland Research Institute, the Soay sheep has been looked at from this point of view. It does not require shearing, appears to have many of the desired characteristics and is capable, when mated with a Down ram, of producing quite acceptable meat from the crossbred progeny (Spedding, 1972b).

The main point is that, should the integrated use of land for agriculture and other purposes reach a large scale, and there are signs that it might, this could also be an important factor in determining the future of meat-producing animals.

## OTHER CONSIDERATIONS DETERMINING THE CHOICE OF SYSTEM

It cannot be stressed too strongly that these issues are exceedingly complex and oversimplification is unhelpful and could be very harmful.

Whilst it is not possible, therefore, even to mention all the other factors that may be of importance, it is important to be aware of their existence. For example, a ruminant at a distance from its market may use particular feed resources in a more desirable fashion than a pig close to its consumer. But the costs, financial and social, of transporting the output of the first enterprise or of sustaining a human population to service it, may be high. Of course, the rejoinder is obvious, that it requires less transport to convey carcasses than to cart the pig's feedstuffs; the point is simply that the argument cannot end there and we are in no position to complete it.

No clear conclusion can be presented, therefore, suggesting that one animal species will survive as a meat-producer and another will disappear; even less can a time-scale be introduced. Nevertheless, it is worth trying to ensure that the major factors and their influence have been considered. One, so far only briefly mentioned, that may be of increasing importance is the attitude of the whole community to the ways in which animals are kept. Much emotional and unreasonable criticism has been directed at animal production methods and undue emphasis has often been placed on the consequences of modern, intensive methods.

It would seem very unwise, however, to ignore the proper concern of the community for the 'welfare' and 'wellbeing' of agricultural animals. All such terms are now rather suspect but it is worth noting that we would not regard either 'welfare' or 'cruelty' in relation to children (or dogs) as something that could not be thought about because we have no objective measures of 'stress' and 'suffering'.

Most people would draw the line at some point and the question is simply where to draw it. It is drawn by most people, without objective measurement, and the community is entitled to a view. The relevance in this discussion is simply that such a view may gather force and thus influence the future of any animal that can only be kept (economically) in ways which the community finds intolerable.

A major reason for ending on this note is that it emphasises the breadth of the view that has to be taken in assessing the potential for meat-producing animals.

## REFERENCES

ARC (1965). *Nutrient requirements of farm livestock,* No. 2, Ruminants
ARC (1966). *Nutrient requirements of farm livestock,* No. 3, Pigs
BAKER, R.D. and BARKER, J.M. (1971). Unpublished data
BLACK, J.N. (1971). *Ann. Appl. Biol.,* 67, 272
BLAXTER, K.L. (1970). *Anim. Prod.,* 12, 351
BRADFIELD, P.G.E. (1967). Unpublished data
CAMPBELL, A.G. (1968). *Span,* 11 (1), 50
CASTLE, M.E. and HOLMES, W. (1960). *J. agric. Sci., Camb.,* 55, 2
CONWAY, A. (1968). *Span,* 11 (1), 47

COOPER, J.P. (1970). *Herb. Abstr.*, 40 (1), 1

DICKERSON, G. (1970). *J. Anim. Sci.*, 30 (6), 849

DUCKHAM, A.N. and LLOYD, D.H. (1966). *Farm Economist*, 11 (2), 95

FAO (1970a). *Production Yearbook*, 24

FÃO (1970b). *Nutr. Stud.*, No. 24

FÉVRIER, R. (1971). *Proc. Int. Cong. Anim. Prod. Versailles*, 381

FOSGATE, O.T. and BABB, M.R. (1972). *J. Dairy Sci.*, 55 (6), 870

HAMMOND, A.L. (1972). *Science*, 177 (4054), 1088

HOLMES, W. (1970). *Proc. Nutr. Soc.*, 29 (2), 237

HOMB, T. and JOSHI, D.C. (1973). 'The biological efficiency of protein production by stall-fed ruminants', in *The Biological Efficiency of Protein Production*. Ed. J.G.W. Jones. CUP

KENNAWAY, E.L. (1943). *Br. Med. J.*, 1, 760

KIVIMÄE, A., WADNE, C. IDEFJELL, C. and HILDINGSTAM, J. (1971). *Lantbrukshögskolans Meddelanden A.*, No. 157

LARGE, R.V. (1965). Unpublished data

LARGE, R.V. (1970). *Anim. Prod.*, 12, 393

LAWTON, J.H. (1973). In *Resources and Population*. Ed. B. Benjamin, P.R. Cox and J. Peel. Academic Press, London

MACFARLANE, W.V., HOWARD, B., HAINES, H., KENNEDY, P.J. and SHARPE, C.M. (1971). *Nature*, 234, 483

MALOIY, G.M.O. and TAYLOR, C.R. (1971). *J. agric. Sci., Camb.*, 77, 203

MLC (1971). *Joint Beef Production Committee. Booklet No. 1*, revised edn.

MLC (1972). *Joint Beef Production Committee. Booklet No. 3*

MORRIS, T.R. (1971). Personal communication

NIX, J. (1972). *Farm Management Pocketbook*, 5th edn. Wye College

NORMAN, L. and COOTE, R.B. (1971). *The Farm Business*. Longman, London

OLSSON, N.O. (1969). 'The Nutrition of the Horse', in *International Encyclopaedia of Food and Nutrition*, 17, part 2. Ed. D. Cuthbertson. Pergamon Press

OWENS, M. (1973). 'Resources under Pressure - Water', in *CICRA Symp. on Intensive Agriculture and the Environment*

PIRIE, N.W. (1971). *Leaf Protein, Its Agronomy, Preparation, Quality and Use*. IBP Handbook No. 20, Blackwell, Oxford

PYNAERT, L. (1951). *Le Manioc,* 2nd edn. La direction
de l'agriculture, Ministère des Colonies, Belgique
ROSSIER, E. (1973). Personal communication
SPEDDING, C.R.W. (1971a). *Grassland Ecology,* OUP
SPEDDING, C.R.W. (1971b). The Future for Grass.
Presented to the BA Meeting, Swansea, Sept. 1971
SPEDDING, C.R.W. (1972a). *Proc. 2nd World Conf. on
Animal Feeding,* IV, 1245
SPEDDING, C.R.W. (1972b). *Ann. Rep. Grassland Research
Inst.,* 129
SPEDDING, C.R.W. (1973). 'The Meaning of Biological
Efficiency', in *The Biological Efficiency of Protein
Production.* Ed. J.G.W. Jones. CUP
TAYLER, J.C. (1970). *J. Br. Grassld. Soc.,* 25 (2), 180
TOMBS, M. (1974). The Significance of Meat Analogues.
*Proc. 21st Easter Sch. in Agric. Sci., Notts.*
TREACHER, T.T. and HOXEY, A.M. (1969). *GRI Internal
Rep.,* 170
TURNER, H.N. (1971). *Outlook on Agriculture,* 6 (6), 254
WALSINGHAM, J.M. (1972a). 'Ecological efficiency
studies - 1. Meat Production from rabbits',
*GRI Technical report No. 12*
WALSINGHAM, J.M. (1972b). Unpublished data
YOUNG, N.E. (1973). Personal communication

# 26

## ALTERNATIVE LIVESTOCK: WITH PARTICULAR REFERENCE TO THE WATER BUFFALO (*BUBALUS BUBALIS*)

W. ROSS COCKRILL

*Animal Production and Health Division, FAO, Rome, Italy*

### INTRODUCTION

World supplies of animal protein are limited and strained. The search for new sources is becoming ever more stringent. Not only orthodox categories are constantly being sifted, but serious consideration is being given to some very unorthodox animal proteins.

The particular type of malnutrition caused and perpetuated by protein deficiency is common and widespread in the developing countries in many of which domestic livestock are few and of poor quality. These countries are unable to produce sufficient meat or milk. Their needs are steadily increasing with the inexorable growth of human populations.

Many such populations depend largely upon traditional sources of animal protein and attempt to meet their needs by drawing on wildlife resources. In parts of Africa, antelopes provide a supply of meat. Where the off-take is carefully controlled a steady and certain harvest can be obtained. Eland, for example, yield quality meat with a low fat content and a dressed carcass percentage of 59%. Growth and reproduction rates compare well with those of cattle (Kyle, 1972).

Many proposals have been made for the industrialisation of game cropping and even for the large scale domestication of wild species. There have been some convincing demonstrations of the feasibility of such projects. However, they meet the needs of a small

proportion of the human population, and their importance
is limited in both the geographic and the demographic
sense.

Wildlife resources comprise a very wide range of wild
creatures.  In many parts of the world all species of
cats, including *Felis domestica*, are eaten.  Such small
mammals as guinea pigs, hares, porcupines, mice, rats
and squirrels are trapped for food.  Monkeys are eaten
in parts of Africa, as are anteaters which are considered
a rare delicacy.  The capybara, the largest living
rodent, is highly prized for the table of certain South
American countries.  Bats and birds are important meat
sources.  A variety of reptiles including tortoises,
turtles, lizards and snakes are eaten while some
insects, certain maggots and giant snails are also
useful protein sources (Asibey, 1972).

All these origins of animal protein, and many more,
are of great and even vital importance to localised
human populations.  For the most part, however, they are
utilised by people who accept their environment.  These
are mainly primitive communities of hunters and
gatherers.  They are not the mass of more or less
civilised peoples who attempt to restructure and control
nature to meet their needs.

Chicken meat and pig meat are well recognised as being
among the more efficiently produced forms of high-quality
protein but, like a host of other foods, they are barred
to millions on grounds of religion or superstition.
Hunger is sometimes as much a matter of deliberate food
avoidances as it is of food shortages or of poverty.

A system of calculated cannibalism could go a long way
to solving simultaneously the problems of protein hunger
and of over-population but might be expected to arouse
aversion.

The world's fishing industry, which is sophisticated
hunting and gathering, contrives to supply a small but
significant proportion of the animal protein consumed by
man.  Less technical fishing with rod and line, wicker
traps, derris extract or explosives provides small
quantities of fish and prawns.  Fishpond farming also
makes a contribution.

The future for meat analogues appears to be promising.
The mass production of concentrated protein by growing
various micro-organisms, such as yeasts, algae, and
certain bacteria on a variety of media ranging from
mineral oils and waxes to grass bagasse, offers
possibilities for increasing the food of man and the feed

ot his animals. This is especially so since flavour
manipulation has emerged as a minor science, and since
food texture has become a parallel study.

This chapter, however, approaches the question of
future prospects for meat from a conventional and
specialised angle. It assumes that meat as an almost
universally popular item of human diet is here to stay.
Eating a balanced ration is not only desirable but is,
or should be, a source of enjoyment to the consumer.
Domestic animals will, at least for some time to come,
continue to provide the major part of this animal
protein factor.

Within the narrow framework of my title 'Alternative
Livestock', this thesis directs attention to the water
buffalo (*Bubalus bubalis*) as a neglected domestic animal
and an important potential source of quality meat.


## THE WATER BUFFALO


At the risk of boring the reader with tedious taxonomous
zoological detail it must be made clear that the buffalo
in question is not the animal which is commonly referred
to as a buffalo in North America, which is not a buffalo
at all but a bison, as its tautological Latin nomenclature
makes doubly clear - *Bison bison*. Nor does this chapter
consider the African wild buffalo, *Syncerus caffer*, which
has never been domesticated. This animal is among the
finest of African big game: shrewd, incredibly fast and
extremely dangerous. Both the bison and the African
buffalo are distant relatives of the domestic buffalo,
the water buffalo of Asia, *Bubalus bubalis*. Their
relationship may be roughly defined as a distant, tribal
kinship, similar to that of the sheep to the goat, or
the ox to the yak.

The water buffalo is an animal of great but still
largely unacknowledged importance in at least 30
countries reaching from the Caribbean across to Italy
and all the way through the Near and Far East to the
USSR and China.

There are at least 18 distinct breeds of water
buffaloes. Most of these are primarily milch animals
and are found in the Indian sub-continent, where almost
half of the total world population of buffaloes is
located. India has approximately 55 million and Pakistan
around 12 million. They include the Murrah (*Figure 26.1*),
The Surti (*Figure 26.2*), the Nili-Ravi (*Figure 26.3*)

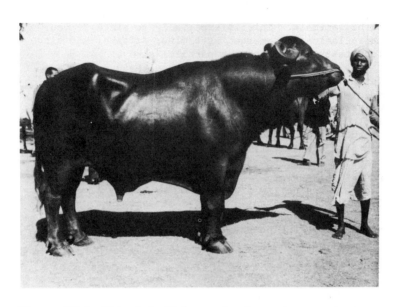

Figure 26.1   Murrah buffalo, male (by courtesy of the Indian Council of Agricultural Research)

Figure 26.2   Surti buffalo, female (by courtesy of the Anand Cooperative, Gujerat, India)

Figure 26.3 Nili-Ravi buffalo, male (by courtesy of Z. Khan)

Figure 26.4 Jafarabadi buffalo, male (by courtesy of Studio Jyoti, Junagadh)

Figure 26.5   Swamp buffalo, Thailand (by courtesy of
B. Rumich)

Figure 26.6   Swamp buffalo in wallow (by courtesy of
A. Welle)

and the Kundi the best of which are outstanding milch
animals. Less well-known breeds include the Jafarabadi
(*Figure 26.4*), the Mehsana, the Nagpuri and the
Pandharpuri.

Buffaloes exist in greatest concentration in under-
developed or developing countries where provisions for
production-oriented research on livestock are usually
not extensive. Science has paid insufficient attention
to the buffalo and especially to its productive capacity.
It is relatively seldom that one encounters such thoughtful
papers as those of Kaleff (1932), Vallegas (1932) and
Macgregor (1939). Even more rare is it to find in the
literature any references to milk production potential,
or to the possibility of increasing work output by
rational breeding, feeding and the improvement of work
implements and vehicles. Rarest of all are publications
on buffalo meat and meat production. It might be
remarked in passing that other neglected livestock which
offer considerable inducements for investment and
development include the camel - especially in the
so-called oil-rich countries - the donkey and the duck.

Until very recently, the world population of water
buffaloes was seriously underassessed, owing to the
widespread practice of lumping cattle and buffaloes
together under the heading of 'cattle' for census
purposes. It is now certain that earlier world census
figures placing the total population figure at around
80 million were grossly in error. The true world
population figure is at least 150 million. This is a
significant figure even compared to the estimated total
world cattle population of 1,140 million, especially
when it is considered that the majority of buffaloes are
productive in terms of milk, work and meat (or any two
of these outputs) whereas a high proportion of the
world's cattle is economically useless.

In all but a very few buffalo countries stocks are
increasing steadily, often despite official apathy or
active discouragement. In some areas the current rate
of increase is phenomenal. In a substantial part of the
Amazon valley, for example, where nucleus stocks were
introduced early in the present century from India,
Guyana and Trinidad, the rate of increase is said to be
10 per cent per annum (Gade, 1970). It seems certain
that the overall world population will increase
substantially during the next decade.

Over 30 years ago, Macgregor (1939) divided buffaloes
into two main groups which he designated swamp and river
types. The Swamp buffalo (*Figure 26.5*) is a breed in

its own right.   It is used mainly for work and there are
wide variations in size and conformation.   It is
widespread throughout the Far East, including China
(where there may be as many as 30 million), Indonesia,
Philippines, Thailand and Indochina.   The river types
include the milking breeds of India and Pakistan, and
some European countries, for example Italy, Bulgaria,
Greece and Yugoslavia.

All breeds present a picture of complete bliss when
immersed in a wallow, whatever its nature (*Figure 26.6*).
They chew the cud with half-closed eyes, and sigh deeply
and characteristically in utter contentment.   The wallow
and the plastering of mud which may result, provides a
protective mechanism against solar radiation and the
unwelcome attentions of biting insects.   The wallow is
not strictly essential, provided that ample shade is
available against the heat of the day.

## PRODUCTION

### MILK AND MEAT

Buffaloes may yield less milk than cows maintained under
identical conditions, but in India the average milking
buffalo gives considerably more milk than the average
milking cow.   The milk, moreover, is much richer than
cows milk both in butterfat and in non-fat solids.   Over
60 per cent of all the milk consumed in India and
Pakistan is buffalo milk.   This figure is steadily
increasing, as is the buffalo population.

Animal production of any sort in India is subject to
all sorts of difficulties not encountered in other parts
of the world.   The sacred cow, which is both a philosophy
and a physical entity, has led to the existence of a large
population of useless and unproductive bovines which yield
no milk, do no work and may not be slaughtered for meat.
In this environment the buffalo is emerging as the true
livestock wealth of India.   In spite of there being no
official support for intensive buffalo breeding programmes
of the type lavished on cattle, the buffalo is supplanting
the milch cow in many areas and more than holding its own
in others.   It is not protected as is the sacred cow;
such protection is among the great agricultural problems
of the world.

Depending upon the breed, the environment and management
practices the daily milk yield of a buffalo ranges from

the 2 to 4 litres produced by an actively working
draught female to the 16 litres or more of the
exceptional dairy animal. Buffaloes which combine work
with milk production naturally give a lower milk yield
than the exclusively milk breeds.

The lactation period is usually 270-300 days. Yields
vary greatly with breed, location, nutrition and management from 500 litres to over 3,000 litres. The gestation
period is approximately 310 days and is commonly quoted
as 'ten months and ten days'. Two calves in three years
is usual, though with good management a calf every 14
months or less is possible from the age of $3\frac{1}{2}$ years to
18 or more.

A distressing feature of buffalo husbandry in India,
as in some other milk-producing countries, is the high
calf mortality. This may approach 80 per cent of all
buffalo calves in the first two months of life. Much of
this is due to deliberate neglect. Thousands of young
calves are allowed to die annually in Bombay, for example,
because the commercial value of the milk is too high to
allow any to be spared for the calf. In addition, the
replacement requirement is low in view of the long
productive life of milch buffaloes and few heifer calves
are reared. It is estimated that in Bombay alone some
10,000 buffalo calves die of starvation each year.
Although the need for animal protein is very great, the
emphasis is on milk and no attempt is made at rearing
the calves for veal.

This picture of wastefulness, thoughtless cruelty and
failure to realise potential is common in the developing
world. It is particularly reprehensible that the buffalo
should be so treated since as a milk and meat producer it
could be among the greatest sources of relief for the
terrible food problems of much of the developing world.

The classic procedure of mating dairy cows with beef
bulls and rearing the calves for early slaughter applies
with much force to buffaloes. The production of buffalo
meat from dairy herds by the use of beef-type males, e.g.
the Swamp breed on such milking breeds as the Murrah,
offers considerable opportunity for development. Results
obtained in Bulgaria (Polikhronov, 1974) and in
Yugoslavia (Ognjanović, 1974) showed daily weight gains
of over 1 kg. Recent work in Italy (de Franciscis, 1971)
has shown that the early weaning of buffalo calves on to
reconstituted milk, and subsequent rearing for beef
production and slaughter at around one year of age with
a liveweight of 300-320 kg is perfectly feasible.

A highly satisfactory yield of top-quality beef is obtained at a much lower cost than with cattle. Dressed carcass percentages range between 50 and 58 per cent and the meat component is around 70 per cent of the meat:bone total.

The popularity of buffalo milk and milk products, and of buffalo meat also, is in no way lessened by a universal belief in their aphrodisiac qualities.


MECHANISATION AND MEAT

In the technically advanced countries we are prone to regard the use of the draught animal as being synonymous with under-development and indicative of a rather primitive agricultural society. However, in very large parts of the world the machine has not yet proved itself to be a satisfactory or economic rival to the work animal. It has been estimated that something like 84 per cent of the world's cultivable land is still tilled by human and animal labour. The rate of mechanisation is very slow and will continue to be so in most if not all of those countries where the buffalo is an important economic factor in food production, and where human labour is plentiful and cheap.

The draught buffalo is not disappearing, nor is it likely to disappear for many decades. It is one of the most efficient work animals, second only to the camel, in terms of work output balanced against maintenance costs and taking into consideration the ease with which it is trained, its docility and its exceptionally long working life.

About half of the world's human population regard rice as a staple and major part of their diet, but the great rice bowl of the East is for the most part cultivated by exactly the same practices as it has always been. In the cultivation of padi rice the water buffalo is the main source of agricultural power. Mechanisation has, as yet, made little impression. In the developing countries it may decline or cease altogether in the aftermath of the series of energy crises which, at the time of writing, seem inevitable. A swing back to the full use of the working animal is likely to be a general phenomenon in the developing countries. This revolution-in-reverse will be seen in areas where there are available not only the actual or potential animal resources, but also ample man power. It will be many years before the importance of the work animal in the developing world will decline.  Work animals, mainly in small-farmer ownership, are generally ignored in development programmes.

There are too many work animals of low calibre in existence:  if they are bred, fed and managed properly the work can be done by fewer animals of greater efficiency.  An important aspect of the revolution-in-reverse is that there can be a steady off-take for meat. The qualities of weight and muscularity required in work animals are those of the productive beef animal, and here the buffalo has a head start.

The working ox has almost disappeared from the West. As the number of working animals diminished, cattle did not disappear from the scene.  Instead they became direct producers of food.  When energy crises are finally and permanently resolved and machines begin at last to make an impression in the developing agricultural world, the buffalo will not vanish from the scene.  The turnover to meat production will accelerate, stimulated by a growing realisation of the attractiveness, acceptability and economic virtues of the end product.  It can be predicted with some confidence that the overall population of water buffaloes is going to increase by at least 50 per cent in the course of the next 30 years and that this increase will be mainly due to much greater emphasis on meat production.  As an off-shoot of the existing buffalo milk industry this has an excellent potential, but buffalo meat production can flourish as an independent industry in its own right.  The animal lends itself to extensive ranching and also to intensive feed-lot operations.


## AVOIDANCES


Foods may be avoided because of lack of palatability, repugnance on the part of the consumer - usually because of unconventionality (none of us likes sitting in the garden eating protein-rich worms) - or a fancied unwholesomeness (stilton, gorgonzola, and hung venison are unwholesome by any definition but are not generally avoided for this reason).  Vegetarians and vegans shun a spectrum of animal foods which to a majority of the human race are wholesome, appetising and desirable. Religious or social usages cause millions of people to regard the flesh of the pig, the cow or the horse with an aversion amounting to horror.

Buffalo meat is acceptable to all consumers except the vegetarians and the vegans. It is not avoided on religious or any other grounds, though it is true that the flesh of the white or albinoid Swamp buffalo is not eaten in Bali and a few other small and remote areas.

When buffalo meat is disparaged the reason is that traditionally it has come from animals at the end of a long working life, or a long life of milk production and calf rearing. It is usually a rather tough and flavourless product as is the flesh of most ageing animals, but nevertheless there is a demand for it. The fact that there is no ban on the slaughter of buffaloes in India and Nepal and a few other countries where cattle are protected means that most of the red meat consumed there comes from buffaloes. There is not yet, however, any move toward a progressive buffalo meat industry in these countries and the most encouraging trends are in Australia and Europe.

No task is more difficult in the advanced countries of the West than to persuade consumers to buy a new or strange, exotic or unusual item of food. 'Buffalo beef', however, could sell without difficulty even in the food-snobbish, selective, wealthy consumer markets. It would be even better received in the meat hungry developing countries, a considerable number of which, with rising standards of living and higher individual incomes are entering consumer markets such as that for red meat.

There is occasionally a danger that the pet food industry may corner valuable supplies of animal protein which otherwise might be used for human consumption. The buffalo meat industry in Australia's Northern Territory was in some danger of a takeover by pet food interests which could offer an attractive price per pound to owners tempted by the lure of the easy fast buck. In some areas buffalo meat as a quality product for human consumption could go the way of whalemeat as a constituent of attractively packaged expensive foods for cats and dogs. Nor is it much consolation that persistent rumour has it that a considerable proportion of such products are, in fact, destined for clandestine human consumption.

## CHARACTERISTICS OF BUFFALO MEAT

There is nothing at all peculiar about the meat of the
water buffalo.  It does not have a 'buffalo flavour';
it is not a second-best to bovine beef.  It is, in fact,
bovine beef, for the buffalo is a bovine animal and its
product differs from cattle meat only in having less fat
and being more tender.

The meat of young buffaloes, reared, managed and fed
for early slaughter can stand on its own merits and
deserves recognition as a quality product fit to rank
with the best of beef for consumption by a meat-hungry
but discriminating human public.

Joksimović (1969), who conducted exhaustive tests on
the physical, chemical and structural characteristics of
buffalo meat and compared it to meat from cattle,
concluded that there were no significant differences and
that the two meats were basically similar in chemical
and physical properties.  He detected some small
differences in water-content, total proteins and mineral
matter (which were somewhat higher in buffalo meat) and
in fat content (which was somewhat less in buffalo meat).
He considered that the lower fat content might be one
of the causes for the darker colouration of buffalo meat.
He noted that from the standpoint of meat chilling
technology there were no differences and that in its
organoleptic properties buffalo meat is the same as beef.

Some of the most interesting current work on buffalo
meat is being done in Australia.  This is a recent
development, since it is only since 1968 that Australia
has awakened to the potential of its feral buffaloes
which until then had been regarded as a pest, to be
cropped as expedient, and eliminated if feasible.  The
work of Tulloch (1974), who studied the behaviour of
feral buffaloes as none had done before, opened the way
to a new industry, small but economically significant,
in the extreme north of Australia's Northern
Territory.

Australian scientists have not been slow to grasp the
full implications of Tulloch's pioneer work in
domesticating the feral buffaloes.  It has been shown
that the Swamp buffalo of Australia produces a high
yielding carcass with a large proportion of excellent
meat which should be acceptable in the major beef
markets of the world (Charles and Johnson, 1972).
Recent work on Swamp buffaloes has shown the muscle, fat

and bone percentages to be 68.2, 10.6 and 17.4
respectively.

Not all the findings of the most dedicated scientists
concerning the nutritional value of buffalo meat – or,
for that matter of any other foodstuff – would be likely
to convince the public, and perhaps especially the
British public, if it could not be amply demonstrated
that the product is not only nutritious, attractive to
the eye and competitive with beef, but is also
thoroughly palatable and acceptable on all counts as a
juicy, tasty, chewable, palatable, digestible and
nutritious meat, devoid of any unpleasant flavour or
after-taste, and completely above suspicion as a gimmicky
meat about to be foisted on a gullible consumer public
as so many doubtful items have been in the past, ranging
from beaver meat, to snoek, and through whale steak to
kangaroo tail soup.

Tests conducted in Trinidad (Wilson, 1961), Yugoslavia
(Ognjanović, 1974) and elsewhere have shown that there
are no differences in palatability characteristics
between the meat of cattle and that of buffaloes regardless of the method of cooking.  It can be concluded that
any prejudice against buffalo meat is without cause,
provided that the animal is raised for efficient
slaughter at an early age and the meat is properly cooked
and attractively served.  It may be added that buffalo
meat is no better and no worse when it comes from old
animals inefficiently butchered and indifferently cooked
than similar meat derived from cattle.  There have been
many instances in palatability testing when buffalo meat
has been accorded preference over cattle meat.  General
consumer preference also often favours buffalo meat.  On
the Northern Territory properties where buffaloes are
now being reared for slaughter, buffalo meat from young
buffaloes slaughtered and dressed on the property is
generally preferred to the alternative of purchased
frozen or chilled cattle meat.  The sparse human
population of the extreme north ranks with the people of
North America and some of the South American countries
as being both the greatest consumers of red meat in the
world and probably the most critical of connoisseurs.

Buffalo meat is making considerable headway as a
popular product in Darwin and other cities both as beef
and as a constituent of that characteristic Australian
delicacy the meat pie.  Australia has a niche in the
export market to the United States where water buffalo
meat is at present catering to a small luxury demand.

## THE FUTURE FOR BUFFALO BEEF

World demand for red meat is steadily increasing and the demand will not lessen for many years, if it ever does. It seems reasonably certain that considerably more buffaloes than ever before will be slaughtered for meat and that a steadily increasing proportion will be reared for early slaughter. There are many countries where the present buffalo stocks could support meat production and where increased buffalo rearing could be profitable under either extensive or intensive systems. The development of a quality meat industry offers the prospect of substantial financial returns.

The possibilities for private investment may be presented geographically as follows:

EUROPE

   (a)   Several of Italy's large milk producing enterprises are among the best examples of modern buffalo dairying. There is a total buffalo population of at least 100,000, and a growing demand for mozzarella cheese, with excellent opportunities for export for example to the USA. The production of beef-type calves from dairy females, and the establishment of a buffalo meat industry in a country where beef is rapidly being priced out by the high cost of the imported product, has decided opportunities.

   (b)   The importation of nucleus breeding stocks into certain advanced countries where buffaloes do not exist at present offers considerable development opportunities in the light of the facilities for research which exist in these countries. Official approval must be obtained, and the difficulties of complicated quarantine requirements will have to be met, for it is an unfortunate fact that the best buffaloes exist in those world areas where the worst livestock diseases are rife. It might be said, however, that in this day and age quarantine should be the servant of the livestock developer, and not his master.

   (c)   Gibraltar has been indicated as a suitable staging point for the processing of chilled or frozen buffalo meat from such countries as Italy, Greece, the Arab Republic of Egypt, and those of the Near and Far East and Oceania. Markets in Europe,

North America and Africa would be unlikely to
refuse good quality canned buffalo beef from such
a centre.


## AUSTRALIA

There are excellent prospects for the domestication of
existing stocks of feral Swamp buffaloes in the extreme
north of the Northern Territory, where land is available
for purchase. The location of range buffalo herds has
proceeded well on numerous properties. If dishorning
and dehorning became general, the establishment of large
feed lots would be feasible. Possibilities exist for
the export of frozen or canned quality buffalo steak.


## THE FAR EAST

(a)  Philippines and Indonesia, both with vast areas
     of land awaiting development, and both with
     indigenous Swamp buffaloes as well as some imported
     milch breeds offer opportunities for investment
     in milk/meat enterprises. The islands of Negros
     and Cebu, Sulawesi and Sumatra might be mentioned
     particularly since it is respective government
     policy to encourage private investment and to
     offer land for development on favourable terms.

(b)  India, Pakistan, and Ceylon could all benefit
     from an infusion of foreign investment and of
     Western methodology for their buffalo stocks.
     The potential is there, but improved management,
     nutrition, breeding and marketing are required.
     The production of beef animals from dairy herds
     could become a profitable venture since there are
     no religious restrictions on slaughter.


## LATIN AMERICA

The prospects for meat production from buffaloes,
ranched on land which otherwise could provide no yield,
are outstanding in Brazil, Colombia, Peru and Venezuela.
There are vast areas where the buffalo is the optimum
animal. Foreign investment is encouraged. Similar
possibilities exist in Mexico, Guatemala and perhaps a
few other Central American countries which at present
have no buffalo stocks. Nucleus stocks of Swamp

buffaloes can be shipped from Australia, where the
disease risk is minimal, at reasonable prices (*c*. $A100
per head for yearling males and heifers 'free on board').

The buffalo has suffered in the past and still suffers
in many areas from an imaginary and wholly mythical
competition with cattle.  Powerful cattle interests,
both national and international are antagonistic to the
buffalo, as they are to any animal which competes with
cattle for land.  In most countries where they co-exist
there is a place for both animals in the economy; they
are not in competition but complement one another.
There are many areas of the world where cattle reign
supreme and will continue to do so.  There are other
areas where the environment is such that the water
buffalo offers the only opportunity of a reasonable
off-take from the land in terms of animal protein and
animal products.  There are many such territories where
the grazing animal provides the only possibility of a
satisfactory return but where cattle, sheep and even
goats are unable to adapt to the harsh conditions and
to heavy seasonal inundation, and only the water buffalo
can survive, multiply and thrive.  Such areas include
substantial parts of Colombia, Venezuela, Peru, the vast
Amazon valley, the marshes of southern Iraq, and the extreme
north of Australia's Northern Territory.  Some such areas
do not at present have a buffalo population: they
include Yucatan and Las Huastécas in Mexico, and the
Petén of Guatemala.

The water buffalo is emerging as an animal with a
great meat-production future which is ripe for development,
not only in tropical and sub-tropical countries but in
those in temperate zones also.

## REFERENCES

ASIBEY, E.O.A. (1972).  'Wildlife as a Source of
   Protein in Africa South of the Sahara', FAO working
   paper presented to the Fourth session of the *Ad hoc*
   Working Party on Wildlife Management, African
   Forestry Commission (Mimeographed)
CHARLES, D.D. and JOHNSON, E.R. (1972).  'Some Carcass
   Characteristics of the Australian Water Buffalo',
   in *The 9th Conf. Australian Soc. of Anim. Prod.,
   Canberra, 1972*
DE FRANCISCIS, G. (1971).  'The Early Weaning of Buffalo
   Calves', Caserta, Associazione Provinciale Allevatori
   (in Italian)

GADE, D.W. (1970). *Américas*, 22 (5), 35

JOKSIMOVIĆ, J. (1969). *J. Sci. Agr. Res. Belgrade*, 22, 78, 110

KALEFF, B. (1932). *Z. Tierzücht. ZüchtBiol.*, 24, 390

KYLE, R. (1972). 'Meat Production in Africa – the Case for New Domestic Species', Published by Bristol Veterinary School

MACGREGOR, R. (1939). 'The Domestic Buffalo', Thesis, Royal College of Veterinary Surgeons, London

OGNJANOVIĆ, A. (1974). 'Meat and Meat Production', in *The Husbandry and Health of the Domestic Buffalo*. Ed. W. Ross Cockrill. FAO, Rome

POLIKHRONOV, D. ST. (1974). 'The Buffaloes of Bulgaria' in *The Husbandry and Health of the Domestic Buffalo*. Ed. W. Ross Cockrill. FAO, Rome

TULLOCH, D. (1974). 'The Feral Swamp Buffaloes of Australia's Northern Territory', in *The Husbandry and Health of the Domestic Buffalo*. Ed. W. Ross Cockrill. FAO, Rome

VILLEGAS, V. (1932). *Carabao Husbandry*, Manila. Oriental Publishing Co.

WILSON, P.N. (1961). *J. Agric. Soc. Trin.*, 61, 457, 459

# 27

# THE SIGNIFICANCE OF MEAT ANALOGUES

M. P. TOMBS

*Unilever Research Laboratory, Colworth House,
Sharnbrook*

## INTRODUCTION

The last thirty years has seen a succession of attempts
to increase the use of non-meat proteins in human foods:
the original impetus for these efforts arose from the
paradox that while a part of the world's population
suffered from a lack of protein a large amount of cheap
protein, mainly soy and groundnut, was available
(Altschul, 1958). The earlier work concentrated on
nutritional aspects, but the outcome was disappointing
in that only a small amount of vegetable proteins was
diverted to direct human feeding. Nutritional
excellence is not enough! Although these earlier
attempts were not successful, we are once again
experiencing a wave of publications and product trials.
What are the reasons for thinking that the chances of
success have improved?

Non-meat proteins are potentially cheaper than other
protein-rich foodstuffs, although they are still likely
to be among the more expensive foods. For the products
themselves to be acceptable they must therefore be seen
by the consumer as foods which fall into the protein-
rich category, and they should have attributes which
products from cheaper carbohydrates cannot easily match.
It is generally recognised that most people like meat
of one sort or another, so the chances are they will like
something with meat-like attributes. This conclusion is
reinforced by the experience of the last thirty years:

meat analogues have a far better chance of success than anything else. The answer to the question posed earlier is, therefore, that we are probably closer to being able to produce acceptable meat analogues. How close is best answered by considering the technical developments which led to the appearance of these products, and possible future developments. We cannot ignore the associated commercial and legal problems, but the rate-limiting problems are still technical.

The technical problems are, of course, enormous: most vegetable proteins are prepared in the form of an off-white powder - a form which could hardly be more different from meat. Meat is fibrous, and although textile fibres have been made from cheap proteins for over a century, it was not until Boyer (1954) and Boyer, Schultz and Schutzman (1969) modified the process that such fibres were suitable for human consumption. Without this achievement the whole approach may well have been rejected as technically impossible: with it an audaciously simple idea could be tried - to take bundles of filaments, bind them together, and add colour and flavour, in the hope that the outcome is sufficiently meat-like to meet requirements. It is not sufficiently emphasised how surprisingly good such products are, even if they are probably not yet good enough.

It should be noted that there need not be an intention of mimicking any particular meat - the aim could simply be to create meat-like attributes and fibrosity, or more generally 'texture'. Methods for texture production other than fibre spinning have also been attempted (Anson and Peder, 1957), and this line of research is still being pursued. Although the term 'meat analogue' is best kept for products based on spun protein filaments, the range available has always included others, generally described as textured vegetable proteins. Whatever the intentions of the manufacturer, and in some cases the intention was to make products like meat, but not imitation meats, the consumer will have the last word, and probably will come to regard meat analogues as an addition to the range of available 'meats', rather than a substitute for any of them.

# PRODUCTS BASED ON VEGETABLE PROTEIN

There are three important classes of vegetable protein products; some of their characteristics are described in *Table 27.1*. Of these products, extended meats are likely to represent the main use of vegetable protein for the next few years. Soy protein has been added at a low level to meat products for a long time (Meyer, 1970) but ways have now been found to add it at a much higher level without obvious ill effects on the product (Tombs, 1972). Addition takes many forms, and all combinations, from the addition of spun filaments to the addition of meal, have been explored. At the moment this kind of product represents the easiest technical option, but, it should be added, is meeting legal problems.

The main advantage of 'textured vegetable protein' is that meals or concentrates can be used as the starting material, rather than isolates. Raw material costs tend to dominate the cost of products based on vegetable protein, and protein isolate is two to three times as expensive as protein meal. Because the soluble soy carbohydrates are undesirable, and lead to the flatulence effect, it is likely that concentrates will be used rather than meals. One product is made from an isolate and a non-soy carbohydrate, though this may be an attempt to use the texture-forming capacity of starch rather than a wish to avoid soy carbohydrates altogether. There is also a group of products, so far unexploited, which rely on carbohydrate gel-forming systems to produce the texture, the protein being inert (Giddey, 1960).

Meat analogues are the most complicated products to make, and require good quality isolate for the fibre production stage. They are in competition with the 'textured vegetable protein' type of product and are likely to be preferred by the consumer. They are, however, more expensive because of the complicated processing required and because protein isolates must be used. There is a place for both kinds of products in different applications, but no one can say which is likely to represent the greater volume.

*Table 27.1* Use of vegetable protein in meat and meat-like products

| | Protein form[1] | Texture method | Content |
|---|---|---|---|
| MEAT ANALOGUE | Isolate | Protein spun to filaments, then bound with further protein, heated | Vegetable protein, egg – albumin (often) |
| TEXTURED VEGETABLE PROTEIN: | | | |
| 1. | Meals, concentrates or isolates | Extrusion of doughs | Vegetable protein and carbohydrate (cellulose) |
| 2. | Isolate | Extrusion of doughs | Vegetable protein and added carbohydrate (starch) |
| 3. | Isolate | Extrusion into setting bath | Vegetable protein and texture forming carbohydrate (e.g. carageenan) |
| EXTENDED MEAT | Protein isolate or meals or concentrates or filaments | Intimate mixing of meat and vegetable protein, then normal meat processing | Variable, up to 50% vegetable protein |

[1] These are generally forms of soy protein. 'Meals' contain about 45% protein, 30% insoluble (cellulose) carbohydrate and 25% soluble carbohydrate. 'Concentrates' have had most of the soluble carbohydrate removed and contain about 55-60% protein. 'Isolates' have had nearly all the carbohydrates removed and contain at least 90% protein. They also contain small amounts of phospholipids, phytate, polyphenolics, saponins and RNA.

All these products may also contain added lipids, flavour, colours and nutritional supplements such as minerals and vitamins.

## PRODUCTION OF MEAT ANALOGUES

The first stage in the manufacture of meat analogues is
the production of filaments or 'fibres'. The term
filaments is best since most, though not all of them, do
not possess the structural characteristics of fibres
(Astbury, 1958). There are now two processes for the
manufacture of filament: versions of that originally
introduced by Boyer (1954), and the more recent
'mesophase' process (Tombs, 1972).


### THE BOYER PROCESS

The Boyer process has undergone many minor variations,
though *Figure 27.1 (a)* shows the main characteristics.
Protein isolate is brought into solution by adding sodium
hydroxide to a slurry; a very high pH (about 11) is
required. The structure of the protein is disrupted and
this increases its ability to undergo protein-protein
interaction later. This solution is then brought to
about pH 10, filtered to remove particulate material and
extruded through platinum or glass spinnerets into a
bath containing high concentrations of sodium chloride
and acetic acid, at pH 3. The protein is insoluble in
this mixture and aggregates in the form of filaments.
The tow (the bundle of filaments) is collected, and
passed over a pair of drums revolving in such a way that
the filaments are stretched to give them greater strength.
In some cases it is possible to detect ordering of the
peptide chains by X-ray diffraction (Lundgren *et al.*,
1948). The filaments are usually made at 60-100 nm in
diameter in bundles of several thousand. They are often
stored in dilute acetic acid.
  Some of the problems of this process are:

1.  Isolate quality varies, and in particular the level
    of insoluble particulate material must be kept low.
    Occasional batches of protein fail to spin for no
    known reason.
2.  The alkaline solutions are unstable and gel
    spontaneously. Interruptions to the process are
    serious.
3.  The filaments are mechanically weak, and the
    necessity of collection on a drum imposes minimum
    mechanical requirements on the filaments.

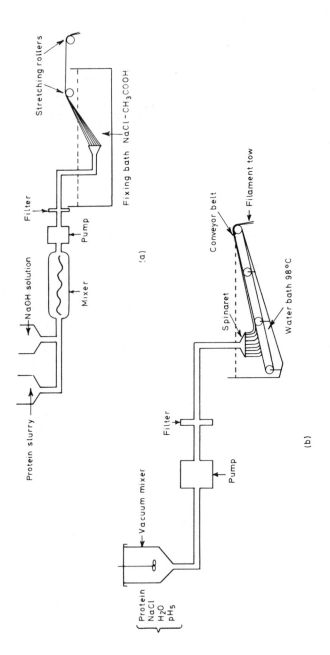

Figure 27.1 Process for producing filament from proteins. (a) The Boyer process (Boyer, 1954) (b) The Mesophase process (Tombs, 1972)

4.  The filaments tend to be unstable, and may redissolve at later stages in processing.
5.  The very high pH means that it is difficult to incorporate into the spinning dope additives such as flavours and colours.
6.  The protein is degraded by high pH - ammonia is lost from amide groups - and very careful control is needed to avoid loss of nutritional value, and production of the toxic lysino-alanine (Woodward, 1973).
7.  The filaments need extensive washing to remove acetic acid before further processing.
8.  It is not very easy to vary the properties of the filaments, though this can be done to some extent by varying the degree of strength.

However, the fact that several products based on this type of filament exist, shows that these problems can be overcome.

The process does have some good points: it is flexible with respect to raw material, and most proteins can be induced to form filaments with minor variation of conditions. It is too soon to say yet whether the filaments have the required properties and in fact filament producers are handicapped because as yet no very clear idea of the required properties exists. For example, the diameter of the filaments is selected to be similar to that of an appropriate meat fibre without any substantial evidence that this is the most suitable.


THE MESOPHASE PROCESS

The mesophase process, although it produces filaments and uses a spinneret, is different in every other way from the Boyer process. In this process (*Figure 27.1 (b)*) the protein is produced in the form of a 'mesophase'. Protein mesophases are perhaps not very familiar, and are best known for β-lactoglobulin (Shaw, Janson and Lineweaver, 1944). Essentially, some proteins under conditions where they would be expected to form a precipitate instead produce a liquid homogeneous phase which can coexist with the solution. Soy proteins, and many other vegetable proteins, form these liquid protein-rich phases providing they have retained their native structure (Tombs, 1972; Tombs, Newsom and Wilding, 1974).

The process consists of mixing protein isolate, sodium chloride and water to form a mesophase. This, when extruded into water forms filaments, owing to the loss of salt. The filaments are weak, however, unless heated, and in its simplest form extrusion is directly into hot water. The process is most conveniently done at the isoelectric point ($\sim$ pH 5).

There is no need to stretch the filaments to give them mechanical strength. In order to avoid the mechanical limitations of collecting on a drum they are simply collected on a conveyor belt. They are washed to remove residual traces of the setting bath water, and may be stored frozen or dehydrated, or in any other convenient way. Flavours, colours and other materials may be incorporated in the spinning dope, and the problems of high pH are avoided. The filament can be made in any required size and, within limits, of any required cross section.

Although this process avoids many of the problems of the Boyer method, it has its own limitations:

1. It requires an essentially undenatured protein isolate, as well as a low level of particulates. This requires care in the isolation process and isolates to this standard are not normally commercially available.
2. With soy protein it is best to use sulphite to control the disulphide interchain linking in the protein, and although the sulphite disappears during spinning, in some countries there may be difficulties about its use.


ANALOGUE PRODUCTION

The next stage in processing is to take the filaments and bind them together with some material which also contains all the other required ingredients. The filaments are, for the most part, produced in aligned bundles, and the main technical consideration is whether this alignment should be maintained through to the final product or not. There is no consensus on this point; some products appear to be highly aligned while others are not. Thus no single description can be given of this part of the overall process and although products depend at least as much on this as on filament production it has attracted much less attention.

Although a process for binding the filaments together by simply impacting them has been described (Dudman, 1957), most processes use a binder protein such as egg albumin. This is expensive and probably was used because no satisfactory alternative was available. In the mesophase process it is possible to use the soy mesophase protein itself as a binder, and to make analogues entirely of vegetable protein. Lipids and colours of appropriate types are also usually included in the binder as an emulsion, and the mixture of binder and fibres is invariably set up by a heating process of some sort. It probably will be necessary to add minerals and vitamins to the binder phase, so that they appear in the final product. Legislation at present under discussion includes such provisions.

## FLAVOURING

Last but not least, appropriate flavours must be added. There is still a shortage of good meat flavours. However, there is another problem: all these products have a tendency to retain some of the 'beany' flavour of the material from which the protein was isolated. In some products this tendency is quite marked, and the problem is not made easier by the fact that undesirable flavours can reappear at any stage up to the final cooking performed by the housewife. A combination of this and the lack of good added flavours makes flavouring the most intractable problem surrounding meat analogues.

## CONCLUSIONS

Although there have been some technical advances in meat analogues it is too soon yet to say whether they are sufficient to overcome the problem of making acceptable products. The problems of immediate concern are predominantly technical, which is not to say that there may not be commercial and legal problems as well. At present these products, which have so far not been an outstanding success, could be better in many respects - in texture, in flavour and in the detailed formulation. Probably the technical problems can be solved, but the time scale is very difficult to predict; at least we can identify the problems more clearly than hitherto.

Meat analogues are only one way in which vegetable proteins and 'unconventional' proteins can be introduced and the underlying impetus remains the utilisation of a wider range of protein for human food, which may yet happen in quite unexpected ways.

## REFERENCES

ALTSCHUL, A. (1958). *Processed Plant Protein Foodstuffs*, Academic Press, New York

ANSON, N. and PEDER, M. (1957). U.S. Patents, 2,813,025 and 2,813,024

ASTBURY, E.T. (1958). *Disc. Farad. Soc.*, 25, 80

BOYER, R.A. (1954). U.S. Patents, 2,682,466 and 2,730,447

BOYER, R.A., SCHULTZ, A.A. and SCHUTZMAN, E.A. (1969). U.S. Patent, 3,468,669

DUDMAN, R.K. (1957). U.S. Patent, 2,785,069

GIDDEY, C. (1960). U.S. Patent, 2,952,542

LUNDGREN, H.P., STEIN, A.M., KOORN, V.M. and O'CONNELL, R.A. (1948). *J. Phys. Chem.*, 52, 180

MEYER, E. (1970). In *Proteins as Human Food*. Ed. R.A. Lawrie. Butterworths, London

SHAW, T.M., JANSON, E.F. and LINEWEAVER, H. (1944). *J. Chem. Phys.*, 12, 439

TOMBS, M.P. (1972). British Patent, 1,265,661

TOMBS, M.P., NEWSOM, B.G. and WILDING, P. (1974). *Int. J. Peptide Protein Res.*, 6, 253

WOODWARD, J.C. (1973). *J. Nutr.*, 103, 569

# 28

# ASSESSMENT OF ALTERNATIVE NUTRIENT SOURCES

W. HOLMES

*Wye College, University of London, Ashford*

## INTRODUCTION

In this chapter an attempt is made to assess the
competition which alternative foods might offer to meat
production and consumption in the future. It therefore
endeavours to place meat consumption in perspective.
Consideration is given to the extent to which normal
vegetable foods might replace meat, the possibility that
vegetable crops might be treated in novel ways to provide
alternatives to meat, the prospects that micro-organisms
might contribute directly or indirectly to meat production and the alternative forms of animal products which
could compete with meat.

Meat is regarded primarily as a source of protein and
is usually compared with other foods on that basis, but
Crawford (1971) has pointed out that it has other
nutritional merits such as the unsaturated fats obtained
from animal tissues.

Hollingsworth (1971) summarised, for the United
Kingdom, the proportion of protein derived from major
food sources since 1909 and the data are reproduced in
*Table 28.1.* There has been only a small rise in average
total daily protein intake from 80 g before 1939 to about

*Presented by Mr. J.B. Moran, as Professor Holmes was
unable to attend.*

Table 28.1 Protein value of supplies moving into consumption in the United Kingdom (after Hollingsworth, 1971)

| | 1909-13 | 1924-28 | 1934-38 | 1943 | 1948 | 1956 | 1962 | 1968 |
|---|---|---|---|---|---|---|---|---|
| Protein (g/head/day) | 81 | 79 | 79 | 85 | 87 | 83 | 87 | 85 |
| % PROTEIN FROM: | | | | | | | | |
| Dairy products | 15 | 16 | 17 | 21 | 20 | 23 | 23 | 23 |
| Meat and poultry | 25 | 25 | 26 | 18 | 17 | 25 | 27 | 28 |
| Fish | 10 | 11 | 7 | 4 | 7 | 5 | 5 | 5 |
| Eggs | 3 | 3 | 5 | 4 | 4 | 5 | 5 | 5 |
| Grain products | 38 | 35 | 35 | 41 | 41 | 30 | 28 | 26 |
| Potatoes | 6 | 6 | 5 | 6 | 6 | 6 | 5 | 5 |
| Other vegetables, fruits nuts, other foods | 4 | 5 | 6 | 5 | 5 | 7 | 7 | 8 |
| % total calories from protein | 12 | 12 | 10 | 12 | 12 | 11 | 11 | 11 |

86 g after 1939. However, in 1968 animal protein accounted for 61% of the total protein, compared with 53% in the period 1909-1913, cereals and potatoes for 31% compared with 44% and vegetables, fruit and other foods for 8% compared with 4% in the period 1909-1913. Over the period reviewed, protein derived from dairy products increased substantially from 15% to 23%, while the supply from all meats moved from 25% pre-war, to 18% in the 1940s and up to only 28% in 1968. The dramatic change in relative prices since 1968 has probably now reduced the proportion of meat in the total diet.

The contribution from fish has fallen from 10% to 5% and the proportion of protein from sea fish is unlikely to rise in view of the primitive and costly methods of harvesting.

The amount of protein which the human body requires for optimal development and well-being is still a matter for debate and is difficult to quantify although there is widespread belief that optimal protein supply is well above the minimum necessary to maintain nitrogen balance. The subject was reviewed by Blaxter (1973). Both for nutritional and social reasons a high level of meat consumption is regarded in many societies as an index of a high living standard. But, as has already been stressed, meat is only one component of the diet, and as will be shown it is costly in resources and in terms of money. It is of interest therefore to consider the alternatives.

Among conventional foods the alternative sources to be considered are dairy products, eggs, grains and vegetable crops. In addition, there is growing awareness of a range of so-called simulated meat and synthetic foods. These have been the subject of a bewildering range of recent publications including symposia edited by Mateles and Tannenbaum (1968), Lawrie (1970) and Jones (1973), a report issued by the Agricultural Development Association (1971) and a wide range of popular articles (e.g. *Farmers Weekly*, 1973).


## CONVENTIONAL FOOD CROPS

Grain products contributed 41% of the average total protein supplies in the UK in the 1940s and although by 1968 this had fallen to 26% (Hollingsworth, 1971) the proportion could rise again if protein foods become scarce and expensive.

Moreover, there has been considerable interest in the improvement of the nutritive value of grain products, either by the breeding and selection of cultivars of higher nutritive value, e.g. Opaque-2 maize, or by the supplementation of cereal diets with synthetic amino acids. The subject has been reviewed by Whitehouse (1973). He stressed the importance of choosing the correct criteria. A cultivar of high protein content which yielded less grain per hectare might be less valuable than a higher yielding grain of normal protein content, or a high yielding variety which could be supplemented readily with amino acids.

Traditionally the pulse crops such as beans and peas are regarded as valuable sources of protein but the yield per hectare is relatively low and the agronomic problems of producing consistently high yields have not been overcome. Field beans (*Vicia faba*), field peas (*Pisum arvense*) and garden peas (*Pisum sativum*) are grown in the UK. Dwarf french beans (*Phaseolus vulgaris*) and Navy beans (*Phaseolus vulgaris*) are more recently introduced field crops. None at present makes a major contribution to protein supplies, although field beans have recently been adapted for a mycoprotein process (Spicer, 1971) and Navy beans may replace imported beans to a small extent.

Green vegetables normally contribute only a small proportion of the dietary protein. Moreover, in the normal processes of farming or market gardening only a proportion of the crop is harvested and then only a proportion of this is actually consumed. With lettuce some 50-90% of the typical crop is harvested. Then only 45% of the purchased vegetable is eaten (McCance and Widdowson, 1960). With the Brussels sprout again the sprouts amount to 30% of the total crop above ground. Only 58% of the sprouts are eaten (McCance and Widdowson, 1960). An indication of the yields of protein per hectare from a selection of the crops mentioned is given in *Table 28.2*. These crops often yield less in practice than the values of 600 kg protein/ha suggested by Schuphan (1965) for kale.

Table 28.2 Protein yields from conventional and processed crops

| | Crop yield (kg/ha) | Amount eaten (%) | Composition of edible portion | | Yield/ha | |
|---|---|---|---|---|---|---|
| | | | Protein (%) | Energy[1] (Mcal/kg) | Protein (kg) | Energy (Mcal) |
| CONVENTIONAL CROPS | | | | | | |
| Wheat | 4000 | 70 | 7.9 | 3.49 | 220 | 9800 |
| Potatoes | 25000 | 86 | 2.1 | 0.87 | 450 | 18700 |
| Field beans | 3300 | 100 | 27.0 | 2.70 | 890 | 8900 |
| Vining peas | 4500 | 100 | 5.8 | 0.64 | 260 | 2900 |
| Navy beans | 2000 | 100 | 21.4 | 2.58 | 430 | 5200 |
| Cabbage | 40000 | 60 | 2.2 | 0.25 | 530 | 6000 |

| | Crop yield (kg/ha) | Protein (%) | Gross yield of protein (kg/ha) | Extraction of protein (%) | Net yield of extracted protein (kg/ha) |
|---|---|---|---|---|---|
| PROCESSED CROPS | | | | | |
| Soya bean | 2500 | 40 | 1000 | 35 | 350 |
| | Dry matter | | | | |
| Lucerne | 8000 | 22 | 1760 | 60 | 1060 |
| Cocksfoot | 9000 | 16 | 1440 | 60 | 860 |

1 Mcal = 4.18 MJ

[1] For human nutrition (McCance and Widdowson, 1960)

## PROCESSED VEGETABLE CROPS

The processing of vegetable crops has received much attention over the last 10 years particularly with reference to soya bean (*Glycine max*) and more recently in the UK with field beans (*Vicia faba*) (Spicer, 1971). The possibility of concentrating leaf protein has also been explored for many years and leaf protein concentrate (LPC) can yield material of high nutritive value (Pirie, 1971).

The technology of soya-bean processing has been well documented (e.g. Meyer, 1970; Holmes and Burke, 1971). A soya-bean protein extract may be readily prepared in the form of dry granules which may be included in processed meat products. In a much more complex process the vegetable proteins are spun into fibres, coloured, flavoured and assembled in bundles, to simulate one of the meat products (textured vegetable protein, TVP). These products are of high nutritive value, much higher in protein but lower in fat content than animal products. Material from the simpler processes is used as a nutritious 'filler' in the food industry, the more sophisticated product can replace meat on the table and is similar in cost to the cheaper cuts of meat. The effective yield of protein from soya-bean processing is not high since only 30-40% of the original crop protein is retained (Meyer, 1970), and the methods adopted require an advanced technology.

A process in which protein from field beans is similarly extracted and spun to form a meat substitute has been introduced in the UK in recent years (*Farmers Weekly*, 1973). The residues can be used for further microbial protein synthesis (Spicer, 1971 and 1974).

## LEAVES AS A SOURCE OF PROTEIN

Since leafy crops which ensure complete ground cover give maximal yields of dry matter and protein, but grazing livestock convert only a relatively small proportion into edible protein, there has been increasing interest in the direct utilisation of leaf crops. A leaf-processing method should ensure that a high proportion of the crop is harvested and that a high proportion of the crop is then extracted and made available in edible form. The subject is fully discussed

in a symposium edited by Pirie (1971). Arkcoll (1971)
summarised recent work, and indicated that the ideal
crop should grow over an extended season, respond well
to fertiliser nitrogen or be leguminous and should
synthesise protein rapidly yet be slow to lignify. He
quotes annual yields in Britain from green cereals
followed by green fodder crops, of up to 2000 kg protein/
ha when 530 kg N/ha were applied, and yields from
leguminous crops of about 1000 kg protein/ha. Alberda
(1973) indicated that maximal yields of organic N
compounds (N × 6.25) may be about 3700 kg/ha from grass
crops receiving 1500 kg N/ha. The proportion of protein
which can be extracted normally ranges from 40 to 70%.
Byers (1971) reported that the amino-acid composition
of leaf proteins compared favourably with the FAO
reference protein. A problem of acceptability of the
material arises because of colour and flavour and it is
normally included in other dishes as a supplement.

Current work at the National Institute for Research
in Dairying and at the Rowett Research Institute and in
other countries is concerned with the extraction of a
leaf protein concentrate for use in non-ruminant feeding;
the more fibrous residue being used by ruminants.
Estimated yields of protein from processed vegetable
material are included in *Table 28.2*.


## SINGLE CELL PROTEINS (SCP)

The general term single cell protein is now applied to a
range of products of microbial origin. Yeasts, algae
and bacteria are all included under this broad heading
(Mateles and Tannenbaum, 1968). Where continuous culture
methods can be practised high growth rates are possible.


### YEASTS

The yeasts (*Candida utilis* and *Saccharomyces* spp) have
been known and used for many years as a minor source of
protein in baker's yeast, and from culture on molasses
or on sulphate liquor as a larger scale food supplement.
The product has a high protein content (*c*. 55%), is
somewhat low in methionine but also contains about 11%
nucleic acid. This limits its safe use for direct
human consumption to about 10 g yeast protein per day.
Worgan (1973) suggests that sugar products and by-products

fermented by yeasts could yield 2800 kg protein/ha of
sugar beet and Bressani (1968) quotes Willcox (1959) as
estimating that sugar cane could provide substrates for
yeasts which could produce 11,000 kg protein/ha.

More recently yeasts have been used in processes for
producing protein from liquid *n*-paraffins (Walker, 1973).
These materials have been tested in pig and chicken
feeding experiments and the evidence is that at up to
10% of the ration 'they can form a satisfactory
alternative to fish/soya bean mixtures in pig and poultry
rations' (Shacklady, 1970). Care must be taken to ensure
that methionine is non-limiting. The subject has been
reviewed by Shacklady (1972).


ALGAE

Remarkable claims have been made for protein production
from algae. Oswald and Golueka (1968) suggest on the
basis of large-scale pilot plant (1 M litre) operations
in California that yields of 50 tonnes/ha of material
containing 50% protein are attainable. Gordon (1970)
shows that *Chlorella pyrenoidosa* contains 50% protein
and could yield 14,000 kg protein/ha while *Spirulina
platensis* may yield 21,700 kg protein/ha. The algae
examined tend to be low in sulphur-containing amino
acids, and although the protein efficiency ratio (PER)
is similar to that of soya flour the material tends to
be of low digestibility, has been reported to cause
gastro-intestinal disturbances in man and is rather
unpalatable. Apart from these doubts on the acceptability
of algal protein there are problems in maintaining
optimum temperatures (22-28°C), ensuring constant
irradiation and organising the harvesting of the
products. Gordon (1970) suggests that these problems
may be overcome in suitable tropical areas. The problems
of human acceptability remain and until these have been
resolved attention may turn to the use of algal protein
as feed for livestock with the attendant losses in
conversion.


BACTERIA

Work has also been conducted on the use of bacteria,
e.g. *Escherischia coli* and *Pseudomonas aeruginosa*
(Worgan, 1973). While growth rate and nutritive value
as indicated by chemical analyses are high, there are

substantial problems in harvesting and in breaking down
cell walls.

However, the use of methanol as a substrate for
bacterial fermentation has now been developed to a
commercial scale, and yields a product of satisfactory
nutritional value for non-ruminant farm animals (Beatty,
1974; Gow et al., 1973).

These processes for the exploitation of single cell
proteins are still in early stages of development but
they show great potential for increasing the supply of
protein for animal feed.  However, the costs and
supplies of fuel and raw materials are bound to influence
their development.

All the products tend to be high in nucleic-acid
content.  It has been established that they can be used
in the diets of non-ruminant animals but much more work
is needed before they could be used for direct human
consumption.


THE CONVERSION OF SINGLE CELL PROTEINS INTO HUMAN FOOD
PROTEIN BY ANIMALS

It can be estimated that crude protein in feed is
converted to edible protein with the following efficiencies
in farm conditions although ceiling levels may be 1.5 to
2 times as high (Wilson, 1973):

| | | |
|---|---|---|
| Pig meat | 12.5% | (Carr, 1972; MLC, 1973a; Holmes, 1971) |
| Broiler chicken | 16% | (Davidson and Mathieson, 1965) |
| Egg chicken | 18% | (Holmes, 1971) |
| Milk production | 23% | (Holmes, 1971) |

It is difficult to express the production of yeast or
algal protein in terms of yield per hectare and
impossible to do so for proteins based on paraffins or
methanol.

However, if one hectare of sugar products yielded
11,000 kg yeast protein this would convert to some
1400-2800 kg animal protein/ha and yields per hectare
of pond algal protein converted to animal protein would
be of a similar order.  In both examples SCP should form
only a proportion (c. 20%) of the total protein in the
diet.

## THE ANIMAL PRODUCTS

Meanwhile livestock continue to supply a major proportion of the protein for human consumption in the Western World. Spedding and Hoxey (in Chapter 25) have given a comprehensive review of the productivity of conventional and unconventional meat animals in relation to various scarce resources. For comparative purposes in this chapter however it is desirable to show how animals compare with the alternative sources of protein.

The data assembled in *Table 28.3* are based as far as possible on recently published information for average recorded farm situations. As for the crops (*Table 28.2*) they endeavour to show the average real life situation, not the maximum attainable (which is often twice the average). The yields per hectare of animal product, of edible protein and of edible energy are shown. In addition the last column gives the yield of protein when it is expressed per hectare used for the growing of concentrates.

For milk the data refer to the average costed milk producing herd (MAFF, 1972) and do not allow for the rearing of young stock. The same basic data have been taken for cheese. The beef, lamb and pig systems are based on average production data published by the Meat and Livestock Commission (MLC, 1973a,b,c,d). The lamb and pig systems account for all feed used by ewes or sows. The 18 month beef and barley beef systems do not allow for the dam since she is normally a dairy cow. Carcass data are based on a number of publications (Everitt, 1972; Leitch and Godden, 1953; Lodge, 1970; Holmes, 1971; Ingram and Rhodes, 1971). They may somewhat exaggerate the edible nutrients supplied since there are great variations both in carcass composition and in the proportion of the carcass which is regarded as acceptable (Everitt, 1972). Broiler data are based on a conversion rate of 2.5 kg feed per kg gain and the analyses of Davidson and Mathieson (1965) for the fastest growing hybrids. All concentrates have been assumed to be composed of barley and field beans in a ratio of 3:1 at an average yield of 3,300 kg (85% DM) per hectare. Where a comparison is possible the results are similar to earlier estimates based on typical populations (Holmes, 1970 and 1971) and the various classes of stock fall in similar rank order. The processing of milk into cheese gives a higher yield than any form of beef production. By-product beef production is twice as

Table 28.3 Protein and energy yields from livestock products

| Product | Requirements per animal[1] | | | Yield of product | Composition of product | | Yield/ha | | | |
| | Forage (ha) | Concentrates (kg) | Total (ha) | (kg) | Protein (%) | Energy (Mcal/kg) | Product (kg) | Protein (kg) | Energy (Mcal) | Protein (kg/ha concentrate) |
|---|---|---|---|---|---|---|---|---|---|---|
| Milk | 0.64 | 1230 | 1.01 | 3760 | 3.4 | 0.66 | 3620 | 126 | 2460 | 340 |
| Cheese (Cheddar) | 0.64 | 1230 | 1.01 | 358 | 25.4 | 4.25 | 355 | 90 | 1510 | 245 |
| Egg chicken (10) | – | 474 | 0.144 | 140 | 10.5 | 1.63 | 972 | 102 | 1580 | 102 |
| | | | | Yield of carcass | | | | | | |
| Barley beef intensive | – | 1880 | 0.57 | 220 | 14.0 | 4.00 | 385 | 54 | 1540 | 54 |
| Beef semi-intensive | 0.31 | 1170 | 0.66 | 265 | 14.0 | 4.30 | 401 | 56 | 1720 | 105 |
| Beef suckler cow, and calf yard fattened | 0.70 | 1070 | 103 | 238 | 14.0 | 4.2 | 231 | 28 | 860 | 89 |
| Lamb from ewe weaning 1.4 lambs | 0.106 | 56 | 0.123 | 24 | 14.0 | 4.00 | 186 | 27 | 790 | 190 |
| Pig | – | 344 | 0.104 | 62 | 11.0 | 3.9 | 596 | 66 | 2320 | 66 |
| Broiler chicken (100) | – | 425 | 0.129 | 110 | 11.3 | 1.33 | 853 | 96 | 1140 | 96 |

[1] Per cow, ewe, sow, beef animal, per 10 egg producing chickens and per 100 broiler chickens.

productive as beef or sheep production from breeding
herds or flocks.

The last column indicates the yield of protein per
hectare devoted to the production of concentrate feeds.
Naturally lamb production, which depends heavily on grass
and to only a small extent on concentrates, is much more
productive by this measure, but milk production again
gives the highest yields of protein per unit of
concentrate feed and even processed into cheese its
protein yield exceeds any of the meats.

Some estimates of maximal production of animal protein
were published following a recent symposium of the New
Zealand Society of Animal Production. Hutton (1970 and
1972) contrasted intensive New Zealand grassland farming
with intensive mid-west USA crop farming for animal feed.
He estimated that grass and arable farming would yield
580 kg and 420 kg protein/ha from milk, respectively,
and 180 kg and 290 kg from beef, respectively. Everitt
(1972) estimated a maximal yield of 102 kg/ha from beef
(excluding maternal costs) on intensive grassland, and
Coop (1972) estimated a yield of 20 kg meat protein and
45 kg wool protein/ha from typical New Zealand sheep
flocks.   The ratio of edible meat protein:wool protein
is about 1.8:1 for the more prolific British breeds.


# FISH


Sea fishing is a form of hunting.  While in total it
yields substantial quantities of high quality protein,
no meaningful estimates of productivity per hectare can
be made.  It has also been pointed out that modern
trawler fishing is very costly in terms of fuel use
(Leach, 1974).

Cultivated fish farming has been practised in some
tropical areas for many years and is becoming more
popular in temperate regions.  Hickling (1962) quotes
yields in Western Europe of 140-450 kg fish/ha which
at an average content of 15% edible protein (McCance and
Widdowson, 1960) indicate protein yields of 21-68 kg
protein/ha.  The higher levels of production depend on
the provision of supplementary feed which may yield
0.25-0.50 kg fish/kg food.

## GAME FARMING

There is considerable interest in game farming based
mainly on East African experience (Talbot *et al.*, 1965).
In Britain the only methods of practical importance are
the management of red deer (*Cervus elaphus*) and of grouse
(*Lagopus scoticus*) in the hills.  Yields of protein per
hectare are low although the financial rewards may be
high.


## THE USE OF ENERGY IN FOOD PRODUCTION

The impending scarcity and high cost of fossil fuels has
stimulated much work in recent years on the energy costs
in food production.
   Black (1971) made some comparisons of primitive and
developed agricultural systems which indicated that if
all fuel sources were considered, so-called primitive
systems were relatively efficient.  Duckham (1973)
pointed out that in a full analysis, attention should
also be paid to the 'upstream' costs (e.g. manufacture
of machinery) and 'downstream' costs (e.g. milling,
baking and distribution of bread).  Where to draw the
line involves complex philosophical judgements.  Slesser
(1973) has published a comprehensive analysis of energy
inputs for food production.  Leach (1974) extended this
form of analysis to include several food crops and
animal production systems.  His data are summarised in
*Table 28.4*.
   As a further exercise the data on animal and crop
production referred to in *Tables 28.2* and *28.3* have been
assessed in terms of their direct energy costs for fuel,
fertilisers and feeds, *Table 28.5*.  The energy network
inputs for fossil fuels and fertilisers are based on the
estimates of Leach and Slesser (1973) and for animal
feeds on Leach (1974).  The resulting coefficients depend
heavily on the assumptions made and no great precision is
claimed.  Moreover, they ignore many less obvious
variable costs and probably include only half the total
support energy used.  However, they demonstrate once
again in a different currency the relative inefficiency
of all the animal industries and show that even the
processes of modern mechanised crop production do not
yield a high return from the use of fossil fuels.  It
was difficult to assess the additional support costs for

Table 28.4  Summary of energy budgets:  preliminary
estimates for average conditions (after Leach, 1974)

| Product | Edible energy / Total support energy | Mcal support energy per kg protein |
|---|---|---|
| Potatoes | 1.0 | 36 |
| White bread | 1.4 | 22 |
| Sugar (beet) | 0.49 | – |
| Eggs | 0.16 | 77 |
| Broiler | 0.11 | 64 |
| Milk | 0.33 | 55 |

Table 28.5  An estimate of the direct energy costs of
food production.  Indirect costs of processing, transport
etc. may double these costs.  (The yield data are based
on Tables 28.2 and 28.3)

| Product | ENI[1] in fuel fertiliser and feed production (Mcal/ha) | Food energy output / Fuel energy input | Fuel energy input Mcal kg protein produced |
|---|---|---|---|
| Wheat | 2200 | 4.4 | 10 |
| Potatoes | 6500 | 2.9 | 14 |
| Field beans | 1400 | 6.4 | 1.5 |
| Navy beans | 1400 | 3.7 | 3.0 |
| Cabbage | 6200 | 1.0 | 12.0 |
| (Mcal/animal unit) | | | |
| Milk | 3700 | 0.67 | 29 |
| 18 month beef | 2900 | 0.40 | 78 |
| Lamb | 70 | 1.0 | 28 |
| Pig | 450 | 0.3 | 64 |
|  |  | (0.5) | (56)[2] |
| Broiler | 5 | 0.3 | 40 |
| Egg | 59 | 0.3 | 42 |

[1]Energy network inputs (Leach and Slesser, 1973).
[2]Data calculated from Wye College Pig Research Section
for 1972/73.

Table 28.6  Farmers prices for protein and energy supplies in UK in December 1973

| Product | Farm price | Edible protein[1] (%) | Price per kg protein (£) | Energy (Mcal/kg) | Price per Mcal (p) |
|---|---|---|---|---|---|
| Milk, | 24p/gallon | 3.4 | 1.5 | 0.60 | 8.6 |
| Cheese | 21p/lb | 25.4 | 1.8 | 4.25 | 10.9 |
| Beef | £18.50/live cwt | 7.7 | 4.7 | 2.50 | 14.6 |
| Lamb | 16.3p/lb liveweight | 5.9 | 6.1 | 1.92 | 18.7 |
| Pig meat | £3.50/20 lb liveweight | 8.6 | 4.5 | 2.69 | 14.3 |
| Broiler chicken | 12p/lb liveweight | 7.4 | 3.7 | 0.87 | 30.6 |
| Eggs | 32p/dozen | 10.5 | 4.5 | 1.63 | 29.2 |
| Beans | £80/tonne | 27.0 | 0.36 | 2.70 | 3.0 |
| Navy beans | £150/tonne | 21.4 | 0.70 | 2.58 | 5.8 |
| Wheat | £60/tonne | 5.6 | 1.1 | 2.44 | 2.4 |

[1]For product as sold.

the processing of vegetable crops or for the production
of SCP but Slesser (1973) indicates that the energy
costs of some processes is high and there is some support
for this conclusion from quoted financial costs (Gordon
1970; Meyer, 1970).

Broadly it can be seen that the crops gave of the order
of 10-fold higher returns than the animal products.
Within crops, the leguminous plants with little need for
fertiliser nitrogen gave the highest returns and the
vegetable crops with high fertiliser, fuel and hand
labour requirements show the lowest return per unit of
fuel energy.

Of the animals, those normally requiring the lowest
input of concentrated feedingstuffs show the highest
return while the non-ruminants, with a high feed input,
yield the lowest return for energy supplied.

Such calculations may overdramatise the situation.
After all, fuel energy costs 0.2-1.0 new pence/Mcal on
average while the energy in animal products is worth
8-30 new pence/Mcal and in crop products from 2-3 new
pence/Mcal (*Table 28.6*). We are unlikely to revert to
a peasant economy immediately because of these high
costs. But they should make us more aware of the heavy
dependence of modern agriculture on fuel, fertilisers
and herbicides which all involve a major drain on fossil
fuels. These factors introduce a further critical
factor in evaluating methods of food production.

REFERENCES

AGRICULTURAL DEVELOPMENT ASSOCIATION (1971). *New Protein
    Foods*. Conference Report, University of Reading
ALBERDA, Th. (1973). 'Potential protein production of
    temperate grasses', in *The Biological Efficiency of
    Protein Production*. Ed. J.G.W. Jones. Cambridge
    University Press, Cambridge, England
ARKCOLL, D.B. (1971). 'Agronomic aspects of leaf protein
    production in Great Britain', in *Leaf Protein, Its
    Agronomy, Preparation, Quality and Use*. Ed. N.W.
    Pirie. IBP Handbook No. 20, Blackwell, Oxford
BEATTY, D.W.J. (1974). ''Fermentation protein',
    *J. Farmer's Club*, 1974
BLACK, J.N. (1971). *Ann. Appl. Biol.*, 67, 272
BLAXTER, K.L. (1973). 'The purpose of protein production
    in *The Biological Efficiency of Protein Production*.
    Ed. J.G.W. Jones. Cambridge University Press, Cambridge
    England

BRESSANI, I.R. (1968). 'The use of yeast in human foods', in *Single Cell Protein*, pp.90-121. Ed. R.I. Mateles and S.R. Tannenbaum. MIT Press, Cambridge, Mass.

BYERS, M. (1971). 'The amino acid composition of some leaf protein preparations', in *Leaf Protein, Its Agronomy, Preparation, Quality and Use*. Ed. N.W. Pirie. IBP Handbook No. 20, Blackwell, London

CARR, J.R. (1972). *Proc. N.Z. Soc. Anim. Prod.*, 32, 204

COOP, I.E. (1972). *Proc. N.Z. Soc. Anim. Prod.*, 32, 178

CRAWFORD, M.A. (1971). 'Current nutritional research: implications of the new proteins', in *Agric. Dev. Assoc. Conf. on New Protein Foods, Univ. of Reading, July, 1971*

DAVIDSON, J. and MATHIESON, J. (1965). *Brit. J. Nutr.*, 19, 353

DUCKHAM, A.N. (1973). *Energy and Food Supply*. Mimeo AND/73/2, Dept. of Agriculture, Univ. of Reading

EVERITT, G.C. (1972). *Proc. N.Z. Soc. Anim. Prod.*, 32, 185

FARMERS WEEKLY (1973). 'A Farmers Weekly Guide to Synthetic Food', Supplement to *Farmers Weekly*, 15 June, 1973. IPC London

GORDON, J.F. (1970). 'Algal proteins and the human diet', in *Proteins as Human Food*. Ed. R.A. Lawrie. Butterworths, London

GOW, J.S., LITTLEHAILES, J.D., SMITH, S.R.L. and WALTER, F.B. (1973). 'Single cell protein from methanol'. *Proc. MIT 2nd Internat. Conf. (on Single Cell Protein)*

HICKLING, C.F. (1962). *Fish Culture*. Faber & Faber, London

HOLLINGSWORTH, D. (1971). 'Protein needs and demands in the United Kingdom', The British Nutrition Foundation Information Bull. No. 6. July, 1971

HOLMES, W. (1970). *Proc. Nutr. Soc.*, 29, 237

HOLMES, W. (1971). 'Efficiency of food production by the animal industries', in *Potential Crop Production*. Ed. P.F. Wareing and J.P. Cooper. Heinemann, London

HOLMES, A.E. and BURKE, C. (1971). 'Simulated meat foods', British Nutrition Foundation Information Bull. No. 6, July, 1971, pp.57-64

HUTTON, J.B. (1970). 'Crops or grasses for efficient low cost livestock production', in *Proc. 11th Int. Grassld Cong.*, A 78-87. Surfers Paradise, Queensland

HUTTON, J.B. (1972). *Proc. N.Z. Soc. Anim. Prod.*, 32, 160

INGRAM, M. and RHODES, D.N. (1971). 'The changing nutritional value of meat', British Nutrition Foundation Information Bull. No. 6, July, 1971, pp.39-56

JONES, J.G.W. (1973). (Ed.) *The Biological Efficiency of Protein Production*. Cambridge University Press, Cambridge, England

LAWRIE, R.A. (1970). (Ed.) *Proteins as Human Food*. Butterworths, London

LEACH, G. (1974). 'The energy costs of food production', in *The Man/Food Equation*. Ed. A. Bourne. Academic Press, London

LEACH, G. and SLESSER, M. (1973). 'Energy equivalents of network inputs to food producing processes'. Mimeo, University of Strathclyde, Glasgow

LEITCH, I. and GODDEN, W. (1953). *Tech. Commonw. Bur. Anim. Nutr.*, 14

LODGE, G.A. (1970). 'Quantitative and qualitative control of proteins in meat animals', in *Proteins as Human Food*. Ed. R.A. Lawrie. Butterworths, London

MCCANCE, R.A. and WIDDOWSON, E.M. (1960). *The Composition of Foods*. HMSO, London

MATELES, R.I. and TANNENBAUM, S.R. (1968). (Eds.) *Single Cell Protein*. MIT Press, Cambridge, Mass.

MEAT AND LIVESTOCK COMMISSION (1973a). *Pig Feed Recording Service. Summary of Results for Period Ending March 1973*, Report No. 11 July 1973. MLC, Bletchley, Milton Keynes

MEAT AND LIVESTOCK COMMISSION (1973b). *Sheep Notes*, No.8, Aug. 1973. Report of Results from Commercial Flocks 1972. Sheep Improvement Services, MLC, Bletchley, Milton Keynes

MEAT AND LIVESTOCK COMMISSION (1973c). *Beef on Target for Profit*. MLC recorded results 1972-73. MLC, Bletchley, Milton Keynes

MEAT AND LIVESTOCK COMMISSION (1973d). Beef improvement services. Newsletter No. 19, Sept. 1973. MLC, Bletchley, Milton Keynes

MINISTRY OF AGRICULTURE, FISHERIES AND FOOD, 1972. *Costs and Efficiency in Milk Production 1968-69*. HMSO, London

MEYER, A.E. (1970). 'Soya protein isolates for food', in *Proteins as Human Food*. Ed. R.A. Lawrie. Butterworths, London

OSWALD, W.J. and GOLUEKA, C.G. (1968). 'Large scale production of algae', in *Single Cell Protein*, pp.271-305. Ed. R.I. Mateles and S.R. Tannenbaum. MIT Press, Cambridge, Mass.

PIRIE, N.W. (1971). *Leaf Protein, Its Agronomy, Preparation, Quality and Use*. IBP Handbook No. 20, Blackwell, Oxford

SCHUPHAN, W. (1965). *Nutritional Values in Crops and Plants*. Faber

SHACKLADY, C.A. (1970). 'Hydrocarbon-grown yeasts in nutrition', in *Proteins as Human Food*. Ed. R.A. Lawrie. Butterworths, London

SHACKLADY, C.A. (1972). 'Yeasts grown on hydrocarbons as new sources of protein', *World Rev. Nutr. and Diet.*, 14, 154. Karger, Basel

SLESSER, M. (1973). *J. Sci. Fd Agric.*, 24, 1193

SPICER, A. (1971). *Vet. Rec.*, 89, 482

SPICER, A. (1974). 'The Biosynthesis of Proteins from Carbohydrates', in *Industrial Aspects of Biochemistry*. Ed. B. Spencer. Fed. Eur. Biochem. Soc.

TALBOT, L.M., PAYNE, W.J.A., LEDGER, H.P., VERDCOURT, L.D. and TALBOT, M.H. (1965). 'The Meat Production Potential of Wild Animals in Africa: A Review of Biological Knowledge', *Comm. Bur. Agric.* Farnham Royal, Bucks

WALKER, T. (1973). 'Protein production by unicellular organisms from hydrocarbon substrates', in *The Biological Efficiency of Protein Production*. Ed. J.G.W. Jones

WHITEHOUSE, R.N.H. (1973). 'The potential of cereal grain crops for protein production', in *The Biological Efficiency of Protein Production*. Ed. J.G.W. Jones. Cambridge University Press, Cambridge, England

WILLCOX, O.W. (1959). *J. Agric. Fd Chem.*, 7, 813

WILSON, P.N. (1973). 'Livestock physiology and nutrition', *Phil. Trans. R. Soc. Lond.*, B267, 101

WORGAN, J.T. (1973). 'Protein production by microorganisms from carbohydrate substrates', in *The Biological Efficiency of Protein Production*. Ed. J.G.W. Jones. Cambridge University Press, Cambridge, England

# LIST OF DELEGATES

ADAMS, I.M.V.  MAFF, Great Westminster House,
  Horseferry Road, London, SW1 2PE
ALLDEN, DR W.G.  Waite Agricultural Research Institute,
  University of Adelaide, Glen Osmond, South Australia
ARNOLD, J.E.  Department of Applied Biochemistry and
  Nutrition, School of Agriculture, University of
  Nottingham
AUXILIA, PROF. M.T.  Istituto Sperimentale per la
  Zootecnia - Sezione Operativa Via Pianezza, 115 - 10151
  Torino, Italy
BACKHOFF, H.P.  Department of Applied Biochemistry and
  Nutrition, School of Agriculture, University of
  Nottingham
BAGUST, DR J.  The Boots Company Ltd., The Priory,
  Thurgarton, Nottinghamshire
BARDSLEY, DR R.G.  Department of Applied Biochemistry and
  Nutrition, School of Agriculture, University of
  Nottingham
BARLOW, P.J.  University of Aston in Birmingham, Gosta
  Green, Birmingham B4 7ET
BARETTA, DR J.W.  Trouw & Co. N.V. International,
  Nijverheidsweg 2, Putten (Gld), The Netherlands
BARTON, N.F.  Ross Poultry Ltd., Sterling Division,
  Wreay, Carlisle, Cumberland
BASSON, D.S.  Department of Food Science, The University,
  Stellenbosch, Republic of South Africa
BEKAERT, H.  National Institute of Animal Nutrition,
  Scheloleweg 12, 9231 Gontrode, Belgium
BELL, DR E.T.  Pedigree Petfoods Ltd., Melton Mowbray,
  Leicestershire

BENDER, PROF. A.E.  Department of Nutrition, Queen
  Elizabeth College, Atkins Building, Campden Hill,
  London W8 7AH
BERG, PROF. R.T.  Department of Animal Science, University
  of Alberta, Edmonton, Alberta, Canada
BERGSTROM, DR P.L.  Institute for Animal Husbandry,
  Driebergse weg 10$^D$, Zeist, The Netherlands
BICHARD, DR M.  Pig Improvement Company Ltd., Fyfield
  Wick, Abingdon, Berkshire OX13 59A
BLANSHARD, J.M.V.  Department of Applied Biochemistry and
  Nutrition, School of Agriculture.  University of
  Nottingham
BOCCARD, DR R.  Station de Recherches sur la Viande,
  INRA - Theix, 63110 Beaumont, France
BOLEDA, A.  Matadero General Frigorifico de Abrera SA,
  Barrio el Rebato, Abrera, Barcelona, Spain
BOND, K.I.  J. Bibby Agriculture Ltd., Farm Products
  Division, Richmond House, Rumford Place, Liverpool
  L3 9QQ
BOWMAN, PROF. J.C.  Department of Agriculture and
  Horticulture, Earley Gate, University of Reading,
  Reading
BRADLEY, R.  Central Veterinary Laboratory, Newhaw,
  Weybridge, Surrey
BRAUDE, DR R.  National Institute for Research in
  Dairying, Shinfield, Reading
BREW, R.D.  Pedigree Petfoods Ltd., Melton Mowbray,
  Leicestershire
BROADBENT, DR P.J.  North of Scotland College of
  Agriculture, 581 King Street, Aberdeen AB9 1UD
BROCKMANN, DR M.C.  Food Laboratory, US Army Natick
  Laboratories, Natick, Massachusetts 01760, USA
BUCHTER, MRS. L.  Danish Meat Research Institute,
  Maglegaardvej 2, DK-4000, Roskilde, Denmark
BUCKLEY, K.  Pedigree Petfoods Ltd., Melton Mowbray,
  Leicestershire
BUCKLEY, J.  Department of Dairy and Food Technology,
  University College, Cork, Ireland
BUTLER, D.A.  Mars Foods Division of Mars Ltd., P.O. Box
  15, Hansa Rd., King's Lynn, Norfolk
BUTTERY, DR P.J.  Department of Applied Biochemistry and
  Nutrition, School of Agriculture, University of
  Nottingham
CARRERAS, P.  Matadero General Frig de Abrerasa, Barrio
  el Rebato, Abrera, Barcelona, Spain
CHADWICK, J.P.  School of Agriculture, University of
  Newcastle upon Tyne

CHIZZOLINI, R.  Department of Applied Biochemistry and
  Nutrition, School of Agriculture, University of
  Nottingham
CLANCY, DR M.J.  The Agricultural Institute, Animal
  Production Research Centre, Dunsinea, Castleknock,
  Co. Dublin, Ireland
CLARK, DR J.B.K.  Cavaghan & Gray Ltd., London Road,
  Carlisle, Cumberland
CLARK, R.D.  Beecham Agricultural Products Division,
  Beecham House, Brentford, Middlesex
CLARKE, V.J.  Department of Agriculture and Horticulture,
  School of Agriculture, University of Nottingham
CLAUS, DR  Technische Hochschule München, Institut für
  Physiologie,  8050 Freising, Munich, West Germany
CLEAVER, M.M.  Pedigree Petfoods Ltd., Melton Mowbray,
  Leicestershire
COCKRILL, DR W.ROSS, Animal Production and Health Division,
  Food and Agriculture Organization of the United Nations,
  Rome, Italy
COLE, DR D.J.A.  Department of Agriculture and Horticulture,
  School of Agriculture, University of Nottingham
COLLINGWOOD, D.  Ross Foods Ltd., Ross House, Grimsby,
  Lincolnshire
COMBEN, N.  Roche Products Ltd., 15 Manchester Square,
  London W1M 6AP
COOK, K.N.  Milk Marketing Board, Thames Ditton, Surrey
  KT7 OEL
COSENTINO, DR E.  Istituto di Produzione, Animale –
  Facolta di Agraria Portici, Naples, Italy
CRAWFORD, DR M.A.  Nuffield Institute of Comparative
  Medicine, Zoological Society of London, Regent's Park,
  London NW1 4RY
CUTHBERTSON, A.  Meat and Livestock Commission, P.O. Box
  44, Queensway House, Bletchley, Milton Keynes, MK2 2EF
DAKIN, J.C.  Parry Ferguson and Dakin, 15 Borough High
  St., London Bridge, London SE1
DALRYMPLE, R.H.  Institut für Chemie und Physik,
  Bundesanstalt für Fleisch Forschung, D-8650 Kulmbach,
  West Germany
DESMOULIN, B.  Station de Recherches l'Elevage des Porcs –
  CNRZ.  78350 Jouy-en-Josas, France
DE VOS, E.  N.V. Zwan, Brechtsebaan 913, B – 2120 Schoten,
  Belgium
DIJKMANN, DR K.E.  Institute of the Science of Food of
  Animal Origin, University of Utrecht, Biltstraat 172,
  Utrecht, The Netherlands
DOBSON, S.  Department of Animal Physiology & Nutrition,
  Agricultural Sciences Building, The University, Leeds

DUCKWORTH, DR J.E.   Meat and Livestock Commission,
Queensway House, Queensway, Bletchley, Milton Keynes
MK2 2EF

DURRANT, R.J.   Nitrovit Limited, Nitrovit House, Dalton,
Thirsk, Yorkshire

EDWARDS, I.E.   Department of Agriculture and Horticulture,
School of Agriculture, University of Nottingham

EVANS, H.T.J.   Ministry of Agriculture, Fisheries and
Food, Plas Crug, Aberystwyth, Cardiganshire

FIELDING, G.E.   c/o Manorcros Agricultural Investments
Ltd., Crosfields House, Clifton, Bristol BS8 3NJ

FISHER, A.V.   Meat Research Institute, Langford, Bristol
BS18 7DY

FISHER, K.   Blackpool College of Technology and Art,
Blackpool, Lancashire

FOLLETT, M.J.   J. Sainsbury Ltd., Stamford Street,
London SE1

FORBES, A.   J. Sainsbury Ltd., Research Laboratories,
18 Blackfriars Road, London SE1

FOSTER, DR W.W.   Central Development Department, Birds
Eye Foods Ltd., South Denes, Gt. Yarmouth, Norfolk

FRANKLIN, J.E.   Spillers Ltd., Research and Technology
Centre, Station Road, Cambridge

FURNIVAL, D.J.   Bollin House Farm, Mobberley, Knutsford,
Cheshire

GAILI, E.S.E.   Meat Research Institute, Langford,
Bristol BS18 7DY

GARWOOD, P.E.   Birmingham College of Food & Domestic
Arts, Summer Row, Birmingham 3

GAULT, N.F.S.   Department of Applied Biochemistry and
Nutrition, School of Agriculture, University of
Nottingham

GEE, D.W.   Pig Improvement Company Ltd., Fyfield Wick,
Abingdon, Berkshire

GILMOUR, DR R.H.   Griffith Laboratories (U.K.) Ltd.,
Cotes Park Farm, Somercotes, Derby DE5 4NN

GLODEK, PROF. P.K.F.   Institut für Tierzucht und
Haustiergenetik, der Universität, 34 Göttingen,West Germa

GOMEZ, P.O.   Instituto Nacional de Tecnologia Agropecuaria
(INTA), C.C.276 Balcarce (Buenos Aires), Argentina

GRACEY, DR J.F.   Veterinary Department, City of Belfast
Meat Plant, Duncrue Pass, Belfast BT5 7EH

GRIFFITHS, A.J. The Boots Company Ltd., The Priory,
Thurgarton, Nottinghamshire

HALE, PROF. N.H.   Department of Animal Industries,
U-40, University of Connecticut, Storre, Connecticut
06268, USA

HALLIDAY, D.A.  Ulster College, Northern Ireland
Polytechnical, Jordanstown, Co. Antrim, N. Ireland
HAMM, PROF. R.  Bundesanstalt für Fleischforschung,
Institut für Chemie und Physik, D-865 Kulmbach,
Blaich 4, West Germany
HANNAN, DR R.S.  The Wall's Meat Company Ltd., Atlas Road,
Willesden, London NW10 6DJ
HANSON, L.G.  Ministry of Agriculture, Fisheries & Food,
Whitehall Place, London SW1
HARDY, DR B.  Sales Advisory Centre, Cooper Nutrition
Products Ltd., Stepfield, Witham, Essex
HARMS, I.J.  Group Technical Services, Associated
Dairies Ltd., ASDA House, Britannia Road, Morley,
Leeds LS27 OBT
HARRIES, J.M.  Meat Research Institute, (Agricultural
Research Council), Langford, Bristol BS18 7DY
HAY, DR J.D.  DCL (Yeast & Food) Ltd., Glenochil Technical
Centre, Menstrie, Clackmannanshire, Scotland
HEIRMAN, W.  N.V. Zwan, Brechtsebaan 913, B - 2120,
Schoten, Belgium
HINKS, DR C.E.  Edinburgh School of Agriculture, King's
Buildings, West Mains Road, Edinburgh EH9 3JG
HINNERGARDT, DR L.  Food Laboratory, U.S. Army Natick
Laboratories, Natick, Massachusetts 01760, USA
HODGKINSON, W.  Meat Technology Section, FMD Department,
Hollings College, Wilmslow Road, Manchester M14
HOLLINGSWORTH, MISS D.F.  The British Nutrition Foundation,
Alembic House, 93 Albert Embankment, London SE1 7TY
HOLLINSON, D.  Central Development Department, Birds
Eye Foods, South Denes, Gt. Yarmouth
HOLMES, PROF. W.  Wye College, Ashford, Kent TN25 5AH
HONKANEN, MISS E.M.  AB Karlshamns Oljefabriker, 29200
Karlshamn, Sweden
HOOD, R.  Westlers Ltd., Amotherby, Malton, Yorkshire
HORTON, G.J.  Griffith Laboratories (UK) Ltd., Cotes
Park Farm, Somercotes, Derby DE5 4NN
HOUSEMAN, DR R.  Rowett Research Institute, Bucksburn,
Aberdeen
HOXEY, MISS A.M.  Grassland Research Institute, Hurley,
Maidenhead, Berkshire
HUBBARD, A.W.  Ministry of Agriculture, Fisheries & Food,
Great Westminster House, Horseferry Road, London
SW1P 2AE
HUFFMAN, PROF. D.L.  Department of Animal and Dairy
Sciences, Auburn University, Auburn, Alabama 36830,
USA
HUGHES, P.E.  Department of Agriculture and Horticulture,
School of Agriculture, University of Nottingham

INGRAM, DR M.   ex. Meat Research Institute, Langford, Nr. Bristol

IVINS, PROF. J.D.   Department of Agriculture and Horticulture, School of Agriculture, University of Nottingham

JENKINSON, T.J.   Harris (Ipswich) Ltd., Hadleigh Road, Ipswich, Suffolk IP2 OHQ

JONAS, DR D.A.   Chemical Products R & D, Central Research Division, Pfizer Ltd., Sandwich, Kent

KANE, M.C.K.   Ross Foods Ltd., Grimsby, Lincolnshire

KARALAZOS, A.   Department of Agriculture and Horticulture, School of Agriculture, University of Nottingham

KAY, DR M.   Rowett Research Institute, Bucksburn, Aberdeen

KEMMISH, B.   T. Lucas & Co. Ltd., Ruskit Mills, Kingswood, Bristol

KEMPSTER, A.J.   Meat & Livestock Commission, P.O. Box 44, Queensway House, Bletchley, Milton Keynes MK2 2EF

KÖRMENDY, DR L.   Hungarian Meat Research Institute, Budapest, IX

KOTTER, PROF. L.   Bereich Hygiene und Technologie der Lebensmittel tierischen Ursprungs, Universität München, 8 München 22, Veterinärstrasse 13, Germany

KROEGER, K.   Mokwa Abattoir, P.O. Box 12, Mokwa, via Jebba, Nigeria

KYLE, DR W.S.A.   Agricultural & Food Chemistry Research Division, N.I. Dept. of Agriculture, Newforge Lane, Belfast BT9 5PX

LAMMING, PROF. G.E.   Department of Physiology and Environmental Studies, School of Agriculture, University of Nottingham

LAWRIE, PROF. R.A.   Department of Applied Biochemistry and Nutrition, School of Agriculture, University of Nottingham

LAWSON, J.B.   Department of Animal Health, Royal (Dick) School of Veterinary Studies, Easter Bush, Roslin, Midlothian, Scotland

LEDWARD, DR D.A.   Department of Applied Biochemistry and Nutrition, School of Agriculture, University of Nottingham

LESLIE, DR R.B.   Unilever Research Laboratory, Colworth House, Sharnbrook, Bedford

LEWIS, PROF. D.   Department of Applied Biochemistry and Nutrition, School of Agriculture, University of Nottingham

LINDHE, DR N.B.H.   Swedish Association for Improvement of Animal Production, S-631 84 Eskilstuna, Sweden

LING, J.R. Department of Applied Biochemistry and
Nutrition, School of Agriculture, University of
Nottingham

LOCKE, R.D. Ministry of Agriculture, Fisheries & Food,
Crown House, Sittingbourne Road, Maidstone, Kent

LOWMDES, H.J.E. Blackpool College of Technology & Art,
Blackpool, Lancashire

LYNCH, P. Ferry House, Lower Mount St., Dublin 2

MACFIE, H.J.H. ARC Meat Research Institute, Langford,
Bristol BS18 7DY

MACLEOD, DR G. Department of Food Science & Nutrition,
Queen Elizabeth College, Campden Hill Road,
London W8 7AH

MADDEN, E. Department of Agriculture & Fisheries for
Scotland, Chesses House, Edinburgh

MARNER, MISS I. Department of Applied Biochemistry and
Nutrition, School of Agriculture, University of
Nottingham

MARSH, PROF. B.B. Muscle Biology Laboratory, University
of Wisconsin, Madison, Wisconsin, USA

MARTIN, W.J.A. Findus Ltd., Humberstone Road, Grimsby,
Lincolnshire

MARTIN-BIRD, C.J. Plateacre Ltd., 23 West Cliff, Preston,
Lancashire

MARTIN-BIRD, MRS. K.M. Plateacre Ltd., 23 West Cliff,
Preston, Lancashire

MAWSON, J.M. Cavaghan & Gray Ltd., London Road,
Carlisle, Cumberland

MCCRACKEN, P.E. Ross Foods Ltd., Grimsby, Lincolnshire

MCKEE, E.C.M. Henry Denny & Sons (Ulster) Ltd., Obins
Street, Portadown, Co. Armagh, N. Ireland

MELROSE, DR D.R. Meat & Livestock Commission, (Head of
Veterinary Services), P.O. Box 44, Queensway, Bletchley,
Milton Keynes, Buckinghamshire

MICHANIE, MISS S.C. c/o Meat Research Institute,
Langford, Bristol

MITCHELL, R. Marks & Spencer Ltd., 47 Baker St.,
London W1

MOERMAN, DR P.C. Central Inst. for Nutrition & Food
Research TNO, Dept. Netherlands Centre for Meat
Technology, Utrechtseweg 48, Zeist, The Netherlands

MONTAGUE, MRS. L. Weston Research Laboratories Ltd.,
644 Bath Road, Taplow, Maidenhead, Berkshire

MORAN, J.B. Department of Agriculture, Wye College,
Ashford, Kent

MORGAN, MISS N.A. Department of Applied Biochemistry
and Nutrition, School of Agriculture, University of
Nottingham

MORRIS, MISS M.L.   Department of Agriculture and
Horticulture, School of Agriculture, University of
Nottingham

NEALE, DR R.J.   Department of Applied Biochemistry and
Nutrition, School of Agriculture, University of
Nottingham

NEŠIC DJUSIC, MRS. I.   Yugoslav Institute of Meat
Technoloqy, 11000 Beograd, Kačauskoe 13, Yugoslavia

NEWBERRY, MISS V.J.   Spillers Ltd., Research &
Technology Centre, Station Road, Cambridge

NEWMAN, D.G.   J. Bibby Agriculture Ltd., Farm Products
Division, Richmond House, Rumford Place, Liverpool
L3 900

NILSSON, DR R.   Swedish Meat Research Centre, 24700
Kävlinge, Sweden

NORMAN, DR G.A.   J. Sainsbury Ltd., Stamford Street,
London, SE1

NORTON, MISS E.   College of Food & Domestic Arts,
Summer Row, Birmingham

OATLEY, D.J.   Unigate Foods Ltd., Glastonbury Road,
Wells, Somerset

OBANU, Z.A.   Faculty of Agricultural Sciences,
University of Nigeria, Nsukka, Nigeria

O'KEEFFE, M.   An Foras Talúntais, Dunsinea, Castleknock,
Co. Dublin, Ireland

OLIVER, J.M.   School of Agriculture, The University,
Newcastle-upon-Tyne NE1 7RU

OLLEY, DR JUNE N.   CSIRO Tasmanian Food Research Unit,
Tasmanian Regional Laboratory, Stowell Avenue, Hobart,
Tasmania

ONO, S.   Meat Research Institute, Langford, Bristol
BS18 7DY

OWEN, DR J.E.   Overseas Development Tropical Products
Institute, 56/62 Grays Inn Rd., London, WC1

PAARDEKOOPER, DR E.   Hendrix' Fabrieken N.V.,
Veerstraat 26, Boxmeer, The Netherlands

PANIN, J.   Yugoslav Institute of Meat Technology,
11000 Beograd, Kacanskog 13, Yugoslavia

PASSMORE, DR SUSAN M.   University of Bristol,
Department of Agriculture & Horticulture, Research
Station, Long Ashton, Bristol BS18 9AF

PATERSON, J.L.   Department of Physiology & Pharmacology,
University of Strathclyde, Glasgow G1 1XW

PATTERSON, DR R.L.S.   ARC Meat Research Institute,
Langford, Bristol BS18 7DY

PAUL, MISS A.A.   Dunn Nutritional Laboratory, University
of Cambridge and Medical Research Council, Milton Road,
Cambridge

PAUL, PROF. PAULINE C.  School of Human Resources and
Family Studies, College of Agriculture, University of
Illinois, Urbana, Illinois 61801

PEARCE, MISS CHRISTINE  Head of Science Department,
Elizabeth Gaskell College, Hathersage Road, Manchester

POMEROY, DR R.W.  Meat Research Institute, Langford,
Bristol BS18 7DY

POULTER, R.G.  Department of Applied Biochemistry and
Nutrition, School of Agriculture, University of
Nottingham

PRESCOTT, DR. J.H.D.  School of Agriculture, The
University, Newcastle upon Tyne

RANGELEY, W.R.D.  Department of Applied Biochemistry and
Nutrition, School of Agriculture, University of
Nottingham

REDHEAD, S.  T. Lucas & Co. Ltd., Ruskit Mills, Kingswood,
Bristol

RIEGER, D.  Butterworths, Borough Green, Sevenoaks, Kent
TN15 8PH

ROBB, DR. J.D.  Ulster Curers' Association, 2 Greenwood
Avenue, Upper Newtownards Road, Belfast 4, Northern
Ireland

ROBERTS, J.C.  Harper Adams College, Newport, Shropshire

ROBERTS, DR. P.C.B.  Contra-Tech Ltd., Durley Lane,
Keynsham, Bristol

ROBERTS, R.W.  Findus Ltd., Humberstone Road, Grimsby

ROBERTON, D.J.  Roche Products Ltd., 15 Manchester Square,
London W1M 6AP

ROBINSON, A.  Ministry of Agriculture, Fisheries & Food
Veterinary Investigation Centre, The Elms, College
Road, Sutton Bonington, Loughborough

ROGERS, C.  Ministry of Agriculture, Fisheries & Food,
3 Whitehall Place, London SW1

ROMITA, A.  Istituto Sperimentale Zootecnia, Via Salaria
31 - Monterotondo, Roma, Italy

RYAN, W.A.  Commercial Branch, Australian High Commission,
Australia House, Strand, London WC2 4LA

RYDER, DR M.L.  ARC Animal Breeding Research
Organisation, Edinburgh

SALERNO, PROF. A.  Istituto di Produzione Animale -
Faculta' di Agraria, Portici, (Napoli), Italy

SARKIN, R.J.  Scot Meat Products, Bletchley, Milton
Keynes, MK1 1DT

SAWYER, R.  Laboratory of the Government Chemist,
Cornwall House, Stamford St., London SE1 9NQ

SCOTT, B.M.  MAFF/ADAS, Burchill Road, Westbury on Trym,
Bristol

SCOWCROFT, D.W.  Ross Foods Ltd., Ross House, Grimsby

SHARMA, P.D.  C/o Ministry of Agriculture & Natural
Resources, PMB 3078, Kano, Nigeria

SHEPHERD, R.A.  Spillers Research & Technology Centre,
Station Road, Cambridge CB1 2JN

SHORTHOSE, DR W.R.  CSIRO, Meat Research Lab., P.O. Box
12, Cannon Hill, Queensland, Australia, 4170

SMITH, D.G.  Unilever Research Laboratory, Colworth
House, Sharnbrook, Bedford

SMITH, DR R.J.  Department of Agriculture & Fisheries
for Scotland, Greyfriars House, Gallowgate, Aberdeen

SMITH, R.J.  Meat & Livestock Commission, Queensway
House, Bletchley

SMITH, DR W.C. Faculty of Agriculture, University of
of Newcastle upon Tyne NE17RU

SOHAIL, DR M.A.  Department of Agriculture and Horticultur
School of Agriculture, University of Nottingham

SPEDDING, PROF. C.R.W.  Grassland Research Institute,
Hurley, Berkshire

STEANE, D.E.  Meat & Livestock Commission, P.O. Box 44,
Queensway House, Bletchley, Milton Keynes MK2 2EF

STEFANSSON, OLAFUR, E.  Agricultural Society of Iceland,
P.O. Box 7080, Reykjavik, Iceland

STROTHER, J.W.  Meat & Livestock Commission, Queensway
House, P.O. Box 44, Bletchley  MK2 2EF

STUBBINS, DR A.G.J.  Ministry of Agriculture, Fisheries
& Food, Government Buildings, Kenton Bar, Newcastle on
Tyne

SWAN, DR H.  Department of Agriculture and Horticulture,
School of Agriculture, University of Nottingham

SWINGLER, G.  Department of Applied Biochemistry and
Nutrition, School of Agriculture, University of
Nottingham

SYMONS, H.W.  Birds Eye Foods Ltd., Walton on Thames,
Surrey

TAYLOR, A.J.  Department of Agriculture,and Horticulture,
School of Agriculture, University of Nottingham

TAYLOR, S.J.  Department of Applied Biochemistry and
Nutrition, School of Agriculture, University of
Nottingham

THOMPSON, R.J.  Findus Ltd., Humberstone Road, Grimsby,
Lincolnshire

THORN, MRS. JANET  Ministry of Agriculture, Fisheries &
Food, Great Westminster House, Horseferry Road,
London SW1P 2AE

TOMBS, DR. M.P.  Unilever Research Laboratory, Colworth
House, Sharnbrook, Bedfordshire

WRIGHT, D.  Fatstock Div. Ministry of Agriculture, R 556, Dundonald House, Belfast BT4 3SB

YOUNG, R.H.  Department of Applied Biochemistry and Nutrition, School of Agriculture, University of Nottingham

YUDKIN, PROF. J.  University of London

YU SWEE YEAN, MISS  Department of Applied Biochemistry and Nutrition, School of Agriculture, University of Nottingham

# INDEX

TOMKINS, T. Nukamel (Wessanen's Koninklijke Fabrieken),
'Winton', Beacon Road, Crowborough, Sussex

VANDEKERGKHOVE, P. Department of Nutrition & Hygiene,
Faculty of Agricultural Science, State University,
Ghent, Bosstraat 1, 9230 Melle, Belgium

VAN DER MEY, DR G.J.W.Instituut, Heidelbergh 2, Utrecht,
Zootechniek, The Netherlands

VAN DER WAL, DR P.G. Institute for Animal Husbandry,
Driebergseweg 10D, Zeist, The Netherlands

VAN HOOF, DR J.B.M. State University Ghent, Fac. of vet.
Medicine: Dept. Food Hygiene & Technology, Wolterslaan
12, 9000 Ghent, Belgium

VARLEY, MISS J.M. Department of Agriculture and
Horticulture, School of Agriculture, University of
Nottingham

VARLEY, M.A. Department of Agriculture and Horticulture,
School of Agriculture, University of Nottingham

VILHJALMSSON, S. Department of Applied Biochemistry and
Nutrition, School of Agriculture, University of
Nottingham

WALKER, DR W.D. Ministry of Agriculture, Fisheries &
Food, Government Offices, Block E, Leatherhead Road,
Chessington, Surrey

WALTERS, DR C.L. BFMIRA, Randalls Road, Leatherhead,
Surrey

WARD, PROF. A.G. Procter Department of Food & Leather
Science, University of Leeds, Leeds LS2 9JT

WARD, L.C. Department of Applied Biochemistry and
Nutrition, School of Agriculture, University of
Nottingham

WARD, P.F. Spillers RTC, Station Road, Cambridge

WATSON, DR W.A. Ministry of Agriculture, Fisheries &
Food, Block 2, Government Buildings, Chalfont Drive,
Nottingham

WEIR, J. Chemistry Division, West of Scotland
Agricultural College, Auchincruive, Ayr

WILKINSON, R. Spillers Ltd., Research & Technology
Centre, Station Road, Cambridge, Cambridgeshire

WILLIAMS, D.R. Meat Research Institute, Langord,
Nr. Bristol

WILLIAMS, E.F. 21 Woodhurst Lane, Oxted, Surrey

WILSON, DR P.N. BOCM Silcock Ltd., Basing View,
Basingstoke, Hampshire

WOOTTON, DR A.E. Loughry College of Agriculture & Food
Technology, Cookstown, Co. Tyrone, N. Ireland

WRAY, PROF. G. Loughborough University of Technology,
Loughborough